CRIKEY !

that's oil

An outback adventure

By Ivan Macmillan -
Last of the old Aussie roughnecks

Ivan Macmillan

Crikey that's oil!

An outback adventure

WA Publishing Ltd.

W A Publishing
Bournemouth
Dorset. United Kingdom

www.wapublishing.co.uk

This edition published by W A Publishing Limited 2006

First published in United Kingdom by W A Publishing

Copyright © Ivan Macmillan 2006

The Author asserts the moral right to be identified as the author of this work.

A catalogue record for this book is available from the British Library.

ISBN 1-905920-00-8

Printed and bound in United Kingdom by
Lightning Source UK Ltd.
Milton Keynes

To My Friends and Family.

Time has created many changes to our normal lives and lifestyles in the period of years spanned by this book. I hope the younger people will see these changes – for better or for worse, by comparing my accounts of life as it was then to their life as it is in the present. This is all part of that change and I make no apologies – my aim has been to tell it just the way it was. The greater the difference that can be seen between life in the old days and life as it is now, then surely the greater is the interest for you.

Innumerable characters and identities shared with me the experiences and adventures woven into the stories on the pages of this book. The influences each one had on my life come in infinite variations, but some rise above the rest like the rocky hills at dawn on a dusty outback landscape.

Lock Horsburgh was a tough and congenial young fitter I met in the outlaw mining town of Wittenoom Gorge in '49. We travelled together in these intervening years sharing high adventure in a staunch and enduring mateship. So aligned were we, we married two sisters.

Harry Murray is the likeable Aussie larrikin who shared with us his MG sports car and his unquenchable sense of fun in our apprentice years. As a skilled marksman with his Lee Enfield .303, Harry bagged fat pigeons for the pot by skimming the shot "just past their ear so I don't ruffle any feathers."

From beginnings in the rough and tumble boys' world of growing up in the '30s my brother Joe was my anchor and my soul mate. As the innocent victim of the exploding bath heater and other incidents he was always stoic and forgiving of my wayward stunts. He showed me the real meaning of unreserved brotherhood.

I am indebted to those friends and relatives who helped me by filling in some of the gaps where recollection was not so clear – and especially to those who tactfully pointed out some of the inaccuracies too.

Lastly but by no means least, I want to thank my wife Beverley. Always ready with useful advice, she gave freely of a lot of sound judgement. She endured the many days and hours as a "book widow" - with only the minimum of complaint, but mostly with patience and understanding.

To my family, I love you all – I wrote this for you.

Ivan - Christmas 2003

Contents -

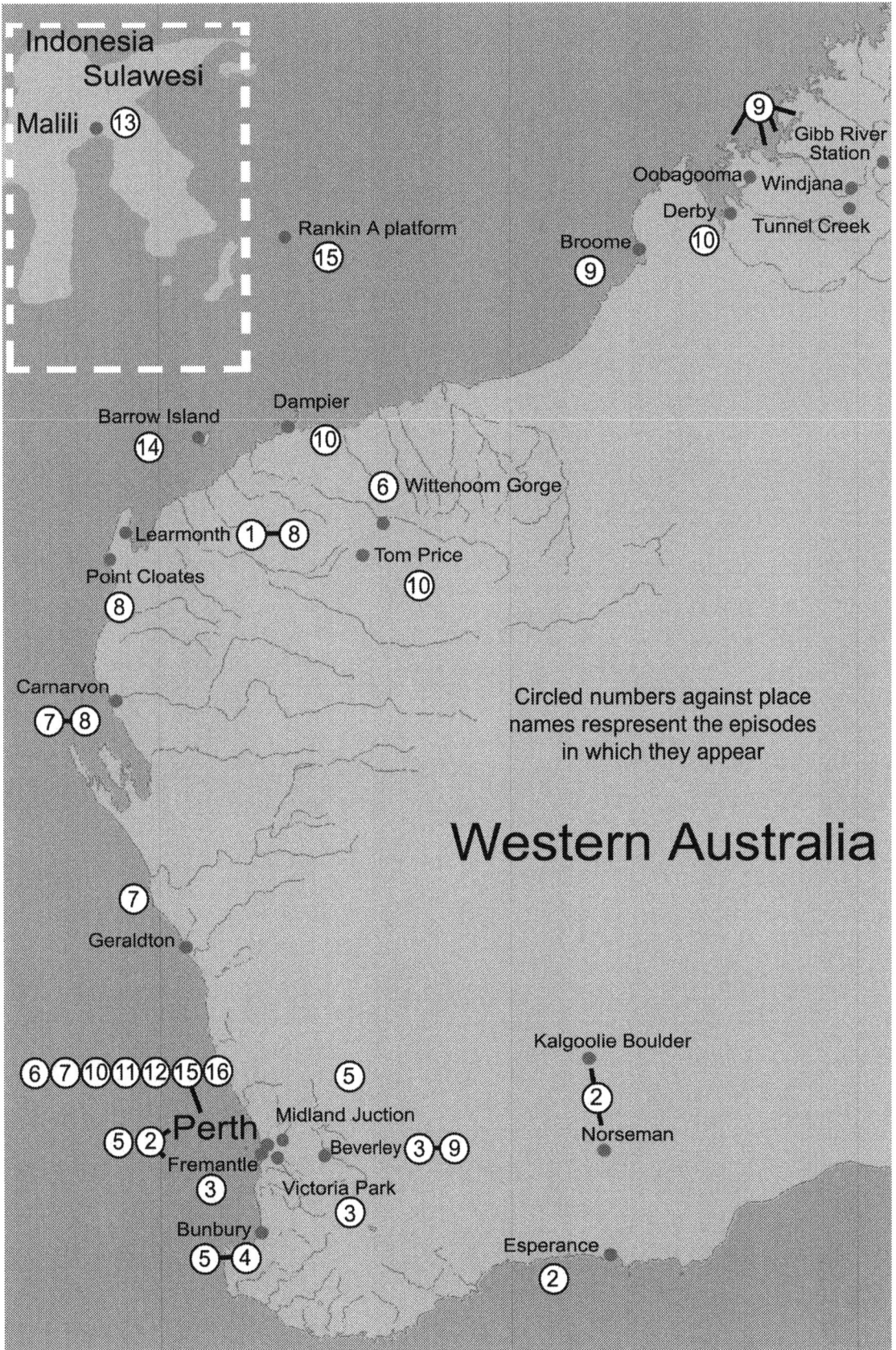

Indonesia
Sulawesi
Malili ● (13)

Rankin A platform
(15)

Oobagooma ● Windjana ●
Derby ●   Tunnel Creek
Broome ●   (10)
(9)

(9)
Gibb River
Station ●

Barrow Island   Dampier
(14) ●   ● (10)

(6) ● Wittenoom Gorge

Learmonth (1)-(8)
Point Cloates   ● Tom Price
(8)   (10)

Carnarvon ●
(7)-(8)

Circled numbers against place
names respresent the episodes
in which they appear

Western Australia

(7)

Geraldton ●

Kalgoolie Boulder
●
(6)(7)(10)(11)(12)(15)(16)   (5)   (2)

(5)(2) Perth   Midland Juction   Norseman ●
Fremantle ●   ● Beverley (3)-(9)
(3)   Victoria Park
(3)

Bunbury ●
(5)-(4)   Esperance ●
(2)

9

Foreword –

This is my life story.

One of my friends humorously remarked when I was planning this project, that the best time to start writing must be when you are old enough to have accumulated a good collection of life experiences, but not so old that you can't remember the best parts. This seems to be that time.

One of the biggest oil exploration and production companies in Australia was holding a special event in our city of Perth to commemorate the first discovery of oil which took place precisely fifty years ago in the remote north west of Western Australia.

My name was on their record books as serving in the role of a roughneck on the drill site and being a witness to the actual event back in 1953. So I was invited to shake out my best blue suit and attend the function. While standing idly by in the press of company officials and very important other people I was listening to the speeches. On hearing my name once or twice I began to take notice. Special mention was being made of the men who pioneered the exploration and especially the drilling of the well which yielded our first flow of crude oil. The significance of this they said, was that the discovery paved the way to the development of what is now an enormously rich oil and gas industry which literally turned the country on its head.

The reason it seems that I was singled out for this special attention was in the revelation that I was at this time the last man standing from that original rig crew. In the quest for survivors the Company searched world wide for others of my crew members who witnessed the first oil on that memorable day, to join them in their celebrations. They failed to locate any of them.

The incentive to write my autobiography was conceived at a happy family get-together when I was spinning one of my adventure stories. In the way of young people who are listening to the same stories for the umpteenth time, one of my sons interjected with the suggestion that I should go away and write it all down and put it in a book. Whether it could be seen as a ploy to gag me or a genuine desire to see it all in print I am not sure, but this is the result.

Episode One

We have Struck Oil!

November days in the remote semi-desert Pilbara region of north-western Australia are stinkers as are most of the days that follow up to about April the following year. November 1953 was true to form though operations were going smoothly at the exploration oil drilling location at Rough Range-1. I was now a sweating member of the rig crew on the National 130 oil rig, which was penetrating deep into the limestone strata and sedimentary sands beneath the range.

This hole was a wildcat, which in the Texas oilman's vocabulary is a leap into the unknown with the first drilling operation aimed at seeing just what was down there. Years of exploration and testing showed it to be a likely spot although in the hot and dusty vastness of the Australian continent, the odds of finding oil seemed just an elusive dream.

The duties of a roughneck are many and varied and I was never idle in the six months in the job. A drilling rig never sleeps, but for many reasons is required to be 'making hole' around the clock, seven days a week. Soon after rigging up, the hills and the gullies of the ranges echo to the roar of its diesel engines and the clang of steel-on-steel throughout the twenty four-hour days. After nightfall the drilling derrick is seen from miles around, bathed in the glare of floodlights and with a chain of lights reaching to the crown at the top of the tall mast. Dawn of each new day reveals the same scene of noise and activities continuing tirelessly as the bit drives deeper into the earth.

I was witness to a bit of drama with a humorous side before I had been there very long. It is a strict rule on the work floor of an oil rig that when all drill pipe is recovered from the well, the opening in the floor is covered immediately with a specially made steel plate. In our case, while drilling at a depth of ten thousand feet, any solid object falling into the well disappears out of sight and plummets straight down all that distance to the bottom. Depending on what the lost object is, the recovery operation to retrieve it can be very costly and time consuming. It is a cause for serious disruption to the drilling program and usually initiates an explosive show of ill temper from our team leader, the driller.

After a six-hour tripping operation one day we had all the drill pipe racked back in the mast and the drill bit out on the deck. No one had covered the aperture in the floor to the well bore. Unfortunately the wet and slippery conditions caused one of the roughnecks to lose his grip on a hammer he was using, which skittered across the floor towards the rotary table. We all watched aghast as it disappeared from sight on its way to the bottom of the deep hole. One of the crew dived for it, but missed. The driller was apoplectic, but there was only one thing to do – trip in again and 'fish' for the hammer.

Twelve sweaty grinding hours later we had retrieved the hammer. It was in the driller's hand. In a fit of bad humour he swung around on the culprit, thrust the tool into his hand and exploded.

"Here's your bloody hammer – so take it and go. You're fired!"

"Oh well," said the chastened roughneck, "I guess I won't need this any more," – and with that comment strolled over to the open hole and dropped it back down the ten thousand feet into the well again!

The job was proving to be tough but obviously not without its lighter moments.

As we Aussies were completely new to this game of searching for oil, I at least was always plying the experienced Americans with questions about oil strikes and the like. Their standard comment on our chances of finding any had quashed my hopes of seeing it happen soon, or ever. They told us of the years spent in similar wildcat drilling in unknown territory in other parts of the world with no result and cautioned us not to get our hopes up as a find in country with no history of oil discovery was a real long shot. On top of all this too was the comment that finding oil in the very first hole drilled was almost unheard of in the history of the oil business.

Hope springs eternal they say and I could not help but feel optimistic.

A memorable day in that year of 1953 was shimmering hot and brassy. All exposed steel on the rig sizzled and we sweated and slipped and cursed as we hauled the drill pipe up from the hole. As each joint of the pipe appeared above the deck, it was secured and then unscrewed, the ninety-foot long detached 'stand' being racked back in the mast. We had a problem with this particular trip though which made our job tougher and very messy. Normally the mud being lifted up inside the pipe drains away back into the hole by gravity and when the connection is broken at the floor level it is mostly free of fluids. This was not to be our day however as we were disconnecting 'wet stands'. This condition is caused by a blockage lower down inside the drill pipe resulting in a violent spray of hot sticky mud which floods across the floor and saturates men and machinery in a coating of slime.

My wiry crew mate Maca had a characteristic habit. His favourite expression was 'crikey.' It was crikey this and crikey that – crikey it's hot, crikey the flies are bad. You could pick him a mile away. My grandfather used to use the expression a lot too when I was just a kid.

We had unscrewed another pipe connection and suffered another hot mud bath, and then suddenly the muttering and the moaning stopped. The jetting stream from the drill pipe looked somehow different. There was a pungent smell like kerosene in the air and the colour of the fluid had changed from its normal muddy grey to a dark green.

Maca and I stared unbelieving and he spun around and called out to the others,-

"Crikey – that's oil!"

While this was a new experience for us Australians, Torchy Burnside, a muscular Texan and our driller had apparently seen it all before. While we stopped and gaped in disbelief, he shut down the operation of the rig and with a superb show of nonchalance strode across the slippery deck. All eyes were on him as he stooped and let the green stuff run over his hand. Having come up from a considerable depth it was quite hot and he grimaced with the pain. Without a word he rubbed it between his fingers and smelled it. He eyed it closely, tasted it and rubbed a patch of it onto the hot steel of the derrick. Turning with a great sense of drama to his goggling audience, he whipped off his helmet, hurled it to the deck and jumped on it and roared, -

"This here is oil boys!"

Maca threw back his head and yelled "Crikey!"

The men in the drilling camp were consumed with joy at our find. With wide grins and much back slapping, we found it hard to control our enthusiasm. The feeling was that we could well be making history at this very moment.

All our hopes and dreams and sweat and toil had now been transformed from a massive gamble into a reality. We had struck oil at last in our State of Western Australia and soon the whole world would hear of our good fortune. One show of oil does not an oilfield make however and we were cautioned that a premature leak of this kind of news could be most damaging to the enterprise especially as the find had not been officially assessed as commercial. Proper testing of the quality and the potential quantity of the reserves was required to determine its value. Samples of our crude oil find would have to be sent to overseas laboratories for detailed analysis and this would take time before we could be confident of success.

At the first flow of crude on the drill floor I found a small jar and collected some. Quite fluid while it was hot straight from the well, I noticed that next morning after it had cooled it was the consistency of a waxy gel.

Then there was the day the press arrived at our drilling site. Word had spread to the newspaper offices in Perth of the initial oil strike and they were hungry for news of further developments. When oil is initially discovered, there still remains a lot of work to do on the well to set it up for testing the flow rates and other qualities of the find. The major newspaper decided that they needed a man on the spot to cover the well testing program and to be in on the scoop when the find was verified as being viable. They actually sent a press team of two men to cover the unfolding story.

The Landrover that rattled up the road to our camp in a cloud of dust contained two personalities well known to West Australians at the time. Kirwan Ward was a prominent journalist and columnist with the Daily News and with him was Paul Rigby who was famous for his cartoons seen every day in the city.

Although we were very much isolated from the general public at the drilling camp and were not up with the latest news stories as were the city people, we had a quiet satisfaction in knowing that rather than just reading the news, we were actually making it. I had no idea as to how widely the word of our strike had travelled, until I received a letter from my aunt who was touring in Scotland. She enclosed a press photo from a Scottish newspaper, which showed an oilman in helmet and overalls lounging on the rail of a big drilling rig above a very significant 'No Smoking' sign. The oilman was unmistakably me! In the way of all aunts she was so chuffed that she showed the photo to half the world I fear.

Under the rules laid down by our company, which was still very sensitive about reports on the testing of the Rough Range-1 well, members of the press were not invited guests at the time and were warned that any information they gathered from the site had to be vetted before it was published. There were to be no photographs taken of the operations at all and to this end the newsmen were banned from approaching the rig.

My drilling superintendent explained to me saying, -

"This is a critical phase in the testing program and any misguided publicity could cause us enormous damage. Your responsibility is to prevent those press blokes getting any photographs of the rig or the operations there."

The two newspapermen had set up camp in the shearing shed of a sheep station a few miles away. I paid them a courtesy visit and explained their position to them and also my responsibility to prevent them recording details of our operations. This was despite the orders from their editors of course who required a flow of good copy and clear photographs for the paper. They appeared to accept my guidelines for their future behaviour with good grace and solemnly promised to obey the rules. The situation was clear - it was their job to obtain photos of our operations, but it was my job to try and stop them.

My boss was livid when he showed me the front page of the major newspaper published within the week with a very good photograph of crews working on the rig floor.

"I thought I told you...!" he exploded - but I had to tell him that I was at a loss as to how they had shot the pictures.

It had been my business to check the perimeter of the rig site and besides, they would have been spotted instantly on the single road leading into the area. It was a case of reconnoitring through the ranges further a field.

My battered old Willys Jeep easily negotiated the maze of tracks lacing the hills around the rig site as I searched for clues to help solve the mystery. It was unlikely that they could have slipped in disguised as oilmen to take pictures or used any other ruse, and in any case their big Speed Graphic cameras would be impossible to conceal. Playing the role of a sleuth really wasn't in my job description as far as I knew, but orders are orders.

It seemed almost time for me to abandon the search, when on one of the sandy tracks I caught the glint of metal in the sun. It was the brass base of a large flashbulb that I had seen. I knew that press photographers used flashbulbs for much of their work. Perhaps their rickety old Landrover had a few holes in it through which they were spilling some of their gear and leaving a trail for me to follow. I was getting warmer. There I spotted another bulb in the dirt and then another, leading me up the back of a scrub-covered hill. Leaving the jeep at a distance I crawled up the slope as noiselessly as possible to take a look.

There hidden in a hollow surrounded by bush was a big black Speed Graphic press camera mounted on a tripod, sporting the largest telephoto lens I had ever seen. My two spies were there too peering at their highly magnified image on the ground glass screen of the drilling rig below. They were discussing the details of the next shot. There was just one word for me to say in this situation and that was in a loud voice, -

"Gotcha!"

The final stages of testing the Rough Range -1 well involved discharging the flow of oil into a flare pit some hundred yards distant and igniting it. Jetting into the pit under its natural flow pressure, it burned with an enormous roiling flame. The towering pall of black smoke was visible for miles and even the master of a ship far out at sea in the Indian Ocean reported a sighting. The flare rumbled and roared, burning steadily like a giant torch for thirty days with the smoke pall turning day into night and the leaping flame turning night into day.

I could see that the stakes were high. On the brink of what could be a multi million dollar enterprise and possibly a major national energy resource for the country, it was understandable that our Company could be a bit twitchy and want to keep a lid on it for a while.

I have to admit to a possible breach of secrecy when I wrote a letter to my brother in Perth. It was prearranged that should we have a strike, a letter from me would give him that information. This was not necessarily to be spelled out, as censorship was a possibility, but would be in the form of an oily thumbprint

Newspaper headlines released the official word that the oil was of commercial grade, on December 4th 1953

appearing on the top of the first page.

Public speculation on the value of our find surged unabated for a month after the strike. We had the makings of a good commercial well - but was the oil of good quality? Without local expertise to analyse the oil, some samples were sent back to the USA for the verdict. There was a rumour spread around during this waiting period to the effect that our product was a worthless sludge. This news served to take the heat out of the share market for a while but the country seemed to be literally holding its breath. I still believed my driller, Torchy Burnside.

Then, on December 4th 1953 the newspaper headlines shouted, -

"OIL FOUND IN WA"

As Churchill was once quoted from a speech during the Second World War, referring to the North African campaign, -

"This is not the beginning of the end, - but the end of the beginning."

The rest of this story can be found in the ongoing history of the massive and spectacularly successful search for oil and gas in Western Australia in the ensuing years.

The author is photographed during a typical work day on the drilling rig.

Episode Two

In the Beginning

      May I assume that I am forgiven for starting the series of stories of my life somewhere in the middle, rather than as is customary - at the beginning? The truth is, the oil strike at Rough Range in the North of Western Australia was not only an historic event in Australian history, but was also a pivotal event in my own life as it was the ultimate goal I cherished from boyhood - to be an oilman. In fact, in my younger years, the more I saw of my grubby Hollywood heroes' antics at the cinema, the more I wished to eventually be one of them.

      From this point onwards I promise to follow a more or less chronological account up to and beyond those heady days when I was one of the first roughnecks hired in this country back in 1952. Now, at the time of writing I believe also that I am the sole survivor from that team of men who worked on that drilling rig when oil was found and important history written. It was my good fortune to be on the spot when the first oil came to the surface which spawned an industry bringing colossal wealth to my country.

      Of that small band of pioneers I am in fact the last man standing to tell the tale.

Giant flame marks the well under test.

Family Tree – Macmillan

If you care to climb back up the family tree with me you will find that siblings Joe, Gwen and I descended from two families - most people do. On our father's side was the line of the Macmillan, while my mother belonged to the family Martin, which was her maiden name.

This episode will show you pictures and also some word pictures of those grand old people, and relate some accounts of their life and times for as far back as I cared to look. As you will have noticed already, it is not a family tree in the strict graphical sense of the term, but more a collection of information, which will help to describe some of their circumstances while living in those days past. There appears to be a few notable characters among them who achieved fame and fortune, some ordinary honest folk, and one rascal - just to make it interesting.

Our Macmillan ancestral roots are in Scotland in the identity of the Clan Macmillan. History tells us they had a pretty hectic lifestyle. This was mainly due to the then King of England sending his troops to subdue the wild Scottish highlanders and trying to bring them to heel. It was in the years circa 1770.

Another clan known as the McDonalds, with whom our lot was pretty friendly, was a thorn in the side of the ruling classes and was holding out for their ongoing independence and refused to sign the papers of submission. The Campbells - another clan, were a powerful force and didn't like the McDonalds one bit and offered to try and wipe out the whole McDonald clan in an act of foul treachery at a place called Glencoe - and they did.

Tower in Knapdale - oldest in Scotland, built by Alexander Macmillan in 1470

Feigning an attitude of friendly cordiality, the Campbells infiltrated the main centres of their communities, wining and dining with them until five the next morning. At this early hour they massed their forces and went on a shock rampage of murder and mayhem, which reduced the numbers of the McDonald clan down to a mere handful.

Our lot, being allies of the victims in this instance and who were also involved in the fighting at Glencoe, were blacklisted by the British and were soon on the run towards the south end of Scotland. During a long period of harassment and persecution, our ancestors were finally divested of their lands and property and wound up on the island of Arran in the Firth of Clyde. From here they dispersed to lands further a field.

Our Buckled Crest insignia.

19

Now there are more Macmillans outside Scotland than within.

My grandfather Joseph first saw the light of day in Northern Ireland at a place called Randallstown where there was a very ancient pile known as Shane's Castle. His father was the head steward at the castle and it was there that grandpa received his schooling.

He was also related to other famous people of the same name - among them being Macmillan the book publisher of Glasgow who was grandpa's cousin. There was also a direct line to Sir Harold Macmillan, the British Prime Minister.

Grandma Laura's father was a British sea captain while her mother was French. When she was seventeen and living in Devon her mother was urging her to marry a rich landowner aged forty. This did not sit well with young Laura so she ran away and sailed to Ireland.

The Russian Ambassador to Ireland in Dublin had two young daughters at the ambassadorial residence and Laura was employed as their governess. One of the little girls was named Olga.

As fate would have it, Grandpa Joseph, who was serving an apprenticeship to a builder in Belfast at this time, used to visit Dublin frequently on business. Also as fate would have it he was a welcome guest at the Russian Ambassador's residence and stayed there often. He was reputedly a favourite of the two little daughters.

They were in the park one day in the care of Laura their new governess, when this handsome and brash young man approached them. He made a great show of trying to pick her up, so Laura left suddenly with her two charges and dashed off home. She was just explaining her predicament of this horrible man who was following her, when he walked into the house as bold as you like. It was explained that he was a part of the family too and this caused much amusement.

They must have shared a mutual attraction stemming from all this. The two lovers eventually emigrated out to Australia and were married on their arrival in Melbourne.

This was not all cut and dried however and there was an element of chance involved that could have changed our family history. While planning to emigrate, Joseph tossed a coin with his brother to see which of them would go to either Australia or America. We are all glad that things turned out the way they did. In the USA his brother eventually rose to be head the Presbyterian Church there and it is reputed that his son was chief engineer on the construction of the Golden Gate Bridge.

It is an interesting thought that grandpa Joseph and all his descendants including me and mine, are here as fate decreed, but with that fate balanced on the flip of an old Irish coin.

It was in the time of the discovery of gold and the subsequent gold rush to West Australian fields. Many hopeful people seeking their fortunes journeyed from the eastern states, heading for Norseman, and were disembarking from sailing ships at the port of Esperance. Joseph and Laura were among them, together with their baby daughter Olga, who was named after one of the two daughters of the Russian Ambassador. Leaving his small family in Esperance, Joseph travelled by horse and cart to Norseman where he secured good government building contracts and started on the family home as well. He built the town post office along with other projects there. This was in 1896.

Grandpa Joseph

Disaster struck when he was established and operating most successfully. A fire destroyed his factory and timber stores and he had to start again from scratch.

Harking back to the Scottish clans' wars and massacres, it was ironic to hear that in Norseman, the family they were closest to were descendants of the Campbells - their sworn enemies in another land and at another time.

My dad was born and grew up in Norseman, but received his education at the Guildford Grammar School in Perth.

Olga, his sister was engaged to be married when she came of age in Norseman but another contender came to town and swept her off her feet. She married him instead. This newcomer was Percy Hodgson. He was a young and brilliant doctor who ultimately played quite a big part in my young life, to say nothing of his influence on so many others. He had entered university in Melbourne at the age of 15 and won his degree in surgery at 21. By 24 he was head surgeon in Melbourne's St. Vincent's Hospital. Percy (Doc) Hodgson was as tough as he was bright and set up several surgery clinics in the Goldfields. From the town of Leonora where he had a practice, it was part of his rounds to travel to Laverton. While not such a feat these days, it has to be explained that he made the 125 km trip over the rough track riding a pushbike.

Doc saved many lives in his career, which you might say was his job anyway, but one of these was my father's when he was just a lad. Dad was aged 12 when he was shot in a hunting accident by one of his cousins. Doc showed up at the house in the nick of time and proceeded to conduct a surgical operation right there on the kitchen table. As Dad had been shot in the right leg, Doc examined the wound but eventually discovered the bullet in the left leg and removed it (the bullet). The wound injury included the bullet grazing Dad's femoral artery so his condition was touch and go. Doc was 27 at the time. In later years he brought another man back from a near-death experience after a tractor had rolled on top of him. The weight of the machine would have killed him instantly were it not for the tin of tobacco in his shirt pocket. This must be one of the rare instances when being a smoker actually saved someone's life.

In the pioneering days of many of the medical procedures we now take for granted, he was always experimenting. Once while working with primitive X-ray equipment, he was badly burned by the radiation and carried the scars for life. In the middle of a diphtheria epidemic while he was inoculating the town population, he contracted the disease too. So that he could keep going he conducted an operation on himself. With the aid of a mirror he performed a tracheotomy and inserted a breathing tube in his throat. One of his successes, for which I will always be grateful, was in pulling me back from the life of an invalid when I was 12.

Grandma Laura

Doc and Olga had three daughters - Mary, Olga and Sheila. This last and youngest cousin was known by the fond family name of Dibs. While I was staying in Beverley recuperating from my alleged illness, and also on wartime evacuation later, the doctor adopted me as his trusted hunting companion. You will understand my feelings when I say that I owe a great deal to that uncle of mine. Among the other successes attributed to his skill, were those with horses, dogs, cats and sheep - he took the role of the local vet as well, to help out the farmers in the district.

Grandpa Macmillan was a general favourite with the youngsters. Always energetic and fun loving even into his autumn years, he was an inspiration to us with his boundless knowledge and skills. We loved to be with him when he crafted fine furniture in the workshop under the house and as he tended his large fruit and vegetable plots in the back yard. Right up until his death at 84 he still retained a thick mane of silvered wavy hair, which was complemented by his perennial flowing white moustache. Also a feature of his profile was the big pipe that he smoked with such obvious enjoyment. We kids found the loading and firing of this instrument to be a constant source of entertainment.

Grandma Laura Macmillan was made of sterner stuff. As you can see in later episodes she was strict and authoritarian in the early Victorian manner. Joe, Gwen and I owe her a great debt of gratitude however. Even though she was at a very advanced age, she took us in and cared for us in those bad times around when Mum died. She continued to look after us in Victoria Park where we attended the Victoria Park State School, and on up until twelve months later, when Dad remarried. We left them then to resume a normal family life with our stepmother Gwen, and latterly, with her four children as well.

My Dad had quite a burden to carry around with him all his life - his name. Christened Joseph Ivan Rabone Macmillan, he answered to the name of Pete. There are no prizes for guessing where my first name came from, but where they got it from in the first place, no one knows.

My father was quite a mixture as to his talents and skills in life. He joined the Bank of Australasia to serve in that capacity for the rest of his working life. If you ask anyone what Dad's feelings were about being a bank officer they will tell you he hated it with a passion. I suspect that there was a lot of parental pressure on him to secure his future in a decent and respectable occupation, as this was what it was seen to be in those days.

His real and natural talent emerged however in the field of mechanics, and he could frequently be found in his workshop repairing and building and tuning all kinds of machinery.

Perhaps contrary to this image of the man who loved to work with nuts and bolts, he had a fine musical talent as well. The violin was another of his passions and so accomplished was he with a deep understanding of classical music, that he gained quite a reputation, and was in demand for the role of first violin in a number of orchestras.

All this dedicated and varied activity came together as I observed it one day, in the matter of the space of a few hours.

The 1927 Fiat tourer swung into the back yard one afternoon as Dad arrived home from the bank. While not exactly running, he wasted no time in diving into the house. He emerged in his customary 'comfortable' work clothes and made a beeline for the workshop. Some mysterious project or other claimed his attention until teatime. Having scrubbed the grease and grime off his hands for the meal, he did not tarry for family chitchat, but pushed back his chair and disappeared into the living room. There was no doubt what he was doing this time, as we could

Grandpa Joseph's birthplace - Randallstown,

hear him tuning his fiddle, and then the strains of a Brahms violin concerto came floating through the house.

Because of his own thirst for knowledge, Dad was a mine of good information, especially for two curious boys. If a 'suitable' blokes' film was showing at the local picture theatre he would take Joe and me along to see it. Sitting one on either side, it wasn't long before we would be plying him with questions about this and that throughout the show, which to his credit he would answer - every one. In later years with sons of my own and in this similar situation, I reflected on this and I can tell you how I felt about it - it would have driven me nuts! He loved hunting and fishing and camping too and would take us every time until he was older and then we would take him. A particularly enjoyable outing with Dad while we were quite small was an air show at the Maylands aerodrome. To demonstrate the latest in fighter aircraft, the Air Force staged a simulated dogfight between two of the latest planes - they did some bombing runs as well.

These fighters were Bristol Bulldogs - canvas-and-wood constructed aircraft that dived and circled each other and fired blank ammunition from their machine guns to the delight of the crowd. The bombing run came after a steep dive, when one of the planes roared across the grass field and the pilot tossed bags of flour over the rim of the cockpit.

There was never any doubt in my mind that Dad was proud of us and encouraged us in all things. Joe and I left Bunbury at age seventeen to start apprenticeships at Midland Junction, and Dad was always keen to know all the details of our work and the skills we were learning. My enduring feelings are that we could not have had a better Dad.

After an otherwise eventful period in my own life, something really fortunate happened to me. I married a girl called Beverley Laker.

We all knew that Dad had been ill for some time with cancer, but he did survive to see us married. We had returned from our honeymoon and I was washing the car at Bev's house, preparing to make a fast trip to Bunbury, when a policeman walked into the yard and asked for me. Our worst fears were confirmed - my Dad had passed away.

Dad and me - Suburban Road house - 1928

Joseph Ivan Rabone Macmillan was joined in holy matrimony with Dorcas Pretoria Martin on February 9th 1927. My birth was recorded as being on December 5th 1927. The timing of my arrival seems to have been a close one - but was really OK as it happened. It is probably good to remember that in those days there were no distractions, like television for instance, with which to while away the evening hours.

Your author first drew breath on that December 5th day in a small private hospital with a nice view, situated in Berwick Street South Perth. Starting life at such an early age has its disadvantages I found, in that I don't remember a thing about the blessed event or anything that followed it for quite a time.

The family home I was whisked away to was in Suburban Road South Perth. They later changed the name to Mill Point Road and that's what the sign on the corner says it is to this day.

With the faculties of sight, hearing, smell and yelling all developing pretty normally, those first three years in South Perth passed quickly. My brother Joe showed up eighteen months after me and seemed to be claiming all the parental care and attention for a while. This was not a good thing as far as I was concerned.

Little Joe was in Noddy Land one day on the sunny front veranda. He was tucked up in a big wicker pram that Mum had carefully and safely parked back in a corner. Probably feeling a little restless and bored and wanting to exercise my stumpy legs, I backed the pram out of its corner and lined it up with the front flight of concrete steps. It is intriguing to see how effortlessly a pram travels when it is given even a slight push - my memory is clear on this. Right on line with the drop, Joe glided along the veranda to disappear completely over the edge. The baby-pram combination bounced quite spectacularly all the way down to the garden path and then capsized. Joe made such a din - but then he was always cranky when he was awakened from a nice nap. In the short space of time after the crash my mind is a blank. That episode of amnesia was probably induced by the body's defence mechanisms, which serve to lock away painful memories.

Dad often took us on walks along the foreshore of the Swan River, which was a great green stretch right opposite the house. These outings were pretty exciting for me as a part of this area was also an aerodrome with planes of all shapes and sizes landing and taking off.

At the low level of the universe as seen by little kids of two or three, my concept of the world around me was rather limited. I remember only the colourful diamond pattern of the linoleum on the floor and Dad's shiny winkle-picker shoes topped by his snake charmer pants. Mum's fluffy pink slippers came and went and there was a ginger cat.

There is also a fuzzy mental image of the times I toddled out to the front gate to meet my father when he came home from work one afternoon. It seems this touching scene must have moved him as he took a photograph of it.

There is not really much you can say about life when you were three is there? That was my age when we packed up and moved down to Fremantle, to where my Dad had been transferred in that branch of his bank.

Family Tree – Martin

The Martin family, the parents of my mother, occupied the other side of the grandparental equation.

We lived in East Fremantle for a good part of my young life and Dad insisted that we visit both sets of grandparents on a regular basis. He obviously thought it important that this connection was maintained.

It was a ritual we followed every weekend, to drive the twelve miles to Perth in the old Fiat tourer and spend the day with first one and then the other of the old folks. Images are still clear of the trips home in the dark - the three of us bundled together under layers of rugs on the long back seat. If we had been to Macmillan's house it was likely that we would be stuffing our faces with one each of grandma's enormous rock cakes, a parting gift that would keep us occupied - and speechless all the way back to Fremantle.

The old bus had side curtains that you put up with snap fasteners to keep out the wind and the rain - or most of it.

The name which comes down through my mother's forebears and which was grandma's maiden name was Bonney. This family name has been traced back through the centuries to its French origins, when it was Bonnet. They were French knights who left their homeland during the Huguenot Wars to show up in England some time in the 1500s. The family tree as it appears from those times to the present reveals some people of note, some ordinary individuals, and one scallywag.

This one Joseph Bonney had apparently upset the authorities in England with some misdemeanour or other and was nicked. He was sentenced to deportation for his crimes in 1812 and found himself on one of His Royal Majesty's ships heading for the colonies in 1814 - as a convict. A record of his conviction at the Old Bailey turned up which described his crime – stealing a roll of carpet.

Further down the line of descendants, a John Bonney married Frances, and then begat another Frances. She in turn married Mr. M J Hinkler. He sired a son, who grew up as Bert Hinkler, the famous Australian Aviator. To give it meaning and put it all in context, this Hinkler fellow was the cousin of my Grandma Martin.

Bert Hinkler racked up a number of daring exploits in those days of wood-and-canvas aircraft and conquered many flying records. It was in 1928 that he flew solo in an Avro Avion from Croydon England to Darwin. He wasn't the first to do this trip, but he beat the existing record by 12 1/2 days. One of his most remarkable achievements was the solo flight in a Puss Moth from New York to England via the South Atlantic. The end of his distinguished flying career came abruptly when he crashed and died in the Italian Alps while trying to break another record.

Another flying Bonney was a mere slip of a girl called Lores. Her justification for being included in the family tree can be attributed to the fact that she was the wife of Harry Bonney, my grandma's brother. While still a teenager she persuaded Bert Hinkler to teach her to be a pilot, after which there was no stopping her. In an age when men ruled the skies, she set off to fly solo from Australia to England in 1933. This feat is listed as one of the great achievements in Australian aviation history. Lores was to be the first ever pilot to fly solo around Australia. Then she added to her laurels by flying a single-engine plane on a solo flight, which took in Brisbane, Asia, North Africa and then down to Cape Town. On this jaunt she covered 29,000 km. The plane she favoured for the long hauls was a Klemm monoplane, which she affectionately called

'My Little Bus'. Her only passengers were two parrots travelling in the plane with her - for company I suppose. She had named them Hitch and Hike.

King George IV awarded Lores an MBE in recognition of her skill and daring.

My maternal grandmother Beatrice Lois Bonney started her life in Bonnievale, a small mining settlement in the Goldfields north of Kalgoorlie. While just seventeen she was starting her career as a seamstress and dressmaker. Romance blossomed for her in the shape of a handsome bloke called Jim Martin. He, aged twenty-six, had a butcher's and bakery business there. Jim's Dad was a hard rock miner in the 'fields and taught him much about prospecting for gold. Jim also taught ballroom dancing there in the town - a man of many parts.

For many men in the goldfields at that time, prospecting was a part-time occupation. No doubt they all thought there was a good chance of striking it rich. As my Dad was then a young lad in Norseman, Jim Martin took him under his wing and taught him the gold prospecting game. Perhaps my father was infected with gold fever too, as in later years he was happiest out alone in the bush with his pick and his dolly pot and his panning dish.

Grandfather Jim Martin

Jim and Beatrice were married in 1896 there in Bonnievale. At the tender age of seventeen she had to get her parents' permission for the nuptials.

We saw very little of our grandfather Jim Martin on our regular visits to their house in Victoria Park while we were young. In his later years he was crippled with Arthritis and generally immobilised. Before this time when the pains were troubling him he would travel to Perth to receive gold injections, which were thought to be a remedy then. He would not stay long however as he needed to be in the drier atmosphere of the goldfields country.

Grandma Martin's skill with the needle and thread was becoming well known and sought after. While living later in Perth, she was employed as the wardrobe mistress by the Eric Edgely stage productions company at the Theatre Royal.

My mother inherited the name Dorcas from one of her forebears, and as a young girl in Norseman excelled in playing the piano and the organ. An accident while playing games with other children caused her to suffer a twisted ankle. Mainly due to

Grandma Martin - nee Bonney

neglect, or lack of appropriate medical care, the leg became affected with gangrene. Her foot could not be saved and after a long and painful journey to Perth she had to suffer an amputation. She was virtually crippled for a time awaiting further medical care to bring back her health and mobility. The church congregation did not want to lose the services of their star organist, so they made arrangements to get her to the church and back for Sunday services. It seems that two strong young men would turn up at the house and linking hands, would carry her down to the church and then return her home afterwards.

She features in my recollections in other parts of this book and I hope that you will appreciate just what a wonderful mother and wife she was under some very trying circumstances. Kind and gentle and with endless patience, she was to us irreplaceable.

Following a long period of illness when she suffered pleurisy and then serious heart problems, she passed away in a Fremantle private hospital on March 25th 1939. She was only thirty-five years old.

She had a sister, Beatrice Merle (Bobby) who was much younger. Bobby at this time is a bright and feisty ninety-year-old who is still with us today. We visit her in her neat little unit in Bayswater where she likes to bring out her memorabilia of the past and regale us with stories of the old days. Bobby maintains that Mum's death was hastened by the death of her beloved father just seven months previously, with whom she had shared a very close relationship.

This seemed to me to be reminiscent of the romantic ages of long ago, when it was often said that a person could die of a broken heart. Knowing her as I did, it would not surprise me to learn that this great loss to her was an important factor leading to her failing health at that

My mother Dorcas Pretoria aged
seventeen

Episode Three

<u>Barefoot in Freo</u>

We made our home in Stratford Street in the port city of Fremantle in 1930, because Dad, who was a teller in the Bank of Australasia, had been transferred to the Fremantle branch of that revered establishment

To help identify the era, I can tell you that it was in the days when parents were christening their girl babies with names like Ruby, Fanny, Gladys, Poppy, Myrtle, Pearl and Daisy. Not to be outdone the boys grew up with names of the day, among which were Albert, Cecil, Syd, Joe, Reg and Walter. From the bible, which was a best seller in those times, came the names of the faithful followers with John, Mark, Luke and those of their fellow disciples.

Very fond memories began to imprint themselves upon my mind in those years and are with me still. On arrival in Fremantle from South Perth, our family comprised my Dad Ray, Mother Dorcas, brother Joe as a baby, and me. Mum didn't wait too long to fill out the family quota of three with the arrival of sister Gwen.

Dad was the sole breadwinner and provider, which was the way things were in those days, while Mum filled the role of full-time wife, mother, housekeeper and nurturer.

Among the many special remembered images of our close family life in those years is one set in the kitchen on a cold winter's day at the Stratford Street house. When the outside temperature dropped low on crisp cool days, the old black Metters wood stove in the kitchen became the gathering point for us. Joe and I were perched on chairs this chilly morning basking in the reflected heat from the bright crackling fire, and toasting thick slices of bread on special hooks with long handles made from wire. As a special treat Mum had lowered the oven door where we rested our feet clad in thick socks to warm them and we chattered like contented parrots. Mum was sitting a little apart and breastfeeding baby Gwen. A more peaceful family scene would be difficult to imagine.

This natural feeding process was fascinating to me and I can still feel the closeness we shared. We turned to look at my mother who was giggling with girlish enjoyment together with the baby's squeals and liquid slurping noises. Gwen had disengaged from the breast and was gurgling and blinking through the stream of milk, which sprayed her chubby face. My fascination had turned to amazement and with Mum setting the mood of the moment we were all soon cackling like fools and rocking back and forth in the glow of the old wood fire.

Joe and me - innocent then

Mum and Dad had both been raised in families where music was very much a part of their way of life. To add to Mum's competence at the piano, she also had a true and sweet singing voice. She knew many of the fine tuneful melodies of the day and the family shared this expression of her enjoyment as she

28

moved from room to room around the house. Especially good memories for me were those of night times. Sitting on the side of the bed after tucking us all in, she would ask for our favourite songs and in that soft, clear voice, sing as we drifted off to sleep. The recollection of a particular one is very special. She must have been a romantic soul and sang this one with much feeling. It was a love song about a little brown bird. I have included the verses in this story so that you can share it with me.

Our Mother

MUM'S SONG
Words – R. Barrie
Music – H. Wood

All through the night there's a little brown bird singing,
Singing in the hush of the darkness and the dew,
Singing in the hush of the darkness and the dew;
Would that his song through the stillness could go winging,
Could go winging to you – to you.

All through the night time my lonely heart is singing,
Sweeter songs of love than the brown bird ever knew,
Sweeter songs of love than the brown bird ever knew;
Would that the song of my heart could go a-winging,
Could go a-winging to you – to you.

All through the night time my lonely heart is singing,
Sweeter songs of love than the brown bird ever knew

What a wonderful way for any child to drift off to sleep. If you are fortunate enough to hear a rendition of the song, I am sure you will appreciate my feelings about it now.

Our family in 1935 - living in Fremantle

There was no doubt in our minds in those early years as to the importance of Dad's role in our little family. In our young and impressionable eyes he was the epitome of the authoritarian father and there were times when we dared not disturb him. I recall him sitting and puffing on his cherry wood pipe while reading the newspaper or listening to the ABC news on our crackling wireless set on the mantelpiece. While he was stern and gruff at times, he could however show us the kind and understanding side to his nature. There was never any doubt in our minds however as to where the centre of power lay in our household.

When dressed in the blue suit and grammar school tie that befitted his job behind the teller's window, he was to us the pinnacle of respect and importance. Joe and I in the way of growing boys must have been obnoxious pests at times to the grownups. We were also reminded that stern discipline and corporal punishment would descend upon us if we tried any stupid stunts, or heaven forbid challenged his position. That dreaded instrument, the symbol of Dad's authority could be found on a hook behind the bathroom door. Razor strops were common in family homes as the man of the house shaved with an open blade razor. The heavy oiled horsehide belt was essential to keep a fine keen edge on the razor. We were captivated by the stropping procedure as Dad slapped the razor back and forth prior to each shave in the mornings – all except Saturday, when he 'gave his face a rest.' This was all before those new-fangled safety blade Gillette razors appeared.

The other role of the razor strop in our household and I suspect in many others, was in showing rambunctious sons the error of their ways. Used folded double so you weren't hit with the metal hook on the end, it could deliver a painful welt on the legs or backside with a meaty crack. This punishment was meted out infrequently, but with a lasting impression on the mind and body of the punished, and was a common method used in boy control in most families then. While leaving us with a sharp stinging reminder of our transgressions and a healthy regard for discipline, it did not appear to create anger or lasting resentment in us -

First family home in East Fremantle - Walter Street - 1935

more a determination not to be caught again doing whatever it was we had been punished for. I assume this was the object of the exercise. One thing that Dad was heard to say when laying on the strop was "This is hurting me more than it is you lad." That always puzzled me at these times, as I was sure that it was I who was the one on the receiving end and therefore feeling most of the pain.

Sometimes our family breakfast was a bit rushed. Dad would lift out his big wind up watch by its chain from his waistcoat pocket and grunt, with a frown, -

"I'm going to be late for work."

Mum had set his man-sized cup of scalding tea in front of him and the scene was set for the little charade that we all enjoyed. Playing to his fascinated audience of four he, with an elaborate flourish and a wink at us, would pour the steaming tea from his cup into the deep saucer. The larger surface area of the tea in a saucer gives you better cooling, he suggested. He liked to give us the technical explanations for things. He must have had a steady hand as he picked up the brimming saucer with a great sense of drama, crooked his little finger and balancing it on his fingertips, raised it to his mouth. The tension around the table increased as he gently began to blow across the surface of the tea without spilling a drop. With a careful tilt he drank from the lip of the saucer. Throwing his head back he drained the saucer and smacked his lips with exaggerated satisfaction to the cheers of the family. There was no shortage of good entertainment in our house.

Joe and I attended two primary schools (in sequence) during our tender years while living in East Fremantle. Our first was the Richmond State School and the other the Bicton Primary later on, when we moved to the Yeovil Crescent house.

An early recollection of my very first class, which was 'first bubs', was the intimidating size of the teachers. They towered above us in our little desks and assumed an even more grand and lofty appearance when standing on the platform at the front of the classroom. It was a bit of a problem to totally respect Mr Fletcher while looking up into his hairy nostrils. Though it did not bother me too much, the booming voice of Mr. Fletcher apparently did disturb others a bit - Mary McNab would sometimes giggle nervously and then wet her pants. Spotting the little stream meandering down the aisle, I usually managed to whisk my lunch off the floor in the nick of time.

Most of us boys attended school in bare feet. This was a great cost saving to the families and they said that you would never catch a cold in winter. Nobody who has worn shoes all the time has experienced the range of sensations you can enjoy through the soles of your feet as you traverse different surfaces in the course of a day. Between hot sand, bitumen, cool grass and concrete I found the smooth worn board floors of the classroom to be the most sensuous. My pedestrian enjoyment took a serious setback one day when I executed a long slide across the room and was speared by a big wood splinter. There were some advantages in having feet with tough horny soles however. The tougher and hornier they became the more impervious they were to rough conditions, though a big patch of double gees could be a worry. Dad put a stop to this wonderful freedom by saying our feet were spreading to the proportions of a Neanderthal's. We did not know what he meant, but assumed the problem was that our normal shoes no longer fitted the modified shape of our spreading feet.

I was rather fascinated by our tools of trade in the school classroom. The new shiny Staedtler pencils had an enamelled smell of their own which changed to cedar wood smell as you sharpened them. One teacher must have recognised my interest in technical tasks, so gave me pencils to sharpen – for the whole class. My favourite subjects in the early years were drawing as number one and stories. The teacher said I was good at making up stories and did it all the time. To me this was as good as a pat on the head.

Most classrooms featured a fireplace in the corner just for the cold grey days in winter. I was very happy to be assigned the task of lighting the fire and keeping it alive during the day. This probably served to satisfy my latent urges as a budding pyromaniac – which was within the bounds of my childish innocence of course.

During the earliest years of school we enjoyed a treat at the morning recess. The clink of bottles outside the room at about half past ten would start me salivating. What I looked forward to was the issue of a half-pint bottle of free milk. It was delicious and cold and with cardboard caps which you popped open with your thumb. No pasteurised, homogenised, or bastardised drink this; it was the real thing straight from contented cows in a green paddock somewhere. Before the cap was opened you could see through the glass, the lovely thick yellow collar of cream on top of the milk inside. This heavenly stuff was extracted with the index finger before drinking the contents. While I knew nothing of the benefits of calcium for young bones and teeth I did know how good it tasted. There were often one or two bottles left over after the distribution to the class. The right to have one of these was strongly contested and it fell to my lot very rarely. It seemed that this privilege was given as a reward for being good, which probably accounted for me usually missing out. It was obvious however that a spare bottle of milk would go to the teacher's pet. I would have done anything to be able to put a hex on him so he would break out in a nasty milk rash or something equally uncomfortable and unsightly.

A walk around the schoolyard at playtime or lunchtime would show that almost every kid was engaged in a game of some kind. Knots of bombastic boys and giggling girls pitted their skills in games of hopscotch, knucklebones, spinning tops or marbles. I remember a stern warning levelled at the girls one day not to draw hopscotch squares on the roadway with chalk as it could frighten the horses. To any casual eavesdropper, to wit an adult, the language spoken between the players during the course of a game would be unintelligible jargon. Marbles for instance were known as alleys or doogs. Individual types of these had the nomenclature of glassies, bots, blood reels, tombolas or stonks. This last despised variety was made of hard clay and lacked the attractive appearance and the performance of the others. The treasured taw with which you did all your shooting was never relinquished or sacrificed even when you lost a game. Mine was a much-prized blood reel. It was pure snowy white with scarlet swirls and it served me faithfully. One boy began to clean up with his taw that was a big heavy ball bearing. So devastating was this secret weapon that we banned it. Every boy carried an alley bag hitched to his belt. This was in the season of course.

One strange phenomenon of boyhood life was the way the different games came and went like the seasons. Cigarette Cards was a popular game where the cards were flicked across the lunch shed to see which could get the closest to the wall. They were also prized as collectors' pieces and were a medium of exchange too. A typical negotiation between two schoolboys might be – three cigarette cards for your blood reel alley, after which a half-eaten Granny Smith apple would be handed over as change. If your Dad or older brother didn't smoke, or rolled his own, the supply was meagre, but this was seldom the case then. Cigarette card games would merge into spinning tops and then into tip cat or kites and then perhaps into alleys. We knew nothing of the hidden forces controlling these changes – we just accepted them. Gings – or catapults - were perennial and most boys would have one in their back pants pocket so there was always something for them to fall back on.

Many of the individual games were abandoned when the word flashed around the school that there was going to be a fight. If the combatants looked at all determined and serious about doing each other bodily harm, the teachers took charge and brought out two pairs of boxing gloves. Honour was served then according to Queensbury rules, and usually the tempers cooled and the junior gladiators shook hands. Many great friendships and alliances started this way. They say that two strong men respect each other don't they? Sometimes a couple of girls would decide to mix it too. Their kind of combat was pretty strange to us boys as it included bashing with school bags and pulling hair with a lot of screaming. They couldn't throw a straight punch either.

As we advanced to higher grades, pencils gave way to pens. Some schools still used slates but these were old hat. This new instrument introduced us to a craft called penmanship, the object of which was to produce lines of copperplate script handwriting.

The pen had a fat wooden handle and the nib was of standard issue steel. Each desk contained a little ceramic inkwell set in the top right hand corner of the top. So the pupil could keep the writing straight and of the correct height, faint blue lines could be found ruled across the special paper. A good example of penmanship featured the correct uniform slope on the characters, lovely fine upstrokes and bold decisive downstrokes and all without blots. These inkblots could be the result of overloading the nib with ink, having your elbow jogged or as a result of trying to impale a resting fly. Impossible to disguise, blots earned you a painful rap over the knuckles with the teacher's big wooden ruler. Handwriting has never been my long suit. I was humiliated one day when the teacher likened my best efforts to the track left by an insect that had just crawled out of the inkwell and walked across the paper. It must have been that his eyesight wasn't too good because sometimes that is just what it was. My writing is still pretty atrocious – it must be a brain thing. One of my classmates got into a heap of trouble and then had to stand up to receive punishment with the cane from his teacher. The little ceramic inkwell he found to be just the thing for drowning flies and conducting other experiments. His major crime, the one that invited the cuts, was when he grabbed the end of one of Polly Kelly's blond plaits hanging enticingly over the back of the seat in front and dipped it in the inkwell too.

School excursions were an enjoyable distraction from schoolwork and I always looked forward to them. We were taken to places that were real eye openers. The teacher made us write an essay on the trip afterwards, which I usually did fairly well. I was deeply impressed with the kids' concerts put on by the Perth Symphony Orchestra under the baton of Bernard Heinz. There was a particular orchestral item, which I have always enjoyed. It is the musical story of Peter and the Wolf, introducing many of the instruments in the orchestra playing solo parts and creating the impressions of the animals that were the characters in the tale. We visited the South Perth Zoo, the Museum and the Art Galleries and the State Library.

By the time you have finished this book you will have picked up on the fact that I am interested in guns of all kinds. They were very common when I was young and nobody took particular notice of them. The military and weapons display in the Perth Museum captured my attention on every visit. On our first school excursion there I was particularly taken with a German military pistol in the small arms display cabinet. On a visit to the museum again shortly after with my Dad, I pointed out to him that the gun was no longer in its place in the case. Dad reported this to the curator and apart from my breathless account to Mum when we got home, we forgot about it.

One day back at school soon after, the headmaster appeared in the doorway accompanied by two big men in suits with stern faces. They asked me to accompany them to the principal's office. They explained that they were investigating the theft of that same pistol from the museum, and it seemed that I was the only one to have noticed it was missing. They asked me to describe it to them and as I had been reading up on it I began.-

"It was a German self-loading pistol known as the Luger model P08. It fired the 9 millimetre parabellum cartridge and..." They said that was fine and I had been a big help in their enquiries and thanked me on behalf of the Police Department. All this attention made me feel useful – even a bit important.

We did not know, nor did we appreciate until later years, just how laborious was the nature of our schoolwork in those days, presumably because we knew nothing different. Many of the tasks we were set and the material we had to learn are so much easier today, and some have even disappeared altogether. It has been my good fortune to be involved in the evolution process that changed the three Rs to simpler forms. Probably the greatest labour-saving changes for kids in schools were in the swing to metrication and decimal currency.

It is not surprising that the old Imperial system of weights and measures was so diabolically complex in its conversion factors. The standard inch was reputed to have its origins in the distance spanned by five dried beans put together. The standard foot was derived from the shoe size of one of the old English monarchs. The furlong (furrow long) was the length a horse could plough in a field before it felt knackered. I rest my case.

Rather than convert up or down in a system using the scale of tens as now, we had to learn that twelve inches made a foot, three of those made a yard and 1760 of these made a mile. The confusion didn't stop there either. To measure weight we divided the ton into hundredweights (20). These were divided into stones (8), which were again made up of pounds (14). The pound weight comprised a number of units called ounces (16). This is not to mention other of the old units, which showed the temperature of boiling water as 212 degrees F. while this stuff froze at 32 degrees in that scale. The units for area and volume were just as complicated too. Then it came down to money – our currency. The guinea comprised twenty-one shillings. The pound was only twenty. The crown was five shillings while the florin had two. The shilling got nasty with twelve pennies, but then the penny was boosted up to threepence and sixpence, but denigrated to two halfpennies. Can you imagine a young shop assistant required to give a customer the change of a ten pound note for a purchase of items totalling three pounds thirteen shillings and seven pence halfpenny, all calculations scratched out on a piece of butcher's paper with a pencil stub?

This reminds me of the local butcher's shop, which with great fanfare installed a brand new cash register. The kids stood around in awe of this mysterious machine that appeared there one day on the shop counter. It was built to look imposing with its intricate cast iron lace all painted in shining silver finish. Many of us would hang around in the shop when it first appeared, scuffing the sawdust on the floor, and waiting for the next cash transaction to take place. First the butcher would bash a series of keys on the machine to total up the sum. This caused little white flags to pop up along the top with the money amounts printed on them. Having totalled the cost to everyone's satisfaction, he grasped a long handle at the side and pulled it down smartly. With the ringing of a bell somewhere in its innards, the machine went 'KACHUNG' and the cash drawer flew open so he could put in the money and count out the change. Not being used to this new-fangled machine initially the butcher kept getting struck in the stomach by the ejecting drawer. His wife, being smaller, got hit in the chest. I noticed that they soon learned to jump back when they pulled the handle.

Fine measurement in my engineering trade was learned using the inch as a unit which was divided up – not into lovely tenths – but into halves, quarters, eighths, sixteenths, thirty seconds, sixty fourths and one hundred and twenty eighths! But remember – there were no electronic pocket calculators, no automatic digital tills in the shops, no ATMs and no EFTPOS.

It is my contention that the kids of the 1940s and before deserve a medal for understanding and being able to use the old systems. There would be many medals awarded to acknowledge their overtaxed brains.

New toys for the youngsters were pretty hard to come by especially in the years around the beginning of the Second World War. We decided that we would have to make our own. A butter box, some scraps of wood, a piece of rope and four pram wheels from the council tip were all you needed to make a good Ginger Meggs hill trolley. Tin canoes, clubhouses, bows and arrows, cranes and trucks were manufactured from scraps – the list was endless. There was as much fun in the making of these as there was in using them. My first bike had its origins in the sum of its parts all scrounged from rubbish dumps. We didn't have to keep up with the Joneses as they were doing a similar thing. We pleaded with Dad to buy us two inner tubes for our bike. He saw the advantages in being able to inflate the tyres. We had them initially stuffed with grass.

We made a range of small toys too for the younger siblings using scraps and unwanted items. They had great fun with cotton reel tractors powered by elastic bands and the jumping beans made from aluminium foil and ball bearings. You didn't have to wind them up the batteries didn't go flat and they could go forever. If there was a little dumb kid hanging around with us, we would drag an old truck tyre to the top of a steep hill and with a bit of cajoling and promises of great rewards he would be convinced to ride the tyre. It sometimes isn't easy to coil up a kid and stuff him into a tyre but when you eventually succeeded, the results proved to be worthwhile. If the hill is steep enough and long enough the kid becomes a blur as he bowls along at high speed, bouncing over obstacles and ending up God-knows-where at the bottom.

It was my first and only firearm offence and was committed in the classroom of grade five. A 'lacky gun' is made from wood and fires the common elastic band at acceptable velocities. Best bands were those Dad brought home from the bank. Our targets in the form of sticks of school chalk, we had standing up in a row on the teacher's table at the front of the room. The objective was to knock these over from ever-increasing ranges. I was totally engrossed in taking a bead from the fourth desk - a difficult shot, when the large hand of Mr. Logan closed on my neck. I honestly had not heard the bell ring to begin the lesson.

Dad brought two of his own boyhood treasures into our young lives. A very large and elaborate Meccano set in a box with its hundreds of nuts and bolts and links and running gear kept us occupied for hours. His most cherished possession, which dated back to 1905, was his stationary hot air engine. Crafted to replicate the great industrial stationary engines of the time it had a big brass cylinder, gleaming connecting rods, a big flywheel and miniature handrails. Joe and I had our first lessons in the proper care of fine machinery as we went through the routines of cleaning and oiling its working parts. I still recall those winter nights sprawled on the carpet in front of the lounge room fire when the engine was brought out and started up. Today it rests on the shelf here in my office. Almost a hundred years from the time my dad's eyes grew wide at the first sight of it, it still has similar appeal for my young grandsons today. It would be nice to think that I will be there when they show the old engine to their sons too.

Purely by accident I found myself in East Fremantle recently (the year 2003) and curiosity took me through the gates of the Richmond State School. A gardener was tending some rose bushes bordering the curved path up to the school buildings. The rose stems were thick and gnarled with age I could see, though they still blossomed profusely. My mind travelled back over the years.

"Did you know", I asked the gardener who was a lad of about forty five or so, "that these rose bushes were planted in about 1933, around seventy years ago?"

"I reckoned they were pretty old," he said, "probably put in before my time. How do you know this?" he asked.

"Because I was about six or seven when I first attended this school and I and my schoolmates dug the holes and stuck them in."

It was a hot day and he seemed interested so we sat on a bench in the shade of an old Jacaranda. I told him of the day when as a class project we were given shovels to dig the ditches and dump the manure and plant the small bushes. He said that hearing this story had made his day.

The holy day of the year to all small boys – and large ones too, was the fifth of November Guy Fawkes Day celebrations. Burned into my memory by rocket trails and coloured fire it signified the ultimate in excitement and fun. I have to admit to having a passion for all things pyrotechnic. I still do. There were 364 days in every year for me. That is the number of days after cracker night one year until cracker night the next.

Guy Fawkes Day was usually a big family get-together with grans and aunts and uncles and cousins and friends descending on us just before dark. They brought lots of food and drinks and depending on their means, substantial quantities of fireworks.

In the weeks leading up to November the fifth we begged for extra pocket money to build up our own stash of fireworks that we had chosen carefully down at Stammers' local grocery shop. Five shillings bought a large Weeties box full which I could only just manage to carry home from the shop one hot afternoon. Our hoard would be spread out on the floor at home and counted and gloated over and counted again until we practically wore them out. The crackers ranged in size from tom thumbs in strings, increasing in size and ferocity up to the big bangers and deafening cane bombs. These latter were renowned among the juvenile tribes of the suburbs to be just the shot for demolishing a standard letterbox. That was what I was led to believe anyway.

Gangs of barelegged boys scoured the bush for bonfire fuel in the preceding weeks. Dragging home enormous leafy branches they stacked them tepee-style to heights of twenty feet and more. These imposing pyres dedicated to Mr. Fawkes himself rose in many of the vacant blocks and bigger back yards around the suburbs. Effigies of the guy who tried to blow up the British houses of parliament were fashioned from old clothing, given life-like faces and stuffed with straw. Lashed to the top of the centre pole of the bonfire, Guy hung there with lolling head, seeming to gaze sadly at his ring of childish tormentors while stoically awaiting his fiery fate.

If you had a really impressive 'bonnie' built, the days leading up to the big night were fraught with worry. Gangs of kids with devilment and destruction in mind roamed the streets and would prematurely put to the torch any bonfire left unattended for long. Many of us organised security teams with military precision for round-the-clock guard duty to fend off these urban terrorists. I remember helping to build many excellent bonfires. We never lost one.

When the sun went down on The Day we were poised for action. On pain of banishment no one was to strike a match until Dad gave the order.

"It has to be really dark," he declared, "to get the maximum effect."

My father had prepared foot-long pieces of hemp rope soaked in saltpetre solution and dried in the sun. These smouldered all night, were ideal for lighting fuses and everybody had one.

The family fireworks kaleidoscope still burns bright in my memory.

There was the whoosh of rockets leaving fiery trails into the night sky and exploding into multi-coloured stars. Big bangers were exploded hanging from the clothesline - for safety reasons and the whole yard was soon a riot of noise, coloured light and drifting smoke. Kids weaved in and out of the crowd trailing sparks and shouting with excitement at every new explosion or burst of coloured fire. There were strings of crackers firing volleys, jumping jacks leaping and banging through screaming gaggles of girl cousins, catherine wheels and roman candles and Mount Vesuvius' too. Not the least enduring of all to me in those memories was the intoxicating smell of burnt gunpowder.

Guy Fawkes atop our bonfire paid the ultimate penalty when he was engulfed in swirling flame and showers of sparks. We howled with delight when the big banger secreted in his trouser pocket exploded and blew him into smoking fragments.

Our supplies of ammunition were exhausted all too soon and then the adults drifted back into the house for a cuppa and a chat. To extend our enjoyment further we kids played out another of the traditions.

We sat in a circle in the dirt around the glowing pile of the bonfire's remains. Long pieces of wire filched from someone's fence came into play combing the hot coals for gastronomic treasures. Big fat potatoes had been pushed into the fringes of the fire before it was lit and the search now began to retrieve them. Black and smoking and covered in ash, the spuds were rolled out of the embers and split open to expose the creamy steaming insides. We reckoned there could be nothing hotter on earth than a coal-fired potato as they were tossed from one hand to another with howls of pain while we blew on them to cool them a little and to dislodge the ash and grime. Happiness, as we all know, comes in many forms. A corroboree of chattering children stuffing steaming potato into grimy faces with dancing eyes in a fire's glow must surely rank among the most joyful of all scenes.

Schooldays

The colloquial term for the port city where we lived during my early school years was 'Freo.' We always referred to it as Freo and when you spoke of your hometown and called it Freo; everyone knew what you were talking about. It's hard to remember the exact day it was when somebody pointed out to me that the name was in fact 'Fremantle.' This didn't really change anything, but at least now I was able to find it on a map.

As an active, growing barefooted boy in the years of primary school in East Fremantle, I would have to rate at the top of my favourite activities – eating.

All regular meals at home were a collective family effort and we were all most helpful both in setting everything up for the table and then cleaning up afterwards. The big table was attended by our family members, all at the same time and was the core - the epicentre of our family life. Besides eating in a most orderly fashion, we communicated with each other. We talked, we joked, we cackled with unrestrained mirth when Dad told one of his outrageous stories and pulled funny faces. The meal table was where we discussed yesterday and today and planned tomorrow. As a child I was imbued with a feeling of caring and being cared for and of belonging.

There were rules on how to behave however, which were passed down by our parents, as their parents before them had handed them on. Correct behaviour in company and consideration for others were paramount in our house and retribution was swift if we jumped over the traces.

With the meal in front of you, no matter what your rate of salivation might be, you did not pick up the eating utensils until your parents had begun. The fork was in the left hand – prongs down except for green peas. Cut what you required into small portions with no slashing, mashing or stirring. Eat with the mouth shut with definitely no slurping. Elbows must never come to rest on the table. Before taking food from the big serving plates, always pass these to others, taking yours last. Take the portion nearest you and definitely not the biggest. No sorting of, picking through, or rooting around in the dish for the choice bits was allowed either.

Normally, you would speak when spoken to, although youthful exuberance often prevailed – and it was never with your mouth full. Also any child forgetting 'please' and 'thank you' would invite a reprimand.

Our youthful appetites frequently prompted the request for second helpings. Mum would click her tongue in mock disapproval while Dad rolled his eyes in feigned horror. I don't recall ever refusing to eat a meal because I didn't like it. Food fads had not yet been invented. The serving dish was always scraped clean at the finish and we vied for this privilege every time.

At the meal's end, the eating utensils would be placed parallel on the plate – never crossed. You asked permission to leave the table too.

The standard meals at the allotted times were seldom sufficient to keep our schoolboy bellies content. After-school snacks were eagerly looked forward to. My favourite on Mondays was a thick slice of brown bread with an equally thick layer of beef dripping scraped on which had come from the Sunday roast. The first one home got to dig deep into the dripping pot to excavate the thick brown meaty gravy in the bottom. Pepper and salt on top made it a heavenly feed. An odd favourite of mine when the dripping ran out was cold porridge. Cut into wedges from the congealed layer sometimes left in the bottom of the saucepan, it helped to fill the empty spaces until teatime. With no dripping and no porridge available, our slice of bread was coated with good old black Treacle or Golden Syrup.

Grandma Macmillan had some funny ideas about what kids could or could not eat. She made a particularly scrumptious boiled fruit cake, with loads of all kinds of fruit and cherries in it too.

My salivary glands would go into overdrive at the sight of this stuff as she cut it into portions. Then she did a remarkable thing, -

"This cake is too rich for children," she declared. Then she layered the delectable stuff between two slices of bread and handed it to us like a cake sandwich. When we complained - in subdued tones of course, she would tell us again her name for this strange combination. For some obscure reason known only to old Welsh grandmothers she called it 'Matrimony.' I was puzzled then how she could come up with that one - I still am.

There were special requirements for good behaviour for the boys in the family. Compliance with these invited a pat on the head from Dad, or a neighbour remarking what a nice boy you were. Females of all age and station were always deferred to, especially if they were family members. We were taught to offer our seat in a bus or tram to females, especially the elderly and to stand in the aisle. Slipping ahead of Aunt Charlotte and opening a door for her, or escorting her across a busy street and other courtesies were what we did and I believe the ladies were appreciative. The fact that many of the origins of these civilised customs were destined to be buried in antiquity and considered no longer relevant just didn't occur to us. We accepted our responsibilities with aplomb.

My class at Richmond school East Fremantle - I am on extreme right, back row

There was a right side and a wrong side on which to walk with a lady, depending on the circumstances. When on a footpath you should stay at her side nearest the roadway. This is so that you cop the geysers of mud thrown up by passing horse – and horseless carriages. In open areas you always took the right side. This was designed to leave your right arm free to draw your sword and defend the lady's honour. I wondered at the confusion that might have resulted with this rule in the old days in the case of a left-handed swordsman.

Jokes aside, we were taught to care for, to protect and even revere all members of the fair sex and I feel that someone is missing out in the changes which are taking place in these later years.

We lads usually possessed three sets of clothing. Best pressed was hung in the wardrobe for church, Sunday school, public appearances and visiting. School togs were a uniform of some sort, especially at high school where pull-up socks, stout shoes, a blazer and a ridiculous cap were the thing. Our knock-about clothes on the other hand were for activities like beachcombing, rock and tree climbing, running errands and gang warfare. These were frequently just hand-me-downs from your father or perhaps from a bigger brother. For reasons of economy, the eldest boy was usually favoured with the new clothes. When these became too tight or too short for him, they were passed to the next boy in line and on to the next and the next in the series. This scheme only worked if the clothes were of such stout materials as to withstand hard service with multiple owners. As the eldest, I recall wearing shirts starched to the texture of ship's canvas, pants with a crease to cut your finger on and wool jumpers, which scratched bare skin like barbed wire.

The populous Murphy family down the road solved their clothing shortage crisis by taking down the window curtains and cutting and sewing a fine range of shorts and shirts and also pretty dresses for the girls. A major advantage of this was the ease with which a family member could be located in a crowd - by the characteristic floral pattern on the clothes. I don't recall one of those kids ever getting lost. The windows were bare for a while but the kids looked decent.

All kids' shoes were 'sensible.' They were either black or brown with leather uppers and leather lowers, all forged from the stoutest hides, with laces of course. Soles frequently worn down paper thin from stunts like skiing behind a pushbike on the bitumen road, kicking cans and other fun would soon be evident to the wearer when the sock poked through. This was no deterrent to Dad who would say that there was plenty of life left in them yet. He would grab the shoes, his hammer and the cast iron boot last, stick a row of tacks in his mouth and go to work installing a new sole. This would be cut from super grade greenhide leather destined for further extreme service in the line of the succession of sons.

I literally hobbled one day on the five-mile hike to school and back in my newly re-soled footwear. Dad mumbled his apologies later and admitted that he had attached the sole with the rows of razor-sharp cut tacks, but had forgotten to hammer down and clinch the points on the inside. It was an amusing thought that an Indian fakir, lying on his bed of nails, was relatively comfortable by comparison.

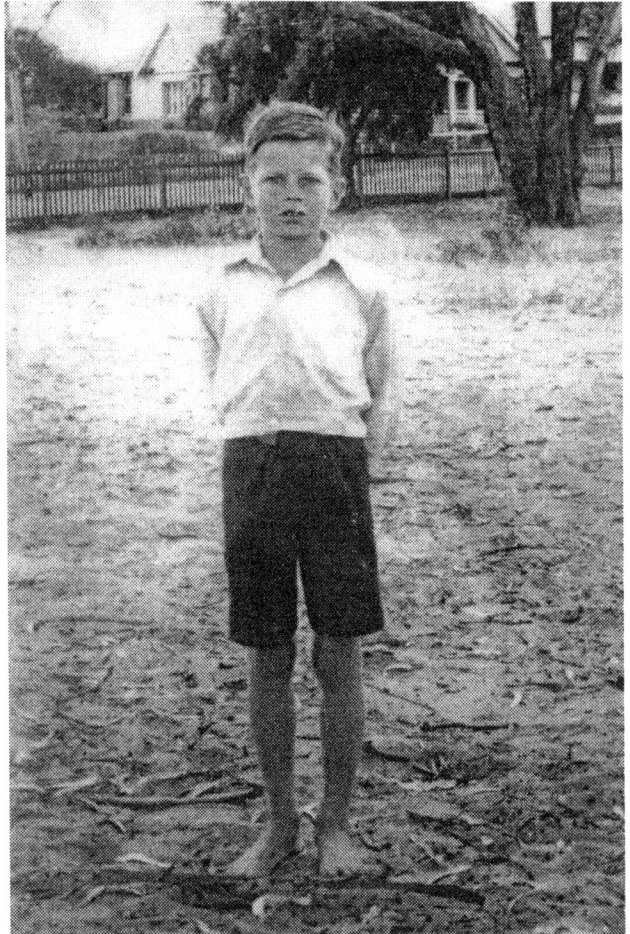

I must have been in first standard at this stage. I did have shoes but they were left at home

39

The war brought shortages of almost all of the family necessities including clothing. If a shirt was ripped, it was sewn up again, or if the seat of your pants was worn through it received a patch. If a button was missing, you sewed it on yourself. Socks were always darned to fill the holes and it seemed that most of Mum's resting time was spent with the sewing machine, the darning mushroom or the knitting needles.

Households of large families created a lot of manual work for everybody and large families were the norm, with several of our neighbours having ten and twelve kids apiece. Dad was the sole breadwinner of course while the burden of bearing and raising the youngsters and looking after them involved heavy manual housework for the wife and mother.

If you are a Sandgroper – to wit a West Australian, you will have had a wealth of experience with buzzing flies. As young boys we waged a protracted war against them. Dad used to say that the only good fly was a dead one. Our defensive weapons against the hordes were crude but provided some interest and activity and skills development for the kids and a pat on the head from the parents if the scores were high.

Usually seen above the kitchen table in almost every home was a descending spiral of sticky flypaper. Pulled out of a little cylinder with a hook on it for hanging on the ceiling light, it was about two feet long, silent and deadly to flies. If a fly landed on its surface it was there with no escape and stayed forever. It was one of my jobs to renew the flypaper when it appeared to be getting overcrowded. It was not a pretty sight with the carcasses of a thousand flies suspended above your lunch, but better there than on the lunch. We had fly sprayers too of the manually operated kind. Ours was painted bright red, which I assumed was a colour feared by flies, and it had a big cylinder pump and handle with a tank underneath. The targets seemed to zoom unconcernedly through the smelly plume of Fly Tox as everyone in the house coughed and sneezed until it had all settled on the furniture. Some experimentation with this contrivance did reveal a more important use for it – it made an excellent flamethrower with the application of a lighted match. Getting right down to basics however was seen to be in the use of a fly swatter. Dad always kept the weapon handy especially at mealtimes. With a swift and very accurate backhand smash he would leave another fly signature on the wall. Sometimes there were two signatures – a double-header no less. With few academic achievements behind him, we had a student at school known by all as Mad Syd. His reputation for catching flies barehanded however was the stuff of legends. With amazing sleight of hand and lightning speed he could capture flies in flight with very high average scores. I paid Syd one day with the going medium of exchange to reveal the secret of his success.

"It's easy," explained Syd, while still keeping an eye on a big blowie droning past, - "You gotta remember that when a fly takes off he jumps backwards. If you allow for that you'll get 'em every time." I remember now, it was my Granny Smith lunch apple I had paid him but it worth every penny.

Monday was a particularly demanding day for Mum. Monday was washday. Our washhouse was an open-sided asbestos-sheeted shed in the back yard. Clothes were boiled in a wood-fired copper and then rinsed by hand in rough concrete troughs. A big ornate hand-cranked wringer squeezed out the water between wooden rollers, after which, the clothes were then pegged out on lines to dry in the breeze. The lines were strung on high crossbars at opposite ends of the yard while a clothes prop held up the sag in the middle. These handy accessories were simply a long thin sapling with a fork at the top end and were sold to householders by bands of wandering aborigines. They had obviously seen the market opportunity in the demand for these essential items. For erasing stubborn stains the sopping clothes were dumped on a corrugated glass washboard wedged into the trough, lathered up with Signal Soap, then rubbed and scrubbed and pounded until they were clean.

Many were the times I saw Mum on a summer day, with the sleeves of her cotton print dress pushed up above her elbows, her hair straggling across a hot sweaty face, turning the handle of the big wringer, or struggling to carry the heavy iron tubs out to the clotheslines. It worried me that she had to work so hard and so long and though still small, I promised that I would help next time she was overcome with work. I couldn't wait for the day when I would be big enough to reach the handle of the wringer and strong enough to carry the tubs for her.

Alas she was gone before I was twelve years old.

Another laborious job was scrubbing and polishing the floors. In between cooking at the hot wood stove, she insisted that the linoleum and the jarrah wood floors must show a beautiful shine at all times, and was often seen on hands and knees with the Relax polish and the rags. Joe and I aided her by skidding the full length of the passageway on bums on mats to heighten the shine. While this may not have added to the lustre, it certainly was fun.

We grew up in an age of friendly door-to-door deliveries and many of our needs were available at the gate. Bread and meat and fish and fruit together with magical remedies from the Rawleighs man were all offered at the front door. Dad said that he was a snake oil salesman, but I never saw any of this stuff among the bottles in his Globite case. The milko plopped his little dipper into his big churn and filled the bright tin billycan left at our door with fresh frothing creamy milk in the early hours of the morning. The milk money left out for him would be scooped up into his leather waist bag and he would jog to his next customer or back to his cart for a refill. Most of the deliveries were from horse-drawn carts with colourful and elaborate sign written displays on the sides. The faithful old horse would plod along the road keeping pace with the driver and it was said the horse knew the route better than the milkman did.

In the food-keeping department, I remember when we graduated from the wet hessian Coolgardie Safe on the back veranda to a brand new ice chest. It featured varnished imitation oak finish with chrome handles on the door and the top. This worked better to stop the butter and jelly from melting so quickly and also provided some entertainment for the youngsters. The iceman was a neighbourhood favourite and delivered every Friday. The ice was in big heavy blocks in his cart and had to be cut to size to fit in your ice chest. He pulled the long gleaming slabs from his cart and chipped around them with his ice pick to get the right size. This created a veritable shower of shards of ice, which were scavenged by the circle of chattering kids and borne off in triumph to the shade of our big veranda. A decent sized ice chip could be sucked for an hour, dribbled water down your chin and turned your hands blue with cold.

There were postal deliveries twice a day in our neighbourhood too. Letters were the main form of communication with friends and relatives and parents always encouraged us in letter writing. Our postman wore a navy uniform with a red stripe and a peaked cap with a red band. There was a whistle hung around his neck and he rode a big, bright red postie bike with a leather bag on the front. The shrill sound of the postie's whistle would see all the kids erupt from the front door, all vying to see who would carry the mail proudly back to the house. There were nice letters from our relatives for Mum and only bills for Dad – at least that's how Dad always described it. If you were really friendly with the postman, or alternately he was by nature a friendly postman, you could give him letters you had just written and he would guarantee to put them in the mail for you. You had to lick and stick on your own twopenny stamp though.

It was my job to pull a hand trolley up and down the streets and with my shovel scoop up manure left behind by the delivery horses. This was used to nourish Mum's vegetable garden by the kitchen door out the back. The big, yellow steaming dollops were most prolific from the dunny man's horse, drawing the wagon loaded with sanitary pans. His was the biggest cart, with the heaviest load, and being the biggest horse, helped our garden the most. This horse was a most prolific and explosive farter too, which was very entertaining for us, but was only natural when you think about it. I had to be up early in the morning to retrieve these choice offerings as he usually came around during the night and I had to beat the kid from across the road.

As I was a boy with a father who could work skilfully with his hands and had taught his sons the use of tools, Mr Logan put me in charge of a special class project. It sounded interesting and there was a chance of missing a few tedious lessons if we played our cards right. There was a long piece of Oregon pine resting on trestles in the school yard, which he said needed dressing down from its rough form to serve as an elegant flag pole for the next Anzac day which was not far off. Gathering a half dozen of my schoolmates, armed with a big hand plane each, we set to with a will to round and smooth the pole. With the piny smell of the timber in our nostrils and the lovely long curl of the wood shavings peeling off, we were enjoying the work immensely.

While not able to recall the length of time we laboured, I do remember vividly the look on Mr. Logan's face when he came to inspect our handiwork some hours later. I can imagine his feelings when surveying the sight of six sweaty, grinning students standing almost knee-deep in wood shavings. The precious piece of prime timber was reduced so much in diameter it would have qualified in the Guinness Book of Records as the biggest toothpick in the world. I doubt it would have had the strength to fly a handkerchief, much less a big Australian flag in a stiff southerly. The lecture he gave us six bow-headed, shuffling boys had something to do with when you are on a good thing you have to know when to stop.

The Fremantle Boys High School in the centre of Fremantle was about six miles from home. It was a forbidding old grey stone building with buttresses, heavy gates and a high, timeworn wall surrounding it. Broken glass was embedded in the top of the wall and its shining shards reminded us of the futility of trying to escape by that route had we wanted to do so.

It wasn't long before I discovered that the school rules and regulations were almost as forbidding as the building. The prison gates were locked when lessons started in the morning, so you had to get in via the headmaster's office if you were running late. If reported late more than once, the principal wielding the cane would give you a painful reminder that you should be more punctual in the future. This was getting the 'cuts,' though fortunately I was not on the receiving end very often. It is probably of interest in these more enlightened days to try and describe this experience. The cane itself was about four feet in length and of solid rattan. The victim stood with arm extended and palm up wards to receive his punishment. This was an ordeal in itself as the schoolboy code forbad any flinching when the cane came whistling downwards. The sound effects were intimidating too, not only to the recipient, but also to anyone in the vicinity. The whistle and the crack were no doubt supposed to send a clear message to all others as a reminder of the power of the Principal. We listened for the howl of pain and anguish from the victim but there never was any as long as I can remember. The number of blows you received was in proportion to the gravity of the alleged offence. You might incur just one for say dropping paper in the schoolyard, while the dreaded sixer was meted out to anyone who challenged the authorities. I don't have to tell you that besides the humiliation (or notoriety) acquired from caning there was also much pain. The hand would burn and sting and swell and be very red for a couple of hours. It would be the left hand so you could still do your schoolwork. The devilish thing about it however was that in a couple of hours all traces of injury would be gone and with no forensic evidence visible, it was useless to complain to anyone, not that grown-ups would be sympathetic anyway. The likely response to the report to your parents was that you must have had it coming. Such is life for a schoolboy in 1937.

The journey into Fremantle along the Canning Highway each day was an exciting experience for this new boy and I soon found out that a certain amount of tram craft had to be learned. The electric tramway followed the highway from the city out to the suburb of Melville. The trams, which rattled and swayed along the tracks, were reminiscent of those depicted by cartoonists and seen in the comic strips. Most were bogey trams running on just four wheels and could be seen grinding up the steep Leopold Hill or bucketing back down on a reckless course on the return trip. They were a comical sight with the trailing arm running on the overhead power cable leaving showers of sparks, platform passengers clinging on for dear life and with the driver stamping on his bell pedal to warn off anyone foolish enough to stray across his path.

There was a particularly successful strategy employed by young boys mostly and it involved the avoidance of paying the tram fare. It was sixpence for the full trip, but sixpence was sixpence and I must admit to operating the scam on more than one occasion on the basis of thinking of all the other things that I could do with sixpence - like buying a cone of potato chips for instance. The plot began to unfold when the tram stopped to pick you up. Firstly, catch sight of the conductor on the tram - he will be working his way from one end to the other collecting fares from seated passengers. His expertise in snapping the tickets off with a grand flourish always intrigued me and I made a mental note to include that job on my list of potential career moves- but I digress.

If the conductor can be seen working his way forward, you swing up on the back platform where you can ride without sitting down. Further on in the journey he will turn at the front of the tram and work his way collecting fares towards the rear. It is at this point that you dropped off at the next convenient stop and ducking out of sight alongside the tram, ran to the front and climbed on there. I must admit to getting the greasy eyeball from the driver occasionally, but found that a sixpence held conspicuously in the hand usually branded you as a potential paying customer. Executed well and with studied nonchalance, I found that you should never have to confront the conductor and could be transported by courtesy of the Fremantle Tramways for a considerable number of trips.

The Second World War was in its early stages in 1940 and was casting a pall over all our lives. Many young men and women had enlisted and were already being shipped overseas. With typical old school pride and patriotism the headmaster had posted a list of all those students and ex-students who had answered the call to arms. Frequent topics in his address to the morning assembly of boys in the big hall were the current state of the conflict in Europe and any news of our boys over there.

The school was buzzing with the news that one of the ex-students who had been flying Spitfires over England was back home and would be talking to us the next day.

He was nineteen or twenty years old and appeared on the dais resplendent in his blue airman's uniform with the flyers' wings on his chest. He also had a bad limp, which he explained by saying it was the result of a bullet wound from a Luftwaffe Messerschmitt fighter he had engaged over the English Channel. He looked so young and yet, in a way quite old and worldly – and weary.

He spoke earnestly and well on the need to display courage and grit and determination in sports and in studies and in fact all of the challenges in our lives. He painted a word picture of the escapades and the daring and the bravery of the flying aces and also gave us glimpses of the dangers and horror in his war. At that point the war, which was to me only a set of black headlines in the morning papers, suddenly became grim reality. I could feel the surge of patriotic fervour through the assembly of boys after this glimpse of the real world out there and then they played the national anthem. With the rest of my school I was caught up and carried along with the spirit of patriotism and determination to serve my country against the aggressors. As it transpired, five years later I turned eighteen. This was when I was eligible to enlist - the war had ended and peace descended a scant three months previously.

We were kept in mind of the toll in lives as the list of servicemen on the notice board in the big hall began to carry the notations of 'Missing,' 'Killed in Action,' or 'Wounded.' The infamous Battle of Britain, which sounded the defeat of German air supremacy, accounted for many of the casualties among the ranks of our Fremantle Boys School representatives. We listened to Winston Churchill's celebrated speech broadcast from England on our wireless one night. These were his famous words, -

"Never in the field of human conflict was so much owed by so many to so few." Those were very stirring and emotional times.

Together with many of the other new students at the beginning of our first year, I had to suffer the indignity and humiliation of the traditional initiation ceremony. We were dragged, still in our new school uniforms, kicking and struggling and flung with flailing arms and legs into a big horse trough. There were tadpoles in the green slime. There appeared to be no purpose in this practice, but it seemed that suffer we must, as it was essential that you were blooded, hopefully not literally, to show you could take it. The masters who were standing by I suspected were there to treat the casualties, and fortunately I was not among them. The stock answer to the question of why new boys were half drowned in the cold depths of the old horse trough every year, seemed to be that it was essential man-making stuff, and anyway, the trough had always been there and so had the tradition.

There was a special door set into the grim grey walls of the schoolyard. It looked very strong, was always locked, and I didn't discover its secrets for quite a time. It transpired that on the other side of that door was a community of extremely attractive, nubile maidens who were the students of our sister school known as Princess May High School. Actually, the big brass padlock was removed and the door opened wide every Tuesday at twelve noon. It signalled the highly anticipated event of our weekly dancing classes. Under the all-seeing eagle eyes of the schoolmistresses, we were singled out by the bravest of the senior girls to join them in the gyrations of the Waltz, the Barn Dance and the Valeta, all to the brassy notes of a big old gramophone in the corner.

The term awkward is a fair description of my performance, as I slipped and stumbled and trod all over my partner's feet time and again - mumbling apologies all the while. In retrospect I must have been suffering from a testosterone overload. The sudden close proximity of a well-formed female person trying to steer me through the dance steps completely unnerved me. I was dizzy from the fresh scrubbed smell of her body and the perfume from her thick curly hair just under my chin. A great bumbling oaf, I was struck dumb with awe and clutched my patient partner with sweaty palms and made a complete mess of everything. Due to these most harrowing conditions, there was little learned in the way of ballroom dancing technique at those times.

We were allowed to walk into town sometimes in the lunch period. Feeling conspicuous in my neat school uniform, I was strolling, munching on a serve of potato scallops - purchased with my deferred tram fare - from the fish shop near the waterfront. A rather interesting scenario was unfolding across the street. There were a lot of coin-operated machines in public places around Fremantle which all accepted a penny in the slot for whatever it was they dispensed. Some of the young local lairs would shave down a leather belt to the exact thickness and width of a penny so they could coax these machines to give them free everything.

I was just in time to see this lad whip off his belt and holding up his slack pants with one hand, feed the belt into the coin slot of a gumball machine. It appeared to work, as the brightly coloured balls poured out and bounced all over the pavement. He pulled on his belt to retrieve it, but it jammed in the slot. Several things were happening simultaneously – his belt stuck, he was rolling around on the escaped gumballs, his pants had fallen down and a policeman had appeared on his beat from around the corner. The ensuing chase and capture rivalled the best seen in the Mack Sennet comedies. There seemed to be little hope of any mitigating circumstances for his ensuing defence case, given the policeman's account of that funny scene to the magistrate.

44

My Invalid Sentence

We were a fairly normal and healthy family - Joe, Gwen and me. While suffering the usual gamut of mumps, measles and other childhood ailments, in the main we gave our parents little cause for real concern.

One exception to this in my case was the painful and embarrassing condition of numerous boils on my person. For reasons of diet, hygiene or other causes possibly unrecognised in those times, boils were a rather common affliction among the younger generation. I was prone to them and they would appear without warning almost anywhere on me. Carrying a particularly large and angry specimen on my bum I found it necessary to find the right book to help with my condition. This was not a medical reference to assist with treatment, but one large enough to sit on. Sliding the book under the unaffected cheek ensured that the boil would not contact the chair when I sat down. There was even a crop of the rotten things on my head all at one time. These were particularly unsightly as due to the presence of a hard skull underneath, they only had one way to grow – which was outwards. I imagined myself to look something like a cross between Old Nick himself and a mature African warthog, such was the effect on my self-esteem – and that was not taking the pain into account!

While living in East Fremantle at about six years of age, I experienced some pain in my leg muscles. The pain was real enough though not debilitating, but I did notice a particular side effect this had. Complaints to my parents attracted generous amounts of sympathy and attention – a most desirable condition. The verdict was that 'Ivan has growing pains but these will pass.' To be truthful I recall using this ploy in times of need, though not frequently enough to cause suspicion. Then something serious was detected.

During the course of a routine medical check, our family doctor dropped a bombshell. Her tests on me indicated, she said, that there was evidence of rheumatic fever and an enlarged heart. My parents were really worried at this news, while I waited to see if I could make any sympathy mileage out of this one.

No, this was not to be taken lightly.

"In order to manage his condition," our doctor said, "Ivan will have to observe a strict set of rules in future." I did not like the sound of this.

"From now on the boy will not be allowed to exert himself. Overstressing his heart could result in dire consequences." Coming from a doctor, this had to be dinkum – doctors are always right aren't they?

As the explanation of my problem and the means to manage it began to sink in, I felt an increasing feeling of dread. No running, jumping, or climbing trees were to be allowed. There will be no competitive sport for me any more. My life was surely going to change - for the worst!

Months passed by in a dreary existence of enforced inactivity. I was doing my best to conform to the new restrictions. There seemed to be little joy in life as I sadly trailed behind the stream of kids pouring out of the school, or watched from the sidelines at the cricket pitch. I felt myself going down hill fast - not in a hill trolley any more – they were banned too!

My general state of dejection and lack of healthy exercise was taking its toll. I became sluggish and overweight. There were no signs yet of my imminent demise but my life was at rock bottom. I felt that I might as well be dead anyway.

"Look at him" said my mother, "the boy needs a change. I think we will send him up to Beverley for a while."

The sheep and wheat town of Beverley was the home of my Auntie Olga. She was a wonderfully kind laughing lady who married Dr. R. P. (Doc - Percy) Hodgson and settled there to develop a thriving medical practice. He was a fine doctor and physician, highly respected in the community and they lived in a large rambling house named 'The Cottage', which had a long gravel drive, lined on both sides by tall sugar gums. Previous visits to Beverley had earned it a place in my heart as the town where boys' dreams are answered. Green farm paddocks swept right up to the lower boundary. All was wide-open spaces with fresh air and creeks for swimming in and lots of rabbits to be hunted. One sound I will always remember in that wonderful place was the calls of whip-birds in the gums.

Doc Hodgson drove a large American Buick sedan in keeping with his important station in the community. He took me on many of his house calls to outlying farm families. A quality he possessed almost above all else in my eyes was his love of fine English shotguns. He had two double-barrelled field guns; one of which he told me had been on order for twelve months in London, being built to his own personal specifications. One was a Boswell and the other a Greener. Doc soon realized he had a kindred spirit in me even at the tender age of eleven and tested me with the heavy kick of firing a twelve gauge. I passed the test and from then on we became firm hunting compatriots.

Doc and Auntie Olga drove the ninety miles to Perth to pick me up for my sojourn in the country. While cruising on the winding road back to Beverley he seemed delighted to accelerate the big car up to ninety miles an hour on the straight stretches. I had begun to feel better already.

Even after I arrived in this exciting country town with all the temptations and opportunities to cut loose, the adults' fingers still wagged at me and the restrictions on my physical activities still persisted.

'Doc' seemed to regard my alleged medical condition with some scepticism and treated it with a certain studied indifference. I felt that were I to seek sympathy from him, none would be forthcoming. Perhaps professional ethics prevented him from becoming involved in another doctor's case – as a doctor who should know all about these things I reckoned he seemed to be biding his time.

Soon after, I was packed off to my Cousin Mary's farm where it was said the country lifestyle would put me back on the road to good health. While feeling an even more urgent desire to break out of this cocoon of care, I was however cosseted and pampered and kept warm and dry with few, if any changes. As far as I was concerned, I still resembled a big white slug. This condition was exacerbated further by a diet which included lots of fresh cow's milk, farm butter and cream.

My mother wrote to Auntie Olga from Fremantle during my stay in Beverley. That letter I have copied verbatim into this episode under the title 'Treasured Letters.'

Back in Beverley at The Cottage, Doc sat with me on a log in the sun one afternoon. He plied me with questions as to how I was feeling, so I unloaded on him my tale of misery.

"Come down to the horse yards with me" he offered, "You can give me a hand if you like." Doc owned a number of racehorses which were kept in the back paddock.

We stood at the railings and he pulled a bundle of hay from the small shed and threw it down at my feet.

"Get under the rail here and into the yard" he said. "I'll pass you the hay and you can carry it over to the feed box on the other side."

Being a city kid I was not familiar with horses. They seemed pretty friendly animals albeit being very big. Staggering across the yard with the armload of feed I failed to notice the big stallion lazing in the corner. His interest was instantly aroused as he spotted what appeared to be his lunch moving away from him and he set off in pursuit of it. The thud of his hooves made me aware of this perceived danger bearing down on me, and the instinct for self-preservation kicked in. Hurling the hay in the general direction of the perplexed animal, I forgot my long-standing restrictions, sprinted wildly for the opposite fence and dived under the rail. Panting hard I sat propped against a post and within seconds my uncle was there dragging a stethoscope from his pocket as he squatted down.

"Lie down and be quiet" - he ordered.

Pulling open my shirt he began a thorough examination of my chest, nodding and grunting as he did so. I went along with this unexpected turn of events. Resistance never occurred to me – I had after all just sprinted fifty yards and had miraculously not died as it had been predicted I would by the first doctor.

Doc's first examination was concluded with a few more nods and grunts. He seemed pleased with his findings.

"Not a word of this to anyone!" he said swearing me to secrecy.

"We will come down here every morning for some more tests - no promises you understand – and don't blab!"

It seemed that someone was on my side at last and understood my problem - and a doctor to boot. Things were certainly looking up.

Two very exciting weeks passed. Doc had me running around outside the fence of the horse yards every morning, while increasing the distance each time. Always at the end of the run he stretched me flat on the ground and did his bit with the stethoscope. Still sworn to secrecy I felt smug in the confidences we shared and I was also feeling a whole lot healthier.

It had been a very cold night. I ran that next morning in bare feet through the crackling frosty grass up the length of the paddock, through the creek and up to the tree line and back. Sucking in the icy air I felt my lungs expand enormously and my heart thumped at a furious rate as our final test was completed. My feet were numb with cold but I was overawed. Doc was a man of few words –

"You are as sound as a bell Ivan and there's not a damn thing wrong with you!"

The difficulty I had speaking in reply, was not just due to my recent exertions over a mile of ploughed paddocks, but also because of overwhelming gratitude I felt towards my uncle for what he had just told me. No more ridicule would I suffer from my mates at school and no more wagging fingers of censure from the grownups. My invalid sentence was at an end, not as just a reprieve but I was to be a free man again.

Doc and I made a joint announcement at the next family meeting around the big dining table at The Cottage. The men congratulated me while my mum and Auntie Olga had a bit of a cry. I almost choked with pleasure and embarrassment.

Grandpa Joe Macmillan squinted through his cloud of pipe smoke.

"So that's what's been going on. I reckoned the boy was looking better every day and pretty pleased with himself. This is great news."

As follow-on to this story, I am happy to report that not a trace of any heart problem has appeared with me since those heady days. (Touch wood) Doctors aren't always right are they? – Well some doctors aren't anyway.

Grandpa's Castle

We were living at our grandparents' house in Victoria Park while Mum was desperately ill and unable to take care of us. Dad had bundled us into the old Fiat tourer with our few prized possessions and dumped the bewildered trio in Grandpa's castle. We knew our grandparents well as we had visited regularly since we were little. Both the old folks welcomed us and reassured us that we had a home there as long as we needed it

Grandpa's house at number twenty-five McMaster Street in Victoria Park was a source of constant intrigue and interest to us. The old fellow had built the place himself, as he was a

My paternal grandparents' house -'Grandpa's Castle'

skilled carpenter and cabinetmaker. Clad in weatherboard timber with an iron roof, it followed the pattern of many of the houses of that era. It squatted on a steeply sloping block, which I imagined, the first time I saw it, to give it the appearance of having its chin on the ground at the front and its bum in the air at the back like a playful dog.

Inside, the house was rich with deep red tapestries, lamp shades, table covers, doyleys and rugs, like a house from Victorian times. Grandpa would give in to my badgering at every visit and lift me on his shoulders to peer at the pictures hung high on the walls in the entrance hallway. I could barely suppress a shudder at the one depicting a bloodied peregrine falcon with a very dead pigeon he had pinned with fierce talons to a rock high on some distant mountain crag. Flanking the ornate carved jarrah hallstand draped in coats and umbrellas were the other two pictures I loved. One entitled The Thin Red Line showed defiant ranks of redcoats standing firm, shoulder-to-shoulder facing a cavalry charge. It was probably the soldiers of the army of the Iron Duke at Waterloo. Smoke drifted across the scene and cannon and musket flashes glinted on the rows of gleaming bayonets. Deeply engrossed, I imagined I could hear the faint cries of the wounded and the scream of injured horses. Pictures of the old ships under full sail usually stir the imagination of most young boys. Showing its red and green port and starboard lights reflecting on the ocean swells, the graceful clipper heeled gently in the shimmering track of the setting sun. The picture was entitled Homeward Bound. No visit to Grandpa's castle was complete without this ritual, perched high on his shoulders and I never tired of listening to the well worn stories he told us every time.

Grandma's kitchen stretched the width of the house at the back and seemed long enough to hold a game of cricket in. The big black-leaded range with two oven doors had a fire in it that never seemed to go out, and mouth-watering aromas puffed and bubbled from the cast iron pots on the top. The largest pot simmered incessantly. It was a mystery concoction and none of us knew exactly what it contained. All scraps of food went into it and that one didn't smell very appetising to me, though as Grandma said,

"It's for the chooks and they reckon it's bonzer."

The old man was retired from the 'fields', as he called the goldfields where they had lived for many years and where my Dad was born and raised. Grandpa had run a thriving building business in Norseman, and he was first registered there in 1903. Responsible for the construction of many of the houses and some of the public buildings in the town, I am sure many of them will still be standing long after we are gone.

He could often be found in his workshop under the rear of the house, which was a kids' wonderland. Interesting smells prickled our young noses where he worked. He made fine strong furniture from Oregon pine using traditional methods and materials. His glue pot bubbled and gave off a meaty pong.

"The glue is made from cows' hooves and horns," he explained. I wondered how they could get a horn or even a hoof into that little pot boiling away there on the bench. He made big water tanks with riveted and soldered joints. He heated the large copper soldering iron with a kerosene blowlamp which had a fat blue flame and roared so loud you couldn't hear yourself speak Wood shavings rustled ankle deep on the floor giving the place a fresh piney smell which blended with those of linseed oil, pure turpentine and the paint he mixed himself. Grandpa's tools were all of the hand driven variety – powered tools were unknown. Chisels and gouges, jackplanes, smoothing planes, rebate planes, handsaws and spokeshaves were all used in his gnarled hands with great skill. I loved to watch him finish off a work piece. After sanding it smooth, shavings from the floor were rubbed over the wood, the surface of which would come up like satin.

Grandpa had me really intrigued one day when I asked what he was going to be doing that day in his little workshop under the house. -

"I'm going to kill some spirits," he declared.

While I was pretty familiar with most of the processes and rituals he conducted down there, this was a new one on me. Somebody with a really vivid imagination might have thought that perhaps he was going to do away with a few ghosts lurking there among the house stumps. It was a mystery we certainly had to unravel, so I was down there with him in a flash. Into an old pickle jar he poured some evil looking yellow liquid that was smoking and gave off a really strong smell as well. With his tin snips he clipped some pieces of sheet iron into it, which caused it to froth and fume and give off a lot of gas.

"Don't strike any matches around here son – that's hydrogen gas." He always had an explanation for me and as this was man's work it was OK for him to divulge the details.

"I'm making my own flux for soldering," he explained. "The liquid is hydrochloric acid – spirits of salts they call it. That's a bit too powerful for soldering so I kill it."

Now it all made sense to me - especially when he explained that the final solution was zinc chloride or killed spirits. That was another useful lesson never to be forgotten.

To us the back yard was a land of discovery. We swarmed like chattering monkeys, red-stained all over with juice and chiacking each other in the branches of the big mulberry tree. Picking strawberries and pulling baby carrots and spring onions from the rich soil in the vegetable garden was beaut fun. Half of these we took to the kitchen and the rest went into our bellies. In our imagination the dense rattling bamboo thicket along the back fence was the haunt of dangerous tigers and leopards and panthers. It was only the big kids who ventured into the gloom down there.

Like a circle of gossiping magpies we would sit on the grass searching for 'puddens'. These were the sweet-tasting seedpods of the shiny green Guildford grass, which Grandma said she hated to see taking over her precious bits of lawn. Oh, and there was watercress, gathered from the fringe of the duck pond down in the corner of the back yard. We munched watercress sandwiches of thick white bread with real butter on sunny Sunday afternoons when our cousins were visiting.

A disappointment for Joe and me was the shortage of male company of our own ages in the family. We had no boy cousins, which meant there was no real competition in many of our games and contests and few common interests we could share with girls in our boys' world. This was partly compensated however by the girl cousins we did have who provided a captivating insight for us into the mysterious female side of a lot of things.

Sheila, or 'Dibs' as she was affectionately known, was a few years older than me and was a frequent visitor to the house in Victoria Park from her home in Beverley. We spent many happy times in Beverley over the years - but I digress. I must admit to being intrigued by her femininity. Dibs was artistic in almost every way and seemed so different and refined compared to us with our rough and tumble boys' lifestyle. She was musical – an accomplished pianist with a sweet voice and a love of music. She wrote poetry and lived in the world of fairies and make-believe and had fine artistic talents. The house was full of fairies in scrapbooks and on odd bits of paper drawn in fine detail with wings and wands and stars. Almost everywhere Dibs had been you would find fairies. She was very theatrical too and staged her own productions at the drop of a magic wand. These penny concerts were held in the shade of the old guava tree in the front yard, with the cast recruited from her cousins, her sister, friends and any passing neighbourhood kid who was unsuspecting enough to be roped in. Her impatience was evident when she tried to coerce Joe and me into roles in her productions. Sprawled on the floor reading the Boys Own Paper, or cleaning Grandpa's .22 rifle (whether it needed it or not) were obviously more important things for us than poncing around trying to be somebody you were not. Also, I didn't take kindly to being ordered around by girls. The Peter Pan role wasn't too bad though because he had a sword and fought battles with Hook, which we made so realistic and bloody that Dibs threw up her hands in disgust and everyone went home. Playing to an empty house is not much fun so we would go to hunt for tadpoles.

My ingrained attitude at the age of twelve was that girls in general were weaker than boys and a bit silly and giggled all the time. This prejudiced attitude was tested one day. While Dibs was older than me by a couple of years, I found myself resorting to the juvenile practice of teasing her mercilessly. She was very spirited though and often gave me back as much as I could dish out, but little did I suspect one day that she had reached her limits of tolerance.

Seated at the table in the long back kitchen I was doing a good imitation of studying, meanwhile keeping up a continuous verbal provocation upon her. Rather than come back at me as she normally did, she fell strangely silent in the bathroom at the end of the room. With no chance to sense danger, in the next second I was knocked off the chair to lie sprawled on the floor. My head seemed to explode and ached so badly and I could feel there was a big lump growing above my ear. Dibs was there in a second, helping me to my feet and apologising profusely and asking was I all right. She had no need to apologise. The missile she had launched was a large and very hard cake of Lifebuoy soap. The range would have been twenty yards or I am a liar, which means she had to allow for trajectory and a raft of other factors like throwing through a doorway to make the perfect strike on my head at that range. While she giggled her profound regrets, I looked at her in awe. She had suddenly assumed a quality of competence in my eyes, which was a complete shock to me and was obviously going to command considerable respect for her in the future. I don't know if any change in attitude towards Dibs was noticed – she didn't say, but from that day forwards I saw her in a totally different light – like an equal – or almost.

The cold bath water made us shiver and break out in a rash of goose pimples. As a regular ritual every Friday night - whether we needed it or not, we sat in the old tin bath and prayed for a bit of warmth. If there was a kettle on the stove we would try to steal it to take the chill off the bath water, but it is scant comfort when one of your feet is defrosted but your bum is still freezing. So you would just scrub your hide with the brush and a bar of Velvet laundry soap and get out quickly just before hypothermia set in. It wasn't that we had a preference for cold baths, but the grandparents were of the opinion that boys should be brought up tough and that was it. The regular nightly cleaning ritual on the other hand involved only hands, face and feet – essential if you don't want mud on the bed sheets. Brushing the teeth was something to be endured too and it was strictly enforced, but not with toothpaste – cooking salt hardens the gums and cleans better they said. I didn't know all of this of course being a young kid, but on reflection I would have given my eye teeth for a tube of Colgates.

Grandma's remedies were apparently inherited from her life back in old England. As active boys we suffered a variety of injuries and ailments so she was often called upon to put her remedies into practice. I soon formed the opinion that for a treatment to work, it either had to be painful or taste disgusting. Purgatives were administered every Saturday morning – like it or not. This was a ritual rather than something born of necessity. It was difficult to decide which of the Saturday morning brews was most obnoxious between castor oil, liquorice powder and Epsom Salts. The castor oil was mixed with orange juice in a small glass to try and prevent an involuntary throw-up. This did little to alleviate the foul taste, but it did put me off oranges for years afterwards. Neat iodine was swabbed onto cuts and abrasions and on naked wounds it burned like fire. I once even had the inside of my throat painted with the stuff. An infection would require the Bates's Salve torture. The thick black waxy stick was held over a candle flame until it melted, to drip bubbling and smoking onto a gauze pad. This was immediately whacked onto the tender affected part and bandaged firmly. Sometimes you had to be held down, though allowing this was considered being a sissy, so you gritted your teeth and waited for the pain to subside. Overall, the grown-ups' excuse that 'it's good for you' didn't carry much weight with us - after all, why should the treatment invariably be more unpleasant than the thing you had wrong with you in the first place.

One of Grandma's proven (and economical) remedies for a sore throat was kerosene. Perhaps the ancient Incas did swear by it, but a spoonful of sugar soaked in lamp fuel down the hatch in one gulp was going a bit too far I reckoned. Given that the cough usually still persisted afterwards, I also think that being close to a naked flame with that kerosene breath would have been a fire hazard.

Then there were the boils. For some reason I was a sufferer and had to stoically bear the home treatments designed for them. It was my lot to have four or five at the one time. The grown-ups watched them ripen to full size and colour and at the appropriate stage disembowelled them by thumb pressure. It was either the squeeze treatment or the Bates' salve – not that I had a choice anyway. A more scientific boil treatment tried on me achieved success by suction. The method, based on well-known laws of physics, involved the use of a standard milk bottle. Filled with steam from a kettle to exclude the air, the mouth of the bottle was centred firmly over the boil and the lot allowed to cool. As any student knows, as steam cools, it condenses and creates an almost perfect vacuum in the bottle. Given that the resultant difference in pressure increases to about 15 psi, the contents of the boil now become the contents of the bottle. This sudden transfer, albeit with considerable pain for the patient, does the trick very effectively and with no mess.

My contemporaries will affectionately remember others of the old medicinal remedies I am sure. Iodex and Zambuck ointments were applied vigorously to tender flesh. Hot Kaloplasm and sugar-and-soap poultices worked on infections like fire pumps. Tonics were popular (with the caregivers only) and the familiar brand names of Clements's Tonic, Cod Liver Oil and good old Bidomac featured in our health and welfare regimens too.

Mostly, we amused ourselves when it came to entertainment. The ornate valve wireless crackled and whistled on the mantelpiece with barely recognisable classical music and of course the hourly news from the ABC. Squeaky tunes could be coaxed from the wind-up gramophone, which had a replaceable steel needle. These needles wore out quite rapidly and had to be replaced frequently lest they damage the Bakelite records spinning at 78 rpm. Drop one of those records and it would shatter into a million bits. We used to collect the used gramophone needles until we had accumulated enough to hammer into a flat wooden board to make a passable pinball game. The overall theme in our world of kids' entertainment was that you created your own fun – the only limits being the scope of your imagination.

Grandpa, in his leisure hours, often sat in the big armchair, his head with the thick wavy grey hair wreathed in fragrant tobacco smoke, nodding in time with some music and puffing contentedly. My Dad smoked a pipe too as many men did – the cigarette was considered a lightweight by comparison with a rich and fragrant pipe. Dad's was the straight Trevor Howard type while Grandpa's resembled the one Sherlock Holmes used to wear. The rituals that surrounded the operation of these instruments intrigued us kids. Apparently you don't just ram tobacco into a pipe bowl and light up. If the pipe is not 'drawing' to the smoker's satisfaction, it is attacked and pulled to bits. The stem is reamed out with a woolly pipe cleaner, which removes the spit, creosote, tar and other odoriferous residues. The bowl is next. The ash is knocked out by rapping on the heel of the boot, followed by a good scraping with a miniature penknife, which was something that every gentleman carried. Grandpa's trick was to put a lump of charcoal in the bottom of the bowl first.

"It acts like a filter and stops the fire going out too," he explained.

Rich, black and fragrant tobacco was shaved from a plug of 'Havelock Best Grade,' using the penknife again and was stuffed into the bowl and rammed home – with the handle of the penknife of course. The cheery glow in the pipe was known to go out frequently. We estimated that Grandpa's expenditure of matches was around one box of matches for every pipe fill.

The best part of all this for us was the test run when the pipe was fired up again. While drawing hard on the stem, the old man would apply the flaring match back and forth across the bowl. The tobacco glowed, he sucked, the match flame danced up and down with the puffs and we all let out a collective sigh as he sank back and relaxed in the obvious pleasure we reckoned only a smoker could feel. Dad had a fashionable technique I saw the hero in a film do once. He would suck in the smoke and then let it escape from the corner of his mouth with soft popping sounds, squinting as he did so. Outside later, we kids could be seen in a crowd, all puffing on pipes fashioned from a hollow grass stem and with a gum nut for the bowl. Nodding sagely at each other, engrossed in deep discussion, we drew on our pipes and squinted and popped away happily, boys and girls together.

Those girl cousins weren't all that bad.

Indelible memories of the old house with the sound of creaking floorboards and the rich ripe smells visit me still. There was a sound peculiar to that house along with the faded images, which is particularly poignant for me.

In the summer time the window in our bedroom was left wide open but was fitted with a flywire screen. The 'Fremantle Doctor', our cool breeze from the ocean, blew steadily through the wire with a mournful keening sound.

My dear mother was seriously ill at that time and was not expected to live much longer. The atmosphere in the old house was hushed and no one laughed or joked any more and we all spoke only in low tones.

I was lying on my bed, daydreaming and sad, alone in the bedroom one day. The breeze was steady and strong and the high fluting note from the window screen was a kind of mournful background to my sombre thoughts.

The telephone rang in the hall and the house was silent as Dad moved to pick up the receiver. I shall never forget his choking response to the message we had all been dreading. Without hearing the spoken words, I knew that Mum had passed away.

We three kids, Joe, Gwen and I suffered greatly at the sudden loss of our beloved mother. Turning to Dad gave us little comfort, as he seemed as hopelessly adrift as we were. It was as though a great chasm had opened up in our lives across which we could reach no one. To go on with life without her songs, her music and the many loving things she said and did to make us feel safe, seemed an intolerable load to bear. What we were feeling was isolation and pain of the worst kind.

While collecting the memories for this book, I have tried to transport myself back over the years to not only remember the events and images in my life, but to actually experience the sensations as well. After the fateful phone message, I recall the members of the family in the house drawing together in mutual grief. The most poignant sensation of that time was that of scalding tears running down my face as we hugged together there in the hallway of the old house. Those tears were shed by Dibs who was trying to say how sorry she was through her sobs.

It was a moment in my memory, which has never faded and the thin fluting note of wind through a window wire screen will always send me back to that sad day in Grandpa's Castle.

After Mum's death in 1939, relatives cared for us for about a year. Then Dad broke the news that he had found someone he wanted to marry and who would be the means of making us into a complete family again. Her name was Gwen Thomas and she was a member of a family that had been closely associated with Dad's for many years. Previously unmarried herself, this suited her as she wanted to have children of her own and we still needed a mother figure in our young lives.

<u>Dibs</u>

My first cousin Sheila was known in our list of fond family titles as 'Dibs'. She featured briefly in the story 'Grandpa's Castle.' It is with a great sense of nostalgia and many warm recollections of her influence on my childhood years, that I dedicate this section to the memory of her.

In the course of collecting material for this book, I visited Dib's sister Peggy who had saved from Dibs' personal effects a number of poems, which were written when she was still quite young. Some of the poems are dated from 1942 when she was just seventeen. There is a passion revealed in the verses, a love of nature and also sadness, which seems to show that parts of her life were not altogether happy or joyful.

It is my hope that you will see what I have seen in these and understand my feelings in reading the poems I have selected from the collection.

Dibs died in 1977 Peggy passed away relatively recently in this year 2002 soon after giving me the poems.

Leaving Australia

She expresses her feelings on the eve of an overseas trip.

Now upon the eve of great adventure,
The time is set, tomorrow to embark –
After the haste, excitement and impatience,
I am prepared to set off alone with my heart,
A coward heart that cries of friends and memories,
That blinds her eyes to tomorrow with tears for the past,
That remembers and longs and sighs for the clinging fingers,
Of things I love, to be brushed aside at last.
Not seeing the need for partings, or understanding,
Why youth will rise from the little place of his birth,
To measure his wit and strength with the world's devices,
And find his own harmony in a discordant earth.
The restless listening night is spent debating,
And if I try to reason we argue more.
I know that when the brave, cold day gleams out of the darkness,
I shall sail on a lonely tide, with my heart on the shore.

Autumn

Grandpa Joe Macmillan tended the family garden. This was a likely inspiration for this verse.

Today in the leaf-strewn garden the autumn wind,
Blew chill from the rainy sky, and the naked trees,
Were huddled in shivering groups and moaning with cold,
In dreary, miserable, leafless companies.
The earth, new-turned, was brown and fragrant and bare,
And the gardener was bending, digging and planting there.

I saw the wrinkled bulbs in his busy hands,
Being safely laid in their quiet beds of earth,
And marvelled to think of the miracle soon to be wrought,
When those dry withered lives will shoot up into wonderful birth.
And the ground will be splendid, with glorious emerald and gold.
As the daffodils wake in the sunlight, and yawn and unfold.

.

The Lily Patch

I remember seeing lilies growing by the creek in the horse paddock behind Dibs' home in Beverley. They were most likely the same ones she refers to in this verse and I believe she had planted them herself years ago.

I have found no words to last,
Past one, remembering,
Nor sung more sweet a melody,
Some other did not sing.

I, who would match the matchless swan,
Have capered through the night,
Where boards are worn by pirouettes,
Spun with steps more light.

Yet I have made a loving mark,
Planted a dynasty,
Next year beside the flooding stream,
Lilies will speak for me.

The envious moon will touch them, strung
Like pearls among the trees.
They'll scatter satin for the sun,
Bewitchment for the bees.

And for the children, ripe for joy,
A harvest of delight,
Of muddy fingers oozing stalks,

Brown cheeks against the white.

Their minds will keep the Lily Patch,
A quiet perfumed place,
Lilies will smile through arid places,
And smooth a wrinkled face.

Nostalgia

Written in Fremantle while on hospital duty and pining for her home in Beverley.

Would I were home, for it is spring again,
I know the hills are green from winter rain,
And everlastings under happy trees,
Would be a silver sea, tossed in the breeze.
I see the river now, so full and clear,
A laughing river and the grasses near,
Must be as thick as grass can be and green,
And flecked with sun, the sheoak trees between,
Well I remember the argent light,
The moon gives in the starry sky at night,
And even now I clearly hear the song,
The birds sing, by my window all day long.
Oh God, if I were home, this day would be,
Complete and perfect, beautiful to me.

Windless Moonlight

One can imagine the feelings of walking home alone on a still moonlight night in the quiet of a sleepy country town.

The silence is so loud, it almost seems,
That all the world is dead, or wrapped in dreams,
So deep that it will never more awake,
There is no breeze amongst the leaves, to shake,
Them from their vacant staring at the night.
No birds or crickets call, the moon is white
And solemn in its soundless steady walk
Across the dark. The river dare not talk,
But like a cat, treads softly through the land,
Without a sound; and grass and flowers stand,
Blankly around. Unearthly silence fills
The night and veils the length of moonlit hills,
And nothing breathes, or speaks and nothing stirs.
I tiptoe past the taut, unmoving firs
And quickly hurry home, e'er even I,
Shall join the trance that joins the sea and sky.

Episode Four

<u>The Invasion of Bunbury</u>

The year of 1940 saw the Macmillans packed into their 1928 Fiat tourer chugging out of Fremantle at a fast clip of forty miles an hour, heading south.

Our family, after Mum's death, now comprised the original members of Dad, Joe, Gwen and me. Now, sitting beside Dad in the front of the car was our new stepmother Gwen. We three were still trying to accept this serious, dark-haired lady whom Dad had married a scarce twelve months after Mum had passed away. She had taken her place as maternal head of the family in a firm and decisive manner and gave everyone the impression that she would stand little nonsense.

The natural deep love and affection we held for our real mum had not died with her. In the readjustment to this new arrangement, mainly at the insistence of our paternal grandmother, we were told that we would have to try and lock away our emotional attachments with her from here on. It would be better they said, if Mum's name was not mentioned in family circles and that we should not speak openly of her again. One thing I do know, the love you have for your mother never fades and though her name was unspoken she was still there inside us. There was and would only ever be one Mum for us, so we decided that our replacement would be known henceforth by the respectful title of Mother – and so it was. To be fair, she did prove to be a great cook and housekeeper and looked after us well.

Joe and I were grumpy on that historical trek to Bunbury. We had heard of the wonderful surf beaches down there and had received a gift from an uncle, which he said could be put to good use when we arrived. It was a surf ski of large proportions and weight. In the style of the day it had a wooden framework that was covered with canvas. While not necessarily a two-man craft, it took two good men to carry it. Dad baulked at the prospect of taking it along with us. Despite all our entreaties his refusal stayed firm. Perhaps the sight of the old Fiat arriving in the town, driven by the new bank officer and balancing a huge surf ski on top which was as long as the car would not have created an appropriate image. After our arrival we soon learned to body surf at the Back Beach, so it was of no consequence really after all.

My first sight of Bunbury from the Te Kianga guesthouse, which was our first stop, buoyed up my spirits no end. A sparkling blue sea dumped white surf on the ocean beach. This was going to be a great place to live I thought, with lots of space to roam and plenty of freedom to enjoy it.

Within a week we moved into a rented house in Beach Road in South Bunbury. Dad haggled over the rent, which was a bit steep he reckoned at thirty shillings a week, but he conceded it was a big place with room to raise a growing family so we settled in there. The location suited us too, with the ocean beach a half-mile away and green scrub covered hills sloping down to the back fence.

On the home front, our original family of three kids began to increase as Mother added her contributions to the roll call. She began by having three girls – Marjorie, Shirley and then Leslie. Ross came along almost as a postscript a few years later. So I found myself the eldest of a family of seven of which number four were girls, including my full sister Gwen. Relationships in stepfamilies are not always predictable, but I am happy to say that we all lived together in harmony and scrubbed along really well. Joe and I enjoyed the roles of the older brothers, taking the others on outings, especially when we became mobile with motorbikes and cars of our own. I have a suspicion that the older girls regarded my rather outrageous behaviour, which passed for fun as we grew up in Bunbury, plus my unsettling revolutionary tendencies, as having the look of the lamb with wool of a darker hue – the black sheep of the family in fact.

Dinner by lamplight - early days at the Mangles St. house.

The war in Europe had been raging for a year and our lifestyle began to feel the effects of it. A great number of our young able-bodied men and women had enlisted in the armed forces and were already being shipped overseas to various theatres of war. There were casualty lists pinned to the door of the post office in town. Mothers and wives and sweethearts were beginning to receive the dreaded telegram from the War Office –'Missing believed Killed in action,' or other messages leaving them bereft and grieving for their men whom they may never see again.

Shortages of life's essentials soon began to bite. The war machine had put enormous demands on the economy of the country. We collected scrap metal to be turned into planes and tanks and bombs. Women - and many men, knitted woollens for the servicemen who were in the cold countries and parcels were prepared for the Red Cross to deliver to our boys overseas. July 1942 saw the beginning of rationing for everybody. Little books were issued with tear-out coupons to be handed over with the money for many purchases. Without the coupons you could not buy petrol, meat, butter, clothing, sugar or tea. We set our minds to alleviating the food shortage as any enterprising young son should do. Growing our own vegetables and having a large fruit orchard took care of the vegetarian side of eating, but many families had no way of supplementing their meat diet. A trip on my bike to the butcher one day resulted in the purchase with the mandatory coupons of only three quarters of a pound of meat to be stretched for a meal to feed our nine family members. With fishing line and net and spear and rifle we boys often provided a supplementary supply of meat for the appetites at our house. Wild rabbits thickly populated the lovely rolling hills beyond Dardanup to the east of Bunbury. We spent many hours stalking and ambushing the bunnies in the bracken and around the warrens and seldom returned with less than two or three for the pot. Early morning and evening would see us casting long lines into the surf and off the jetty for good feeds of skippy, tailer and whiting. Big blue crabs were there for the taking in the shallow estuary with the occasional fat black cobbler lurking in the weed to be speared before he speared us.

It is certain that my love of the bush and hunting which I still enjoy, had its roots in those days when we were the family providers helping to boost the food supply when many essential items were in such short supply.

There is no doubt that we lads did engage in a lot of mischief and nefarious escapades during the war years in Bunbury. I want to say here at the outset that none of these were intended to provoke or hurt anyone at any time. As young lads are wont to be adventurous and possibly take risks, I would like to think that nobody suffered as a result of our pranks. We were not apprehended or charged by the law – more by luck than by good management I suspect. It was quite a buzz really to embarrass the police from time to time, but they proved to be pretty good-natured about minor infractions. Being caught in the act of doing something stupid would usually invite a cuff around the ear from the policeman and then being frogmarched home to your father. He in turn, usually seemed more than willing to show you the error of your ways as well - that was never one of my experiences however – more by good luck etc.

In the company of our school mates, the best time to see fair and honourable behaviour was when a disagreement occurred between two boys. Most of these occurring in the school grounds were settled in a makeshift boxing ring with the gloves on and refereed by a male teacher. Honour was usually served and they shook hands afterwards. An impromptu scrap somewhere else was fought one on one with no interference until one or the other gave in. The ring of spectators would to see to it that a contestant was never hit while he was down. To commit this, or put the boots in, or use any kind of weapon was definitely frowned upon. To break these rules of fair play was to invite a lot more lumps than you had originally expected. Most boys were taught to defend themselves by their fathers or older brothers. My Dad taught brother Joe and me to wrestle, with the contention that an opponent trapped in a good hammerlock or a half nelson was not likely to do you much damage. This didn't always work for me on the few occasions I tried it, but I'm sure he was right. We had sports clubs, boys' clubs, gymnastics, boy scouts and a number of other organizations vying to ensure we grew up clean, strong and capable and above all fair-minded and chivalrous. There was the Temperance League too, which I attended once or twice. Intended presumably for the purpose of steering us away from the demon drink, I still remember one of their very original slogans, - 'It is better to be neat and tidy than tight and needy.' - Pretty inspirational stuff.

Any beneficial effects that exposure to this group may have had on my future lifestyle are still in contention and have been sorely tested many times.

Young sons were never idle. Seriously interfering with our busy outdoor life was the responsibility of work around the house. Leftovers from the kitchen were not an insurmountable disposal problem as these were usually fed to the kids – bubble and squeak was a favourite disguise for the remnants of previous meals. After they had been tempted, the leftovers were passed down to the dog, the cat or the chooks – not necessarily in that order. The stout brown paper bags used to bring home the groceries were great fire starters so were utilised. Waste not eaten or burned ended up being dumped into a large 'rubbish hole' somewhere in the back yard. It was an early version of mobile landfill. As the boys made one excavation and it was used up and then backfilled, another would be started further on. If you lived on an acre block this could go on for ages – generations even. You can see that between chopping firewood, running errands on the pushbike, digging holes, washing dishes, making the bed, trimming grass with a boy-powered hand mower, helping Mum on washdays and off-siding to Dad on the many repair and building projects, there was little time left even to moan about everything.

One of the major ongoing projects we were roped into was house maintenance. Gallons of old black engine oil was mixed with linseed oil and working off ladders we slapped it on the weatherboard cladding so it looked decent. Alone at home one day I was part way through painting our tin roof a nice red colour. It was a cool day but up on top of the house the sunshine was seductively warm and made me feel dozy. It must have been only a short time that I had nodded off, but was awakened by a minor civil disturbance in our front yard. Apparently a passer-by had panicked when she saw a recumbent and motionless body, possibly electrocuted or something, draped over the peak of the ridge capping next to the chimney. I sought to reassure my onlookers by coming to life and resuming the job, but the paint on the brush had gone hard.

The family home in Mangles Street, initially had no electric power. As well as having to operate everything manually around the place there was the matter of lighting. Though smelly, the big kerosene lamps with the tall glass chimneys did a fair job and brought the family together into their rings of light. There was one technical breakthrough I won't forget when the wick on our main table lamp was fitted with an incandescent mantle. Previously softly lit, my pimples stood out like a moonscape. Venturing outside this comfort zone, like going to bed or to the dunny for instance, was fraught with danger, as you tended to fall over things in the dark. Speaking of lamps reminds me of an amusing incident involving one of my sisters, which she found difficult to forget, as her brothers constantly reminded her of the gaffe. Soon after we had electricity surging through the house powering everything, the unfortunate sibling was asked to turn off the room light. Going straight to the wall and just as force of habit dictated, she took a deep breath and blew on the switch to extinguish it. Every member of the family was there and we all fell about with laughter.

Myself and Joe at the Mangles Street house. I did get a larger jacket later on

Having been brought up to have great respect and care for the females in the family I did try, but sometimes they show some amazing qualities. Handing my sister my camera at a family outing I asked her to take a series of pictures of us all fooling around. She clicked away a whole roll of film while we posed and postured at great length. As it transpired, the set of prints would have brought great joy to an ear, nose and throat specialist, as she had taken twelve perfectly exposed pictures up her own nostrils. In mitigation for her faux pas it is only fair to admit that it wasn't easy to identify on that camera which was the lens and which was the viewfinder. She threatened sudden death to anyone who showed the pictures around or even told anyone – so I haven't.

The perennial pushbike was our main mode of transport. My first request for a bike was met with the suggestion that if I wanted one I had better go out and find one for myself. The council rubbish tip was a treasure trove of all kinds of things useful, although in those lean times people threw out little that was of any value. It took me a couple of months to accumulate all the parts for one bike. Stripped and then assembled and painted with great care, no one would have known that the origins of my bike might be traced back to a hundred previous owners and at least a dozen different brand names.

It is my belief that happiness and security are based on having a lifestyle with the minimum amount of stress and worry for the safety of you and your family and property. Our community in those days however did have the great fear of the powerful enemy sweeping towards our shores from the north. We did not consider there was any great threat within our own society. This blissful condition seems unfortunately to have changed to a great degree.

60

I rode into town one day – a trip of about five miles – and in a daydream, walked all the way home leaving my bike leaning against a post in the shopping centre. The first chance to go back and retrieve it was three days later. The bike was still there where I had left it. Dad's car had two buttons on the dashboard – a black one to start the engine and a red one to stop it. It had been built without keys, as these were considered unnecessary. Our house was never locked when we were absent. There was the arrival of unexpected visitors when we were not at home one day. They came inside, made a cup of tea and washed the dishes. Girls walked home in the dark from the pictures or a dance alone and in safety. Lost property was seldom at risk as it was almost invariably returned intact. Petty crime was a rarity – serious crime almost unheard of. The only felony we witnessed was seen on the screen at the pictures on Saturday night and that was highly censored and sanitised - for our protection I presume.

On the shadier side of family life topics such as natural bodily functions were not openly discussed or even alluded to, especially around the meal table. In this we were no different to other families of the time. The place to go, when one had to go, was the lavatory, deep in the wilds of the long back yard. Standing at a respectful distance from the house it complemented the green bulk of the mulberry tree and received some shade from it at the right time of the day. It was a place of rest and relaxation too away from the bustle of life and was the place to look for any one of the family members when they could not be found elsewhere.

Known simply by the title of lav or dunny or dub, it employed the pan system for weekly collection and disposal of the unmentionable. A fairly typical standard design was the model featuring timber-frame construction, clad in sun-curled weatherboard with a gabled tin roof. It was rumoured that people in the posh suburbs in the city built them out of brick, to match the construction of their houses. Our house was jarrah weatherboard too, so we must have been conforming to the latest trends in dunny design. The door sagged a bit which caused it to jam so it required a solid kick by a visitor to open it. This wasn't a bad thing as it gave any dozing occupant the chance to shout a warning. In this situation the newcomer could just go and sit in the green shade under the mulberry tree for a bit, or even talk through the cracks in the wall if they wanted to. These lavatories all had a flap at the back so the dunny man could pull out the full pan and slide in a new empty one - every Friday.

A trip to the dunny was usually an experience to be enjoyed - but for the faint-hearted, usually girls, to be sometimes filled with dread. Spiders lived there in the shadows in abundance and you could imagine an entire family of redbacks and a lone Huntsman scuttling for cover under the seat as soon as someone rattled the door. It was my regular job to ensure that there was a constant supply of paper in the dunny for the use of the squatters. It seemed that girls used a lot more paper than boys did, which was a mystery to me then, but with four sisters in the family it became quite obvious that this was certainly the case. The best and the cheapest paper was sourced from the pages of the West Australian or the Daily News and there was just enough forthcoming to meet our needs in that department. Pages from the glossy magazines didn't work well at all, as they had no grip, which didn't matter really, as we could not afford to buy them anyway.

There were the friends' houses I visited, where you just tore a suitable sized piece from the whole newspaper provided, but I decided to turn our paper dispensing system into an art form. A full sheet was folded and cut into six-inch squares, hole-punched, threaded with string and hung on a nail on the back of the door. Many a happy hour was spent in this retreat with a book, reading the dunny paper or just daydreaming. You could not get too involved in reading our newsprint supply due to lack of continuity. No sooner did you become interested in an article or were going down a column, than you would reach the torn edge. This was very frustrating to be sure, and a search for the piece where the story continued would always end in failure. I was tempted to put my brain to work on a solution and thought perhaps that having the paper in rolls would work, but rejected the idea - it would never come to that I was certain.

There was always plenty of room on the wide dunny seat to put down whatever it was you were carrying when you were called away. Putting a ball on the seat was hazardous, as the slope would cause it to roll of its own accord and disappear down the hole before you got your pants down. A sandwich or a bunch of grapes didn't suffer from the same problem however. A tin of ashes from the kitchen fire was there too, with a little shovel together with a bottle of good old Phenyle. Both of these, used in conjunction after you had 'been' were found to be effective aids in keeping the dunny habitable, especially on hot summer days.

Night-time sorties were especially scary on a dark winter's night with a howling wind and fitful moonlight and a failing torch or a yellow flickering hurricane lamp. This condition of course only applied to the girls in the family. When seated there on cold nights the jets and eddies of freezing draughts through the cracks in the boards were acute goose-pimply discomforts, reminding one of the saying involving brass monkeys. There is nothing that feels more vulnerable than having your tender naked posterior hanging in space under such conditions. One consolation was something I had read about our Redback and Huntsman spiders. They are supposed to become quite torpid, even lethargic in the cold weather. This was a comforting thought.

To try and avoid the hazards and discomforts of the outside dunny visits at night, especially in winter, an alternative was used which kept the whole business in-house so to speak. Hidden away under the beds in homes all over the country was the answer to any urgent nocturnal emergency. Known as chamber pots, jerries, potties or simply pos, these handy china or enamelware receptacles were ready at all hours. They were necessarily very sturdy in order to support the weight of the heaviest posterior with a safety factor of extra strength built in. Depending on the make and model, a jerry could boast one and even two handles for extra peace of mind while it was being carried out the next morning. Some of them were rare and handsome works of art. I believe that Henry the Eighth was reputed to have used the most ornate and the strongest, in the kingdom. One I recall in our family was a fine example of the china painting art with a colourful design of floral wreaths and cherubim chasing each other around the rim. The familiar Blue Willow pattern was popular too.

Me, with two mates on the beach.

You needed to be careful though, blundering around in an unfamiliar bedroom on a winter's morning. Good observation skills were needed so that you didn't step in one lurking by the bed, or in one carelessly left on the floor by somebody after a midnight sitting.

The local shire councils engaged men with special talents to drive the night carts and change the family pans each week. The earliest conveyance I remember which served this purpose was a horse-drawn cart with a capacity of forty dunny pans. There is an old joke that was doing the rounds in those days, which has little relevance in the present. It asks, "What has forty cylinders and flies?" There will be no prizes for getting that one.

The dunny man, as he was known, was one of a sorry lot with much to put up with. His standard dress was blue overalls, a leather shoulder pad and a leather hat with a flap to protest the ears and neck from spills. Arriving at the gate, usually in the evening, the dunny man sprang off his perch on the cart, shouldered a clean, tarred pan and headed down into your back yard at a swift trot. His return journey was often fraught with danger. The sheer weight of a pan at week's-end from a large family was considerable. Regardless, he would enter the minefield of a dark backyard at a canter and then later head for the street and back to the cart. One night I witnessed a dunny-man's waterloo as he took an uncharted shortcut through our yard, falling foul - literally, of a low-hanging clothesline. The taut wire caught the full pan on his shoulder and he was swept clean off his feet in the encounter. Both man and pan crashed into the shrubbery, the former losing his cool and the latter losing its lid and load. It is guaranteed that the garden still flourishes at that spot to the present day.

Small boys were the anathema of the dunny man. They taunted him with most undeserved ridicule and played practical jokes on him too. The worst I ever heard was when a pit was dug in his intended path and camouflaged so that in the darkness he would not see it and therefore would become a victim. He did not and he did.

There were three good years spent attending the Bunbury High School, leaving at the end of year ten to begin working. Those student days hold some very special memories.

It was the middle of winter when I sat with my classmates, warm and comfortable, with a big crackling fire in the fireplace in the maths class in room 3F. The winter wind whipping up off the sea gusted around the building and rainsqualls sheeted and rattled against the big windows. The roar of the surf from the back beach filtered faintly into the room as the bell sounded for class change. Physical training was a segregated activity where we did boys' things and the girls did whatever girls do somewhere else. Our Mr. Colgan was a teacher who held the firm and popular belief that boys should be brought up to have strength and stamina. This usually involved suffering a certain amount of pain and strain and this day promised to be no exception. It was almost a mile down to the beach through the sand dunes. Stripped to bathing trunks and with heads down into the bitter wind we all set off Indian file style. The wet sand was hard going, but we soon fanned out across the beach, ploughed through the belt of foam and disappeared into the surf. Several days of stormy weather had whipped the waves up into monsters. A pleasant surprise that greets anyone venturing into the sea on a cold and windy winter's day like this, is the warmth of the seawater. All is relevant I suppose and there is no doubt that the water was cold, but after being so chilled by exposure to the elements, the swim is seductively pleasant and you don't want to come out to face the weather again. We surfed and sported and spouted for half an hour until the teacher's whistle signalled the time for return to classes. In the gym showers after the mile run back up the hill to the school, we all shared the sheer enjoyment of being young and strong and glowing with good health.

When Japan entered the war and her armies began to invade southwards engulfing the Pacific countries, a wave of deep apprehension ran through our community. The possibility of an enemy invasion became stronger by the day, as the news on the wireless spoke of the grim future we may have to face. In February 1942 the newspaper headlines were black with the accounts of Japanese air attacks on Darwin and our own towns in the North West of the state. This was all so close to home - and then Singapore fell. It was a severe blow to our sense of security and contentment, and then the rumours suggested that we could be next in line to succumb to this seemingly unbeatable enemy.

Our town swung over to a wartime footing with the whole population uniting in preparation for possible air attacks. Bunbury was a main regional centre and important seaport – the kind of target we thought the Japs would want to knock out.

Dad took this all very seriously and decreed that for the safety of the family we would have to construct an air raid shelter. Joe and I stripped down, grabbed shovels and soon had a cavernous hole excavated in the top end of the back yard. Heavy corrugated steel sections formed the underground compartment, while the roof was topped with railway sleepers and sandbags. A heavy steel hatch opened to a ladder leading down into the dark interior.

"It's OK for all but a direct hit," proclaimed my dad, so we all felt safer after that.

To give the air raid shelter its due, it made a fine clubhouse where we could huddle and hatch nefarious plots of devilment. I didn't know the meaning of dark until we all piled in and closed the hatch. At least you couldn't see the spiders, until someone lit the hurricane lamp.

My mates and I entered into the spirit of the moment, having been caught up in the patriotic fervour and feverish activity. We volunteered for everything. School was attended on a daylight-saving scheme, which meant we only had to show up for a half of each day. A shortage of teachers required that students were divided into two groups and we attended school in shifts. A half-day holiday every day was fine by us, giving more time to spend the long summer afternoons surfing, hunting, and beachcombing, or engaged in other patriotic pursuits.

The wartime footing at our school had its lighter side. If life becomes dull and boring with little fun or excitement, kids will usually create a diversion to change things. We took advantage of a development that altered class routines - it involved holding emergency evacuation drills. This was to prepare for the possibility of bombing raids - during the maths period for instance - and what better time could there be to have one.

Teachers organised teams of boys to dig deep slit trenches at various points in the school grounds. We looked upon these not so much from a lifesaving aspect, but with a view to having a bit of fun.

There was a state of controlled chaos when the air raid siren sounded the drill through the corridors of learning. We students all jostled and jived our way down the stairs to fan out in the direction of the slit trenches. The tight pack of hot young bodies wedged one against the other in the trench promised to allow little space to manoeuvre. This seemed to present a heaven-sent opportunity to be close up and personal with the girl you had been mooning over all term. The school dancing classes that we enjoyed with our female classmates were fine and were the source of some pleasant stimulation. There was little scope for making opportunistic moves on a partner however under the steely glare of the senior mistress who was supervising.

After one or two air raid drills, the boys became more devious. All the girls would flee ahead to the trenches and remembering our training, where it was always 'ladies first', we would lag a bit behind without being too obvious and only to the trained eye, appeared rather selective as to our points of entry to the underground. The plain and the beautiful, the thin and the buxom, had all jumped unsuspecting into the dark confines of the trenches. Our strategy was to be seen as trying to protect the fair sex against the possibility of bomb blasts – a chivalrous and self-sacrificing act indeed.

A teacher patrolled the trenches this particular day - oblivious to the risk from enemy action, to assess the effectiveness of the exercise in protecting his charges. He must have noticed the uneven distribution of arms and legs and faces peering up from below, with higher concentrations of student flesh in some areas. The main press of male bodies seemed to be in places where the most desirable females had landed. This could have been put down to the natural protective instincts of the boys, although the waves of giggling and shrieking that came up from the earth seemed to suggest there were other more ulterior motives at work.

In and around the town in 1941 and 1942, the streets rang to the sound of marching columns of soldiers and the whine and clatter of Bren gun carriers of the light horse battalions stationed in the town. Also, miles of ugly barbed wire entanglements had appeared overnight as a blot on our pristine beaches and blackout restrictions were imposed in the dark hours. We all crowded the second story railings of the school one day to watch a flypast of Spitfire and Kittyhawk fighter planes, which thundered over the town roofs. I recall looking down into the cockpit of one of them as he flashed by and waving back to the helmeted figure at the controls.

There must have been a spate of accidents in town involving cars running into each other in the dark, mainly due to the requirement for all headlights to be blacked out to mere slits. Another edict required motorists to paint a broad white stripe completely around the car, so presumably one would have a better idea of just who or what it was that one was colliding with. The trouble was, all cars of the day came in the darker shades of blue, green and black. A white car, while easier to see, would have been considered a kind of albino oddity then.

We lads felt very strong in the desire to serve our community in its time of need in every way possible. With no chance of joining the army at age fourteen, although some tried, we volunteered for service with the ARP. This fine body of citizens, mostly quite young or quite old, with few in the middle, donned tin hats and slipped on armbands and shouldered gasmasks and patrolled the streets, especially at night. Air raid precautions were their business. They jumped on any householder who carelessly allowed a chink of light to escape from their house during air raid drills. They were trained also in the techniques of fire fighting, rescue and first aid and other duties that could be required in a community under bombing attack. I was excited about these possibilities but dreaded the thought of the real thing nevertheless.

Dad was a community-spirited chap and joined the ARP immediately. I soon followed in the role as messenger.

"We need boys on bikes as dispatch riders to maintain vital communications during an air raid" said Dad.

The incentive for joining up was the issue of a tin helmet, a gas mask and a white armband with a big M on it. To look the part was very important in these circumstances of course. That was enough for me. With visions of pedalling through fires and explosions, skirting bomb craters and collapsing buildings to get the vital messages through, I sprang into action. The front tyre on my old bike had been flat for a week so I headed outside to mend it. I learned that in an air raid a brave and speedy messenger on a bike is a most reliable form of communication between the emergency teams and headquarters, especially when all telephone lines are down. To me, the value in this was immediately obvious and I reasoned that smoke signals for instance, as an alternative, would be most confusing under those prevailing conditions.

While my issue gas mask was never used in anger in an enemy attack, I did find it useful at two am one night at home. It was warm summer weather with mosquitos a worry. I woke to the smell of smoke and the sounds of pandemonium in the house. Dad was groping his way down the hallway in his pyjamas wheezing and spluttering something about a fire in his bedroom - as indeed there was. This was a potentially serious situation living in a timber house, but we were not able to get to the fire because of the thick and choking smoke. As if a light bulb had switched on in my head, I snatched my gas mask from its peg in the hall, crammed it on and entered the bedroom. I was just able to make out the mattress on the bed, which by this time was a cheery glowing mass.

All the family, still in pyjamas looked with glum faces at the sodden, smoking mess on the front lawn where we had extinguished the flaming mattress with the garden hose. The pat on the back from Dad was good though. It seems that the cause of the conflagration was that the mosquito coil smouldering on the bed head for the purpose of deterring the mozzies had fallen down due to some vibration or other and ignited the connubial mattress. Using the smoke from a mosquito coil to kill mosquitoes was one thing but this was going a bit too far I reckoned.

My designated responsibilities in the ARP included operation of the gigantic howling air raid siren on the roof of Barney Hay's service station in South Bunbury. As we were told that an air raid could occur at any hour and night attacks were common, I took great delight in cycling down in the dark and firing up the siren. The effect on the whole town was most gratifying. Nobody - but nobody could sleep through the blasts of ear-splitting noise. Also, I had developed a hate relationship with the volunteer fire crews and it was a source for great satisfaction to me to have these characters ripped from their warm beds in the cold early hours to do their duty. There was no one who twigged to just who was hitting the switch on the siren. Ivan would have been the most unpopular kid in town had they known. I certainly wasn't going to tell anyone.

Also, a duty dear to my heart was that of lighting big fires around the town. This is probably a confession to the need to satisfy a hidden desire I harboured – that of a pyromaniac. This should not be seen in the criminal sense of course, but as a strong, though harmless fascination for fires and explosions of all kinds.

In covert conspiracy with the ARP chief one day I was given sealed orders containing the time for the big siren to sound and also the locations for the sites of simulated 'incendiary bomb' strikes. Each of these was to appear as a nice big fire – which I myself had set, to be flaming away on a roadway, behind someone's dunny or on the lawn in the council park. Given all this heady responsibility I lusted for more power to create chaos and disruption – just like what you would expect in a real air raid. Realism was the key.

Having noted with some disappointment, the ease with which the fire teams put out my fires of wood and kerosene and old tyres, and then to see them pack their gear and walk away in contempt, I decided to try and do better. My ARP training revealed that a burning Japanese incendiary bomb could not be extinguished with just a piddling stream of water. It was time for a bit of the real McCoy.

An 'incendiary bomb' had landed one night in the middle of the intersection of Beach Road and Spencer Street. This was MY bomb and it was going to work the way it would in the real situation. I quietly observed the fire fighting efforts on this one from the shadow of a large power pole on the corner of Beach Road.

Thinking themselves by now expert at killing my fires without difficulty, the fire team hit this one with jets of water from stirrup pumps from all angles. They were working in close, chattering away, and no doubt thinking of home and a warm bed when they had it under control.

Then all hell broke loose.

With only a birthday candle-worth of flame remaining, the sodden fire erupted with a whoosh and jetted scorching red flames in all directions. Gas hissed and flared and lit up the dark street like Gloucester Park on a Saturday night. Black silhouettes danced around my conflagration aiming ineffectual squirts, which were turned to clouds of steam in an instant. My secret was buried in the heart of the fire in the form of a large tin with nail holes in the lid and packed full of calcium carbide. With perfect timing the water they were applying had trickled in onto the chemical, which then turned into a seething mass of acetylene gas. And then the lid blew off. With a wonderful potential for explosion and flames, this stuff is also impossible to extinguish with water – just like a real Jap incendiary bomb in fact.

Calcium carbide, my secret ingredient, is a grey rock-like substance, which generates great volumes of highly flammable acetylene gas when it comes in contact with water. In earlier times before electricity, it was used for street lighting and in many other kinds of lamps and for welding.

In the wash-up after that exercise I preened under the praise and commendations of the chief, whom I suspected was a bit of a firebug himself. I had to tell him in confidence of course of my resounding success in the search for realism and also of the consternation that had been inflicted on the cocky fire crew.

Future fires I decided would be booby-trapped like that one, though not all, so they wouldn't know which one was going to flare up and scorch their contemptuous attitude

Bath Night

One of my dad's morning rituals, which were pretty interesting for us to watch, was shaving. This was sacred men's business. The shaving routine required the careful preparation and attention to detail you might find in a Japanese tea making ceremony.

The big wicked blade of the cutthroat razor was to me the ultimate symbol of masculine maturity. The razor strop was retrieved from a hook on the back of the door to begin this rite. Crafted from heavy horsehide, this wide belt had a dual purpose. My bum twitched when he picked it up, as it was also the instrument used for taming insubordinate and rebellious boys. The pot of Imperial Leather shaving cream was placed on the shelf under the mirror alongside the ornate china shaving mug. You could get this in a stick as well. The shaving brush, which applied the thick foamy lather to the face, completed the equipment list. There was one other important item however, which everyone who used a blade razor should have and that was a styptic pencil. It was applied to staunch the flow of blood from minor nicks and cuts. It was one of my essentials when I started shaving with the big blade and I remember it would sting like hell. Larger accidental lacerations were treated by the application of a piece torn from a Rizlax rollyerown cigarette paper. While very effective in the more severe cases it also provided some amusement for the kids at the breakfast table. Dad sometimes had to be reminded to remove the little white paper flags glued on his face with blood before he left for the bank. We had to cover our ears if he peeled off a piece of cigarette paper and started haemorrhaging again on his white shirt collar. It wasn't just "Oh dear – oh dear" or "Blast" or "Bother"- especially if he was running late.

Many were the tales that Dad would recount about his early life in the goldfields. In those days the town barber was apparently a man of some account who commanded respect. Wielding a cutthroat razor when shaving recumbent customers surely must have given him a psychological advantage no matter how tough was the man in the chair. These blokes wielded the cutthroat razor and the scissors and the hand clippers with masculine authority.

It seems that one day while divesting a hard rock miner of his matted beard, a goldfields barber took offence at the show of bad temper and the poor attitude of his client. With only his head protruding through the hole in the big white sheet, the miner was an easy target. The barber decided to repay his victim's lack of manners with a sharp reminder of just who was in charge. Tilting back the bloke's head to expose the vulnerable throat, the barber swiftly drew the wicked blade from ear to ear across his jugular. Ejecting from the chair entangled in his shroud, the victim allegedly clutched his throat and yelled, -

"I'm done for!" - Or with an anguished cry in that same vein.

To all intents and purposes the miner had just had his throat cut, but was in fact on the receiving end of a barbers' trick. Having steeped it first in boiling water, he reversed the wicked blade to the blunt side and then applied it to the miner's neck with a slashing motion. I shall never forget that story as Dad related it to us about thirty seven times or thereabouts and as it was with his other yarns, we never tired of hearing it.

In the Beach Road house in Bunbury our family boasted a very handy appliance in the bathroom – a chip bath heater. With a swing door where you stoked the fire, and a tray at the bottom to collect the ash, it was the latest thing. It rose like a green lighthouse to where the flue disappeared through the bathroom ceiling. As the name suggests, it was used for heating the bathwater, which was a big step from Grandpa's Castle and the frigid conditions there. You light a fire with paper and sticks inside it, feed it with wood chips - blackboy is very good, and as the water is warmed in the casing, you turn on the tap for a lovely hot tub. Our particular bath heater led a pretty rough life as it was forced to do things for which it was not originally designed. We lads frequently had the need for a good heat source in our experimental projects from time to time. In order to heat pieces of metal to a rosy glow for blacksmithing and to melt lead, the fire had to burn very hot. Wood chips would just not do the job, so solid fuels were substituted. I recall seeing the upper section of the flue turn bright red with heat as the fire roared like a blast furnace in a steel mill. Eventually the knob on the firebox door sagged, then melted and fell off. These desirable conditions were attributable to the dry jarrah blocks we stoked the fire with, although by far the best were the mallee roots.

After a period of stressful service our chip heater became tired and started to leak water from its sprung seams down into the firebox. It accumulated there and wet the fuel, which made the fire difficult to light and keep going.

In the interests of responsible water conservation of course, my brother Joe and I shared the bath. To be honest, we often doubled up in the big tub for some high jinks and also to conduct some serious scientific experiments. Like most young lads, especially siblings, we were highly competitive. The underwater marathon was a favourite and determined which of us could stay submerged the longest. Plunging our heads into the turbid water, we stayed under until the lungs were almost bursting. The first to abandon the game was chicken. Mother became worried one day at the ominous silence during one of the contests. The unaccustomed break in the usual sounds of hilarity and unruly activity made her think that something more serious had occurred.

One day it did!

Joe and I were in the old tin bath this day when the stream of water from the heater went cold. On peering into the firebox it appeared to me that the fire was completely dead, having been extinguished by the water leaking in. I wrapped my naked body in a towel and went to search for a more effective fuel. Experience had taught me that the chemical calcium carbide when immersed in water generates a flammable gas, which we already knew from the air raid exercises. What better fire-starter I reasoned, could you have for a wet fire?

The hissing noise we heard after tossing a handful of the stuff in through the fire door indicated that it had begun to work and was generating volumes of acetylene gas up inside the heater. Joe was still complaining that his bathwater was getting colder but my matches were wet too and I had to search for some that worked. Confident that we would soon have some action, I had ignition at last.

What followed is a bit indistinct and blurry in my mind so I will just try to describe the bathroom scene after the explosion. The flue at the top of the bath heater had split all the way to the ceiling and opened up like a big black smutty flower. I could take in the whole scene at a glance as I was now sitting naked on the floor in the far corner, albeit in a state of shock and confusion. Not only were my ears ringing, but also there was a strange unnatural pain in my groin. The ash drawer was resting in my lap after being ejected first, then striking me in the gonads. The black hole in the heater indicated where the fire door used to be, this having been blown off too, and was now resting against the bathroom door. Water gurgled from the split casing of the bath heater and eddied across the floor.

While these changes in the arrangement inside the bathroom gradually became clearer, it was apparent that the most interesting effect was due to the fine white ash blown from the firebox. It filled the air in the small room lending a white, almost Christmassy look to everything as it slowly settled. It was really no time for laughing, but the sight of Joe in the bath really tickled me, which was a distraction from the ache in my nuts. To be honest my mirth was more like a bout of nervous hysteria. He sat there unmoving as the snowy shroud quietly descended upon him. The surface of the bathwater had a layer of rippling ash on it and Joe just sat and blinked at me and shrugged his shoulders in mute resignation. Two little mounds trickled ash off his shoulders and like Pompeii after the eruption of Etna, he was completely covered in the stuff. The powder on top of his head lent a comical touch to the picture and it seemed he had aged about fifty years as he glared at me from under his white eyebrows.

Dad was in the house, which in a way was unfortunate. We could hear his running footfalls and he was soon trying to push the door open. This was difficult as the heater door was wedged against it. When he finally got to survey the scene I swear his eyebrows disappeared into his hairline, which was quite a feat as this was such a long way back.

On the subject of the scientific experiments we held in the bathroom, this really all stems from Dad's insistence that we should never accept that things just happen in the world, but we should try and find the reasons why they do so. There are a lot of natural laws, which control certain phenomena he said, and we should always try and apply these to discover the reasons why various events happen in the ways that they do. This was the main reason I suspect that we grew up with inquiring minds and liked to experiment - putting theories to the test.

Basic bodily functions have always held a fascination for young boys, to the despair of parents and the disgust of sisters who call them gross. Seldom did our bath night go by without the passing of copious wind in the suds. Apart from the obvious, which are the sound effects and the bursting bubbles, there are deeper theories involved that we felt should be explored.

There are the scientific principles involved with the gas in the first instance having to overcome the pressure of water at depth before it can escape. This can be calculated. The reduction of pressure on a bubble of gas as it rises to the surface causes it to increase in size, which can be seen if the water is reasonably clear. Finally the role played by surface tension of the suds as the bubble expands and bursts on the surface is also of interest. The nature of the laws governing this sequence of events then changes from hydraulics to the study of gas behaviour at the surface of the bathwater.

The composition of the gas can be determined by the olfactory senses, which suggest that it is predominantly methane with a component of boiled cabbage. The value of methane gas as a source of energy in industry can also be demonstrated when a lighted match is applied to the big bubble. This results in a colourful flash and a minor explosion giving off heat. It is also necessary to understand that there is a requirement for the gas to be diluted in the correct proportions with oxygen before combustion can take place. This can be researched from a table of the flammable limits of different gases. Too much, or in fact too little gas in the air mixture for that particular gas will result in ignition failure and ultimate disappointment.

In summary I hope you will appreciate that while igniting farts in the bath can be an entertaining and colourful pastime, it should also be understood that this would not be possible were it not for the laws of nature at work. Learning can take place in the humblest of settings but can be so much more valuable when young minds are at work seeking out the reasons why.

While our parents seemed to accept with some suspicion the sounds of hilarity issuing from the bathroom, especially while scientific experiments were in progress, they did question the reasons for the logjam of spent matches blocking the plughole afterwards.

I remember now, the replacement bath heater we got was a blue one and the threats from Dad regarding any future misuse ensured that it did not meet the same fate as the old green one.

The Reluctant Sportsman

Ours was neither a football family, nor a cricketing family nor a soccer family either. As a growing lad I tended to follow the interests of my parents – or rather did not follow the interests that they did not, especially in the early years of my life. Dad was not a barracker of any of the major sporting teams at all. Living in the country as we did, our state of relatively tranquil existence can be understood and appreciated I think.

There was however the test cricket broadcasts that Grandpa used to bend an ear to, coming from the staticy old mantel wireless set we had. There was also the occasional program to be seen on the newsreels at the local picture gardens on Saturday night. They were black and white, grainy flickering images, which seemed to be taken from a spot miles away. There were no streakers on any of these either, not that you would see much under the circumstances anyway. So that you could distinguish the players more clearly I suspect, they all wore white - except for the squashy cap. They were very gentlemanly and restrained too when applauding a good over or a fine show with the bat - regardless of which side it was they were applauding.

A lot of Dad's free time was taken up with his music. He played first violin in a number of big orchestras and showed a real passion for it. His other passion was satisfied in his backyard workshop. That interest was one which Joe and I shared with him and was the source of a good mutual father-and-sons relationship. Dad always said that a man had to be handy with tools. He had a broad knowledge of all things mechanical and could also turn his hand to cabinet making and other skills as well. He taught us a lot of useful stuff in those early years.

In retrospect I was probably a disappointment to him in his efforts to arouse my interest in music. We embarked on some violin lessons, which for me had to compete with bike riding and fishing and other more important activities. He said I had a good ear for music too. Not being too sure what that meant I could still not be persuaded to knuckle down and learn. Whatever latent ability or talent I might have had just died on the vine.

There was very strong emphasis placed on physical fitness in my primary school years and this carried through to high school as well. Everyone was required to participate in sporting activities – it was not optional and there were no slackers. There was a compulsory session called drill. Out in the quadrangle with shirts off and shivery, we did a series of physical jerks like knee bends, waist bends, arm movements, sit-ups and push-ups. You don't stay cold for long doing that lot at a brisk pace. All the major sports were played too with the inclusion of boxing and gymnastics. Joe and I played in the same side in cricket at one of the schools we attended where we had the opposing team a bit confused. Joe would take the run up and deliver the ball right-handed while in the crease he was a left-hand bat. I on the other hand was a conventional right-hand bat but bowled as a mollydooker. We were in this way a bit unconventional.

Any sporting aspirations I might have had at the time were suddenly blown away when I was twelve. Our family doctor declared that I had contracted rheumatic fever and as a consequence now suffered the disability of an enlarged heart. From that day forwards I was banned from joining in any strenuous activity, which found me sitting very unhappily on the fence while all the others enjoyed their games. This condition I endured for about a year and I believe it had a significant effect on my normal physical development for that period and perhaps also affected my interest in sports generally.

There was hope for me yet however in my miserable condition, as my Uncle Doc. Hodgson checked my health and saved my bacon. It had apparently been mis-diagnosed and it transpired that I was completely healthy after all. The account of my resurrection and return to a normal life can be found in a previous episode.

At the Bunbury High School which Joe and I attended, it seemed my performance in most sports was better than some, but not as good as many others of my contemporaries. While I was fit enough, I probably could have done much better. In retrospect, what was lacking was the strong will to win – the killer instinct if you like, or I was just lazy.

While generally turning in average though unremarkable performances in most team sports, I did manage to pull a couple of feats out of the hat.

Our cricket team was contesting a grudge match with a rival school. It was very hot where I stood in the sun in the outfield near the fence. I had little more to do than watch the other team's star bat, who was the last man in, knock our bowling all over the paddock. I was unimpressed by the heat and the inactivity, and an old song kept running through my brain. It was something about Mad Dogs and Englishmen. I presumed it related equally to Aussies.

The cocky batsman was having a field day racking up a big score with fours and sixes, which was bringing the game close to a defeat for us.

He belted another one towards the boundary, which apparently would have just about sealed the victory for them. My dozy state was interrupted when I saw the ball coming like a mortar bomb straight at me. It came high intended to lob over the fence and being the only one in the target area, I decided I had better to do something about it. That ball hurt like hell when it smacked into my palm at the top of an involuntary and uncoordinated leap, but I brought it down for the catch. It was a good feeling when the team chaired me shoulder high off the field at the end of the game, but I was secretly hoping not to be called on to go through all that again.

In most of the track events I was content to stay with the main bunch in the middle of the field – for company no doubt and let the stars and the heroes and the legends take the tape. An exception to this was the day I led the field and won the mile. We had to run the whole of the race in soft sand with bare feet. Nobody was more surprised than I. They said after the performance that I was a stayer, which is something to think about after all.

More a novelty event than a regular crowd pleaser at the school sports was the 'cricket ball throw.' All the cricket stars and some other hopefuls lined up to pelt a cricket ball down the hundred-yard track as far as they could. All efforts were carefully measured for distance.

Harking back to my junior years before this at the Richmond State School, I found myself one day in a tie in an impromptu rock-throwing contest. My major opponent was the erstwhile legendary state footballer Jack Sheedy, though I didn't know this at the time and was therefore not intimidated by it. I beat him that day by a good margin and did notice afterwards that he was not very amused. It was good to hear though that in later years this early defeat did not seem to adversely affect his successful sporting career too much.

My main rival in the cricket ball throw was a lanky fifth year with arms like an orang-utan. At well over six feet in height, I remember he had a very impressive Adam's apple, which bobbed up and down as he was slinging off at me - not his fault of course but it was a bit of a distraction. There's little chance for me here I was thinking, though we soon had the field whittled down to just the two of us.

You may think - correctly, that I would not be going to the trouble of writing this if the other sport had won the contest, but read on anyway.

We made three attempts each in a tense throw-off and it seemed that he was not pleased either as the results came back on the final outcome. It was decided that I had beaten him with a throw of seventy-eight yards. This was not going to get my name on the honour roll in the sportsmen's hall of fame, or an inscription on the 'Cricket Ball Throw Cup' but it was a good feeling for this left-handed lower school reluctant sportsman.

Probably aimed at the development of good coordination and balance and fitness were the compulsory sessions we attended in the school gymnasium. I looked forward to these with great anticipation when we filed into the big echoing school hall. There were springboards,

71

vaulting horses, parallel bars and thick floor mats and a lot of dust. I made a fair showing in most of the exercises but my favourite was the flying neck roll. Pounding up to and launching off the springboard you dive over and across the high vaulting horse and finish with a rolling summersault on the floor mat on the other side. While it gave me a big thrill to do this one, as not everyone was game to try it, perhaps in retrospect I know now the cause of my neck troubles in later life. Maybe these twinges have their origins in crashing on the back of my head so many times. There is a saying I have heard relating to this. – 'Where there is no sense...!'

One team sport I did play and enjoy for a number of years around the age of eighteen was Lacrosse. The history of this game can be traced back to the native Red Indian tribes of North America and Canada. Possibly devised as a way of reducing the high mortality rate they suffered in tribal wars, it was decided to substitute these hostilities with a competitive field game - hopefully without so much of a reduction of the population. Field boundaries were not considered to be necessary, as all they required for a game was the nearest bit of prairie. They swapped their tomahawks and bows and arrows for the crosse, which is still used to play the game in modern times. As I understand it, the rules were pretty lax, so it didn't change the mortality rate by all that much. The crosse is a stout stick about five feet in length made of tough hickory timber. The business end is curved across in a kind of flat blade and laced with thongs to form a net or pocket behind it. The ball can be skimmed off the ground with the blade or caught in the net from the air and carried without slowing the pace. The small heavy ball achieves considerable speed when thrown from the stick, which acts in a way similar to a woomera or spear thrower and gives it greater distances when compared with other field sports with balls.

This does not automatically exclude ladies' competition judging by the number of girls' lacrosse teams in the competition.

The boundaries of a lacrosse field are not clearly defined as in most other team sports, which require frequent stoppages when a throw goes over the line. The ball is actually still in play until it travels over someone's fence or across an adjoining highway. They termed this 'natural boundaries' and were probably adopted to try and emulate the Redskins' games of yore. To further maintain continuous play, the goals are set in from the ends of the field and the action can continue beyond and behind the goal line. The power and speed of a good throw enables a skilled player to score a goal from the goal at the other end of the field. There is never any need for a player to feel left out in a lacrosse match, as everyone is busy most of the time. There is also a curious rule that on second thoughts appears to be very fair. It is intended to weed out a bruiser who as a result of his rough or illegal tactics would be suspended from the game. He must go off the field together with the player he has injured, and he may not return until the injured player recovers and is able to resume. There were some comical situations where the perpetrator sat next to his victim on the bench at the side of the field, obviously eager to get out again and back into the game. The groggy victim was shaken and slapped around the face to revive him sufficiently so they could both get back into the action.

Here there is a confession to make in this regard. I was playing a position known as point. Standing in front of your own goal you are supposed to stop any flying forwards from getting through to score. It is permissible to slow or stop an opponent with a body check, which turns you into a kind of punching bag for the opposition. You are not allowed to whack 'em with your stick. Playing my part with stolid brute force and ignorance I moved to stop this flying missile with the red jumper and we collided with a considerable impact. Getting up from the ground feeling pretty whizzy, I saw the medics racing out with a stretcher. This all seemed unnecessary as I was on my feet again, so I called for the medivac team to go back. They rushed straight past me which was a let down. It was the other bloke they were interested in, who was flat out unconscious and had suffered a broken collarbone

Urban Terrorists

There was one valued facet of my young life and the lives of my fellows and that was the quality of the freedom we enjoyed. I know we did not appreciate it at the time though life for us all, especially during the years of the war was austere and frugal in many ways. Our youthful activities and pursuits however were bounded only by the limits of our imagination and resources. We spent all spare time in the open, at the beaches or in the countryside or working in the yard. There were just not enough hours in the day to satisfy our thirst for exploration and new experiences. We ate plain food, of which there never seemed to be enough, and ran ourselves ragged every day. As the eldest in our family of seven children I found that there were many laborious household jobs to see to as well. These were completed dutifully, though rather reluctantly at times. We got up to a few pranks of course and I think perhaps in this present day, some of the high jinks we were involved in would now be seen as delinquent behaviour or even illegal – we called it fun.

The story 'Invasion of Bunbury' suggests that my youthful character had tendencies bordering on pyromania and that I derived satisfaction from creating explosions of varying types. This is undoubtedly true, but in mitigation I want to remind you that my dad allegedly displayed similar leanings in his youth and related many of his experiences to us. It must be in the genes.

In the war years beginning in 1939 the citizens of Bunbury and in fact the whole nation had the threat of enemy attack uppermost on their minds. Military vehicles ran in the streets, barbed wire festooned the beaches and we all had to get used to the sounds of preparations for war. The rattle of gunfire and the detonation of grenades and demolition charges became commonplace as our soldiers played their war games around the town.

Any odd additional sounds of explosions around the countryside we felt sure would go largely unnoticed.

It must be said in all honesty that the case of gelignite had not actually fallen off the back of a truck – such an event would have entailed a high risk factor anyway. After our great find, the necessary ancillary items were collected over time to ensure that it would all function satisfactorily. Our team did not plan the use of this stuff with ignorance of the potential consequences, as some of my older relatives had been miners. The knowledge gained from them ensured that we handled the explosives with a reasonable amount of care and respect.

Gang wars were the stuff of the vigorous and spirited youths of our ilk. Confrontations between groups became a part of our social entertainment and a way of letting off steam.

While we had never met them formally, there was a gang of ferals from the other side of town that crossed our path one day. The first contact occurred in the sand hills behind the high school. Lined up facing each other at a range of fifty yards or so there was a lively exchange of threats and abuse and propaganda across a gully. Then a fairly vigorous rock fight ensued. This was relatively tame stuff and when we tired of it all went home.

By some mysterious form of boy communication we found ourselves again at the same place at the same time a week later, facing each other across the same gully in the sand hills. There was a change however which we soon detected when hostilities began. It seems that some of their number were now using gings - shanghais, catapults, slingshots or whatever, to turn the tables on us. Resenting this sneaky tactical advantage, which had increased their effective range and firepower considerably, we took cover but continued the engagement, resolving to try and match their weaponry at the next opportunity.

The following week found our two factions at it again, with my lot feeling confident of a level playing field as it were. We were soon aware of the fact that air rifle pellets were now whizzing around us. Gings are no match for airguns of course, although a lucky ging shot did connect with one air rifleman, causing him to retire from the fray. Despite this brief advantage our forces had to fall back defeated again.

The war council prior to the next sally into no-man's land resolved that this continual treacherous escalation of firepower by the opposing forces was unfair and unreasonable. Though we were fairly well matched now, I felt that to save face, we would have to raise the ante to provide a tactical advantage in our favour for once.

With the advantage of a good throwing arm, I sent over our piece de resistance in the latter part of the next skirmish. The trump card was a half plug of gelignite. Trailing smoke from the spluttering fuse and arriving unnoticed, it plopped into the soft sand at the top of their hill. It seems pointless to tell you that we prevailed in that battle. A sudden mysterious hush and an immediate cessation of hostilities followed the flash and the bang and the eruption of sand. The enemy retired behind their hill to reserve positions – like back home, and we didn't meet them there again – ever.

A young policeman had moved into a house just a few doors from us in Beach Road. We had previously exchanged greetings in the street and he seemed a nice enough bloke – for a cop. Normally the anathema of young lads, this one did not seem to constitute a threat to us just yet.

We were roaming in the bush at the rear of his property looking for the potential for some harmless fun one day, and found ourselves peering over the fence of his chook yard. The house itself appeared to be unoccupied, and assuming that the policeman was away on duty, it was decided to disrupt the peace of his poultry.

After preparing our surprise for the chickens we retired behind trees nearby to observe the results. The main focus of interest was the empty four-gallon drum upended in the middle of the chook run. From under its rim there issued a thin wisp of telltale smoke.

The raiding party was still debating the length of time it was taking for the fuse to burn down to the gelignite charge under the drum - when it did.

As seen by the large cloud of feathers mixed with the dust after the explosion it seemed likely that the blast had had the effect of defoliating some of the chooks. We were not out of the woods yet though as the rear door of the house flew open and the copper, after a few long strides was seen standing in his short strides - his underpants, right at the bombsite searching for clues.

It was actually fairly still and quiet after the echoes had died away and the dust and the leghorn feathers were settling, but I noticed that there was no sign of the empty drum. Glancing skywards I froze. The tattered container had been blasted hundreds of feet in the air and now, having passed its zenith, was plummeting earthwards at a great rate.

Unable to make a run for it from behind the cover of the trees, as the irate policeman would have spotted us, we did have a short reprieve, which came literally like a bolt from the blue. The shattered drum had completed its sub-orbital flight and crashed to earth a scant few feet behind the policeman. It was obvious to see that his training had kicked in as he jumped and spun around to face the new menace.

We took advantage of this timely distraction and melted back into the landscape.

There was a pretty exciting game we devised which might have suggested the title 'surf roulette'. We spent much of our free time at the beaches swimming and surfing and acquiring mahogany suntans. As these are enjoyable though tame pastimes, it was decided to liven them up a bit and add a dash of stimulation.

The detonation of a half plug of gelignite is quite loud and spectacular. With some lateral thinking it was considered that we could involve some of this effect in an interesting game that would involve a certain degree of guts on the part of the players. This was to be a test that was likely to sort the men from the boys as they say.

From experience we found that standing in water close to a submarine explosion of this magnitude is not too risky, although the pressure wave can give you a bit of a belt if you are really close to it. To spice things up even more we invited other kids to come swimming with us this day, mostly selecting those who had never played the game before.

You need about half a dozen charges with long fuses to make it interesting. These are lit and then tossed into the waves as they break out in front of you and everyone stands still in waist-deep water to await developments. Moving around was considered unwise as you might step on one. We were very aware that the charges were being washed along the bottom towards us in the breaking surf, with the unknown factor being where they were, as it was impossible to see them in the churned water.

A heavy thump and a high waterspout revealed the position of the first one - and that is coincidentally the point when the panic is likely to set in among the new chums. The remaining underwater detonations go off in succession. The rules of the game stated that you must stand still while this is going on - or going off, and the players who break the line and dash for the beach are declared chicken and roundly heckled.

Without wanting to titillate you further with many more accounts of our fun times – and there were many, I will introduce our grand finale – our swansong.

A really old large dead tree had jutted from the skyline for many, many years on the top of a looming sand hill overlooking the Bunbury townsite. We felt sure that this old relic of a bygone day had no aesthetic, intrinsic, botanical, or historical or heritage value and was therefore in our eyes, worthless. Our team had earmarked it as a solution to a current problem. We had only one detonator left in our inventory but there was still a large quantity of gelignite remaining. My recollection of the actual amount is vague, but it would be true to say, it was a considerable lot to explode all at one go.

It was quite dark when we sneaked up the hill and loaded up the hollow trunk and lit the long fuse, and then took the shrewd precaution of melting back down the hill again.

What I can say about the state of affairs after that explosion is this. The top of the sand hill had suddenly become relatively flat and that solid old tree was reduced to a mix of splinters and smoke and befuddled termites. It no longer played the role of the ancient sentinel on the skyline above the town.

While we witnessed the spectacular event ourselves that night it is always good to hear encouraging reports later from other people about your efforts. The most favourable account came from a school chum the next morning. He seemed to be quite excited when he told his story. It appears that the lad was quietly fishing from the end of the town jetty, which was about a mile from ground zero. He was leaning out to free his fishing line when the shock wave from the explosion swept across the harbour and caught up with him. While it may be hard to believe, he swore that the concussion bumped him off the deck of the jetty and into the sea. His comment was that he thought the army had gone a bit too far this time. There was an exchange of knowing smiles amongst members of the demolition crew on hearing this piece of inspiring news.

Oh yes, there was a bit of an anticlimax to our series of noisy incidents. It occurred at school the following day. Six of us were summoned from class to front up at the headmaster's office. He lined us up like penitent soldiers and then walked down in front of the assembled ranks glaring at each in turn. I have absolutely no idea what prompted his next move. He stopped right in front of me and growled, -

"I have been given disturbing reports that some of our high school students are responsible for explosions around Bunbury recently."

Stabbing me in the chest with his chalky finger he added,-

"And I suspect that you might know something about this Macmillan!"

Though normally a truthful person, it seemed that discretion was called for in this event, and with a show of genuine concern I protested innocence. A stern lecture followed from the stern headmaster and after all the prisoners had declared that they were unable to help him any further, they filed out of the office.

As it was a chilly day I was wearing one of my dad's cast-off suit jackets with inside breast pockets. In a conspiratorial huddle around the corner of the building later, we discussed these recent disturbing developments. I reminded them of the point when the headmaster stabbed me in the chest with his finger - and then unbuttoned the jacket to show them the contents of the inside pocket. My comment was that it was just as well he hadn't jabbed me too hard or he might have blown his hand off. My friends then saw the three sticks of gelignite in the pocket and all made a pretty good show of fainting away.

With our stocks of high explosive finally depleted, culminating in the vaporisation of the old tree stump, we turned to other home made exotic formulas for making effective explosives and pyrotechnics from basic chemicals.

To protect young minds from corruption and temptation, it will not be revealed here the names of the ingredients or the formulas. One of the essential reactive chemicals, which will have to be mentioned however, for the sake of the story, is nitric acid. In its concentrated form it will bubble and fume and eat its way through almost anything it touches. It is consequently regarded as dangerous stuff to handle.

My good friend and ally John and I were strolling with feigned nonchalance down the school drive late one afternoon towards the main gates. Our delay in leaving the school building was due to the commissioning of an inside job in the chemistry laboratory. We had reduced the stock of concentrated nitric acid by a small amount and this powerful stuff was in a corked bottle in John's pants pocket.

DK Wheeler was our dedicated English subjects' master at the high school. He was a stickler for the use of good spoken English but very friendly nevertheless. DK came through the gate at a brisk walk with his academic gown flowing out behind him looking like Superman in flight. We were naturally quite anxious to clear the school grounds at this time and get home safely with our contraband, but the master bailed us up and began to talk. Using the correct use of Common English of course, we chatted for a while.

John suddenly became highly agitated, shifting from one foot to the other with a look of pain and mild panic on his face. He spluttered a terse goodbye and loped with a very strange gait down the path, across the road and into the thick scrub on the other side. With no idea as to the cause for his hurried departure, I shrugged and mumbled something about "when you've gotta go – you've gotta go." DK smiled knowingly, nodded and strode off as I followed the trail of my friend into the shrubbery.

The reason that John had his pants down and was hopping around in dire distress was not as a result of his need to answer the call of nature at all. It was all to do with the effect of the concentrated nitric acid eating into his groin. The cork falling out of the bottle in his pocket and the fiery stuff starting to attack his most tender parts was what triggered his sudden disappearance from our group dialogue with the teacher.

We went into damage control at a nearby water tap and relieved his anguish. His relief on closer inspection, which showed that he had not damaged the family jewels, was heart-warming stuff.

The same friend, whose home was in Busselton, was boarding in Bunbury with a retired schoolmarm known by her students as Ghostie. She was a kind soul and seemed to dote on this handsome lad who lived in the small flat at the back of her house.

With the use of my Dad's metal turning lathe I had constructed a miniature cannon, which proved to be quite powerful, using homemade gunpowder. We had the gun in John's flat where he said he doubted its effectiveness and offered that I point it at the door and bet me that it was just a fizzer. This was an unwarranted and unkind comment to which I took great exception.

A little time later our ears were still ringing from the shot fired in the small room. A shaft of sunlight beamed through the smoke, its source of entry being the ragged hole where the gun's load had splintered the door. There was no doubt that I was very pleased with this performance of course and also for having won the bet.

On meeting John the next day it was a surprise to see him looking as cheerful as he did. Ghostie was fortunately out shopping when the door suffered the load from the cannon, so John promptly set to repair the damage. The appearance of the door after the plugging, sanding and painting was a credit to him and it showed not a trace of the ugly hole. Ghostie was very pleased at the efforts of her fine young boarder who had painted her door out of the goodness of his heart. She loved the colour too. After thanking him profusely for his kindness, she cooked his favourite dinner as a reward.

Ian was one of my closest friends and lived a few blocks away down Beach Road. Together we had quite a few hairy experiences and tempted fate a few times. Ian had cranky neighbours. The father was a doctor and his teenage daughter considered herself rather superior to, and unquestionably a cut above our gaggle of youths in the area.

One of our cronies delivered newspapers every morning on his pushbike, hurling the rolled paper up on each lawn as he rode past. The doctor's house however was somewhat palatial and the lawn was the longest in the street.

The doctor's snotty daughter was waiting for our newsboy at the front gate one morning and gave him an earful for not lobbing their paper up closer to the house. Eager to please and placate the customers, he reckoned he would have to try and increase the range of the paper at this house on the next time around.

Hopping off his bike at the front gate next day he proceeded to dissect the paper. The sheets were now wrapped around a good-sized rock, which he considered would have a greater range than just the rolled newspaper. It did!

Ian, who saw it all, described the outcome of this bold experiment to me. Apparently, loose pages of the disintegrating newspaper peeling off the speeding rock, marked a trail up the front lawn all the way to the house. The rock now free of all encumbrances and obeying the laws of physics completed its trajectory and punched a big hole in the largest pane of the imposing bay windows.

Ian's house was pretty ancient and in the yard were two almost identical, very old and very tall Norfolk Pines. It was a great buzz to climb to the top of these monsters and lob pinecones down on the opposing groups of feral kids from further up the road. As you climb higher in one of these trees the trunk becomes thinner and more flexible. It seems less capable of bearing your weight and begins to bend alarmingly as you near the uppermost branches.

Bunbury experiences very high winds in the course of winter storms and being near to the coast, we often felt their full force. Ian and I would stage a contest, which was best enacted at the height of a good gale, when the wind force would cause the trees to whip back and forth a considerable amount at the top. They were equal in height so taking one tree each we would start to climb. The object of the exercise was to see who had the guts to climb the highest. The wind seemed intent on plucking me off the trunk on the way up and I could see that the tip was doing mad gyrations against the sky.

My mate was higher in his tree than I was in mine. His weight combined with the wind force was bending it like a drawn bow. Watching in disbelief I saw the top of the tree curve over like a shepherd's crook with Ian clinging on for dear life, his legs dangling in space. He was laughing like a crazy man and yelling, -

"I beat ya!"

There are some relatively small but very funny incidents still remembered. This Ian was a big fellow with wide shoulders and a lantern jaw. In keeping with his height, so too was the size of his bike. Standing alongside it you could comfortably tuck the seat under your armpit.

My pal had hurt his ankle, on one of our escapades presumably, and it had a long bandage wound around it. Riding along behind him I noticed the end of the bandage was free and flapping in the wind. The situation started to get interesting when the loose end caught and began to wind itself around the pedal crank and the further he rode, the tighter it bound his foot to the bike. Waking up to his predicament at last, he found the tension of the bandage caused him to slow down and then force him to stop. Normally able to put his foot on the ground to stay upright, he found this impossible on the tethered side. While trying to keep his balance he was teetering one way and then the other. I could see that his centre of gravity was slowly transferring to the wrong side and I braced myself for the inevitable impact. As if in slow motion the big bloke, still fighting to keep his balance, tottered like a big tree being felled. Though generally quite reserved he showed his mettle as he toppled and crashed to the roadway. Throwing back his head he yelled,-

"Timberrrrrr!"

Dad always said that the most effective way of learning is in the school of experience. I have found many opportunities to see that theory in action. We boys were aquatic animals, spending much of our free time in or around the sea. We had decided to explore the bottom of the Bunbury harbour, but realised that to do this required some improvisation.

Searching on the gently sloping beach near the jetty I picked up and nursed a very big rock. Fastened to my belt was a long piece of garden hose with one end clamped between my teeth and the free end secured to a block of pine to act as a float. It was all engineered to allow me to walk down into the harbour, submerge and continue on down into deeper water. The weight of the rock would serve to keep me on the bottom while the float was towed along the surface and I breathed through the hose. The theory seemed watertight.

The science teacher at school next day - I didn't drown, explained why I had not gone very far before having to drop the rock, spit out the hose and head for the surface. It had a lot to do with the pressure of the water on my chest, which increases gradually with the greater depth and prevents the lungs from inflating.

It was just as Dad had said, -

"The best school in which to learn son, is the school of experience."

My First Job

My father seated at the head of the dinner table one night proposed to the family that as Ivan was the eldest son, though still at school, he should look for a part-time job. This would provide a contribution to the family budget and help to offset the cost of satisfying his enormous appetite for food. In addition he declared, he would benefit from the experience, which would help to expand his worldly education.

The whole family agreed and I had to concede that I was soon to join the ranks of the working classes - part time.

Big Barney Hay cut an impressive figure on the driveway of his Highway Service Station. My Dad seemed rather diminutive by comparison as he negotiated with Barney on the terms and conditions of my proposed employment.

My pride knew no bounds when Barney's big hand crushed my shoulder.

"I reckon he'll do. Start Saturday seven o'clock, - that's in the morning, son."

My duties as a service station driveway attendant were varied and almost always pretty strenuous. While muscular development appeared to be taking place on my lean frame already, it was hoped this would be accelerated when pitting my strength against the demands of this new job.

Fuel from underground tanks was dispensed by means of petrol bowsers to the vehicles in the driveway. These tall structures had a large glass bowl on top marked in gallons. Petrol had to be pumped up to this height from the underground tanks by muscle power. With the bowl filled to the right mark a valve was lifted to direct the measured amount gurgling down the hose and into the customer's car.

Looking after one or two clients at the same time was not too bad, but on a busy Saturday morning it could be chaotic. Barney further explained to me – more than once, -

"Good driveway service Ivan, is the key to good garage business and keeps the customers happy."

With this piece of wisdom firmly in mind I did my utmost to attend with a smile to the demands for attention to petrol, oil, radiators, tyres, batteries and windscreens. The omission of any one of these I found was enough to bring a sharp rebuke for the new lad – 'Who had better pull his socks up!'

My dad was by day a bank teller, or as he insisted – 'a bank officer.' Beneath the suit and tie of this typical bank official beat the heart of a frustrated engineer. He possessed a passionate love of gears and cams and piston rods, in fact all devices mechanical. His genes passed down to me ensured that this mechanical aptitude blossomed also in me and I found that the mysteries of such things as valve timing and steering geometry soon became an open book.

Barney brought me unceremoniously back to earth.

"I'll be away for a bit. When you lock up the station, make damn sure you bring in all the oils!"

'The oils,' in one and two-pint glass dispensing bottles of a variety of grades and brands were displayed along the driveway in colourful wire racks with handles. To my mind the display declared irrefutably that this was a service station and not perhaps a flower shop or funeral director.

Carrying out the racks of oils, lining them up on the driveway and then later lumping them back inside the station at closing time signified the decisive beginning and the end of the service station day. So it was burned into my brain with letters of fire that I should never forget this important ritual.

Those school holidays rolled by and merged into a delightful sweaty time of frenetic activity, while I inhaled the pungent odours of petrol, oil and exhaust smoke. The pride and satisfaction I experienced in this my first paying job knew no bounds. The feel of aching limbs at the end of the week was tempered by the jingle of a few florins in my pocket and it seemed this was reward enough for a job well done.

You could always find me in the chair at the end of the meal table opposite my father. From here I regaled my captive audience with vivid and lengthy accounts of my day at the service station. My sisters fidgeted and rolled their eyes with boredom but I usually persisted regardless.

With a view to possibly cutting short one of my verbal deliveries, Dad suddenly announced a family outing. Would we all like to go and see a film at the local picture gardens? The younger siblings of course could not contain their excitement and were soon scrambling, chattering and laughing into our Fiat family car for the drive into town.

My entreaties to drive slowly past *my* service station were met with Dad's grumbling compliance and we approached just as dusk was falling. I demanded full attention from the whole family with expansive gestures as to where they should look as we approached.

Was this a nightmare? There in the dusk were the serried ranks guarding the driveway still! The lights of passing cars winked and flashed on the bottles of Shell and Castrol and Texaco etched against the dark doorway of the locked building. My look of disbelief was mocked by the rows of bottles with the unspoken message, –

'You forgot to bring in the oils!'

At that painful moment I resolved to take stock of my life. It was clear that my worldly education had scarcely begun and that it would be necessary to strive to do much better in the future. While I did not know it then, there were to be many more and much harder lessons to be learned in my life that would have this one remain as just an amusing faded memory.

Sister Gwen meanwhile at sweet sixteen was the belle of the ball.

The Awakening

As with most young men past and present, I experienced the chaotic physical hormonal changes that grab you by the scruff and propel you into adulthood. Few would disagree that this change of life is a heady time of strange and powerful emotions, dreams and desires, when one is drawn inexorably towards members of the fair sex. Normally quite content with my own company, or having fun and competing with my schoolmates and getting up to mischief, I felt there must be more. Perhaps I was by nature not group-oriented – or whatever the hip term is for that condition. Frequently going out alone, it always felt right for me.

My male metamorphosis seemed to begin in earnest around the age of fourteen with the onset of the physical changes we normally experience around that time. By the age of fifteen the surge of chemicals charging my system had really begun to take effect. While acutely aware of the presence of any lissom female within sight, my approach to such a person however was in a manner less than urbane and sophisticated. I wondered later what the girls, who were the object of my fleeting liaisons, thought of my mumbling, bumbling red-faced attempts at conversation. These first efforts at seduction were most unremarkable and the pain of embarrassment during the first meetings I found to be agonizing. There was a teenager's birthday party that I attended. The young guests were having a marvellous time playing spin the bottle. Everyone sat in a circle and you had to go over and kiss the girl the bottle was pointing at when it came to rest. I was in a cold sweat all night at the thought and put a brave face on it, but to no avail. A little later you could have seen me groping my way down through the back yard in the dark and losing my supper all over the petunias.

Things soon began to look up however. Seated alone in the third row of the Mayfair picture theatre in Bunbury one Saturday night I became aware of a subliminal mating signal coming from a point four seats away to my right. My mother had taught me that it is rude to stare at people but it was impossible to take my eyes away from this lovely creature whose bare legs were crossed enticingly. While she appeared oblivious to my existence initially, it was just for a fleeting moment that I thought I caught her looking in my direction. Her red painted mouth curved in a mysterious Mona Lisa smile and my heart raced.

The following Saturday night could not come soon enough for me and there I was in the picture theatre again – same time, same row and same seat and in a fever of anticipation. Everyone rose loyally to their feet as the strains of God Save the King filled the theatre, but when the lights dimmed, I felt crushed, as *her* seat had remained empty this time. From the corner of my eye later, in the flickering glow from a Popeye cartoon, there the familiar figure had appeared again – but now she was closer, having moved one seat towards me. There was however still a yawning gap of three seats between us. You may not believe this as it seemed incredible to me too, but after the short space of only three weeks during which time she moved one more seat closer each time, I finally sensed more than saw her actually sitting next to me.

Intermission came as the first film finished – we called it 'interval' then, as there were always two full-length screenings with interval in between. This was the opportunity to escort my newfound companion outside and into the milk bar next to the theatre. In a state of happy and breathless intoxication, I asked what she would like to drink. Her order was for a spider - not a hairy insect, but the favoured drink bought by the blokes who were really serious about impressing their escort. The outlay for this severely depleted my finances so I gracefully declined having one myself and just fondly watched her consume hers. A spider is a tall glass filled with a soda fruit drink into which is floated a large dollop of Peters ice cream. She was pure poetry as she manoeuvred the long spoon trying to stab the elusive ice cream blob, and then slurped the drink right to the bottom with the straw. It was no surprise to me and I understood completely when she asked for a cream-between to follow. This is a kind of ice cream wafer sandwich, also off the top shelf, and then some chewies to take back inside. I said that she had made a good choice when she asked for Wrigley's Juicy Fruit, which is really a girls' chewie. I preferred Spearmint or the more masculine PKs, although I could not afford these for myself as I had by this time squandered my emergency bus fare as well.

In a small soft voice she asked my name and looked up at me from under long mascaraed eyelashes. I said nothing, being incapable of coherent speech at that time though my stomach was rumbling. During the second film after interval while staring fixedly at the screen but really seeing nothing, I experienced a feeling like an electric shock. A small soft hand had crept into my sweating palm while a cloud of hair brushed my face as her head rested gently on my shoulder. Like a summer garden in full bloom her perfume overpowered my senses. If this was what heaven is like I thought – I will die happy.

Outside after the show in the cold night, my lovely friend asked if I would see her home. This was customary under the circumstances I had heard, and young bucks, eager to impress, thought nothing of walking a girl to her house and then tramping all the way home again. The actual distance travelled on these expeditions did not seem to matter, though it was better if it wasn't raining. My walk this night in the cause of true love was going to be for a distance of nine miles – that is five miles there and four miles back home. This is of course nothing for a young buck who is fit and strong and just happens to be floating with his feet at some distance above the ground.

In the pool of yellow light cast by a solitary street lamp she turned towards me, arched her back and offered her ruby lips to be kissed. Overcome with passion and with more than a little confusion, I missed the target by a fair margin on that first attempt, but was determined to do better next time. Events were taking a very exciting turn and my confidence had begun to soar.

With no heated motorcars available and little public transport either in those days, in the wintertime and at that time of night, we had to do a lot of walking to get around. This was a good example of a situation when we had to do that. Consequently everyone you saw out and about would be rugged up with overcoats and scarves and even gloves to keep the chilblains at bay. My recently acquired pride and joy was the long warm overcoat that I was wearing. In a fashionable light tan colour, which the label on the collar claimed to be 'camel', it was tailored in a material that allegedly was mohair. While usually interested in learning about most things, I was mystified as to the origins of this fabric as I didn't know what a Mo was anyway. It could have come from a camel for all I knew.

Further along the dark roadway heading out of town, that small sweet voice coming from beneath my right armpit was complaining that her hands were cold. She asked could she please put one of them in my pocket to thaw it out. This was OK by me of course, so while experiencing a surge of chivalry I agreed with as gruff and as manly a voice as I could muster. It was only a split second later that I was stopped in my tracks with the shock of the next development.

That tiny questing hand had disappeared into my overcoat pocket like a rat up a drainpipe. It slithered in through the pocket vent, down into my trousers and then further down in a flash to seize the family jewels. The busy fingers latched on to my manly tackle like the tentacles of an eager octopus, or a groper - and they were freezing!

My euphoric and romantic dream world disintegrated and I came back to earth with a thud. This sudden invasion of the very core of my manhood was most unwelcome and not to be tolerated. While I had been brought up to respect and even revere females of all ages, shapes and sizes, this was unprecedented in all my experience. My sudden and violent reaction to this unseemly move was in a way understandable - it was an instinctive act of self-preservation in response to the sneaky attack.

It would not have surprised me to learn that her little arm had been fractured where I delivered the hard karate chop, followed by being dragged out and away from its target. Leaping back like a startled rabbit, I hurriedly readjusted my kit. Giving my opportunistic lover a withering glare, I just stomped off into the night without a word or a backward glance.

My lost love was spotted just the next day in town with some other colt, and I wondered if he realised what might lie ahead of him and the kind of risks that he was taking. It can be said that in my more mature years I became more amenable and slower to take offence in similar situations. Experience tells me that perhaps I made an error of judgement and treated her unfairly on that cold and frosty night.

It is good to reflect sometimes on the number of missed or bungled opportunities of this kind, which occurred during my young life. That was my first, and while I felt shame for what was done to a defenceless young girl, the recollection returns to me with more than a tinge of regret. Had I been older and more mature and experienced – like being seventeen for instance, those events may have progressed to a totally different conclusion and without all that unnecessary drama and acrimony.

We live and we learn.

I discovered that there is safety in numbers

Episode Five

<u>Family Transport</u>

Like most boys at about my age of twelve - and older, the motorcar fascinated me. It was 1939, the war years were upon us and to buy and run a car was beyond the reach of many families. The no-car families in those times relied on public transport and could be seen climbing aboard the clattering boneshaker buses, which ran a service from town out to our neighbourhood in South Bunbury. Bicycles were everywhere and seemed to serve as the next prime mode of transport for all ages. Many were the modifications to the bikes you could see on the streets, some kind of trailer being most popular, which was great for doing the weekly shopping and carting odds and ends around. My sea-going tin canoe was fitted with wheels and was towed this way to the beaches and to the estuary.

In contrast to his daytime job as a bank officer, Dad was also pretty clever in the mechanical sense, which was a decided advantage if you owned a car. Most owners had to be passable home mechanics and able to do their own repairs to keep their vehicle going. Although simple in design, the cars of the day were pretty quirky and temperamental and generally responded best to those who knew their foibles and idiosyncrasies. Our first vehicle was a 1924 Fiat tourer with shiny black fabric top and tall wood-spoked wheels. It boasted a set of side curtains for wet weather touring which kept out most of the rain and freezing winds in the winter, but which flapped incessantly at speeds in excess of thirty miles an hour.

The spokes on the road wheels must have had a close similarity to lamp posts either in size, smell or colour because the dogs loved them. I have seen a canine ambling past on more than one occasion and spotting the wheels of the Fiat make a sharp turn and a beeline for it and cock his leg to make his signature. This fatal attraction for dogs I suspect was because the old bus would carry around the odours of many dogs to various parts of the metropolis with an open invitation for all the rest to sign the book. Dad, upon seeing this delinquent act or spotting the result trickling down over the wheel would be enraged. All within earshot would hear the standard explosion of anger.-

"Damn dogs are fizzing all over my tyres again!"

The car, or more precisely *this* car, required the attention of a driver with good mechanical aptitude to coax it to start, especially on a cold morning. Situated on the top of each engine cylinder was a little shiny brass cup with a tap under it. These all had to be 'primed' with a little petrol. Having done this, the aspiring driver went to the driver's seat to adjust two levers on the steering column to the starting positions. These were the throttle and spark control levers and if not set correctly would cause the engine, as Dad put it, to "boot your head off." Back at the front of the car again, one would reach over the big bumper and swing the starting handle to crank the engine. You didn't have to worry about turning on the ignition as the magneto was always ready for action and as starter motor technology was still very basic at the time, it seldom worked well enough to start the engine. This was assuming that one of these new-fangled devices had actually been fitted in the first place.

The car owner had now reached the tricky part. If the levers were incorrectly set initially or the wrong cranking technique was used, the engine could viciously kick the cranker and cause a serious injury. A sprained or broken wrist could be the consequence of your carelessness.

It was a great day for me, when under Dad's tuition I strained my youthful muscles and coaxed the Fiat into life. After that first successful effort, I was nominated Chief Starter of the family car.

This car had buggy-type springs and mechanical brakes operating on the back two wheels only. In the next model the brakes worked on all four wheels and this great leap forward in braking technology raised fears that in stopping so decisively, a motorist behind could run into the back of you. This risk was overcome by mounting a little red warning triangle on the back bumper with a cautionary sign, 'Danger Four Wheel Brakes.' This proved effective in practice as nobody ever did test the back bumper. The steering wheel was of necessity very large in diameter to provide the leverage required for manoeuvring the car, and the gear lever and brake stuck up from the floor on the right hand side of the driver's seat. It was necessary to clamber over these when getting behind the wheel. Considering the generally poor stopping and steering ability of the old bus, it was probably fortunate that our top speed on the flat was only forty-five miles per hour. All this uncertainty however enriched our enjoyment on family outings when travelling to our destination - and hopefully home again.

My first real paying job was at Barney Hay's garage and service station in South Bunbury. Helping the mechanics in the workshop out the back fed my love of engines and all things mechanical, but I will never forget my involvement with the gas producers.

Born out of necessity by the severe shortage of petrol due to wartime rationing, this Australian invention kept a lot of cars going on the roads. A car with a gas producer is actually fuelled by the gases given off from burning charcoal.

The burner itself, about the size of a forty-four gallon drum, usually sat over the rear bumper. Packed with good quality charcoal and ignited, it generated large quantities of carbon monoxide gas, which was fuel for the engine. One of the commercial travellers in Bunbury claimed that he could drive the two hundred and forty miles to Perth and back on a bag of white gum charcoal, at a total cost of four shillings a bag!

The downside of these contraptions apart from their comical appearance was the requirement to clean out the gas filters periodically. They contained an oily mess of black slime and servicing them must have been one of the dirtiest jobs on earth. As I was on the bottom of the totem pole in the garage at the time, it became my privilege. I frequently arrived home with two eyes blinking out of an Al Jolson face – filthy but happy.

There is eighteen months' difference in the ages of my brother Joe and myself, with me as the eldest. At the age of seventeen we both secured apprenticeships in the Government Railway workshops in Midland Junction. My job was that of an aspiring mechanical fitter, mainly building and repairing locomotives. Joe's trade was that of a patternmaker. These tradesmen made very precise patterns from wood, which were used to form the moulds for metal castings in the iron foundry.

With very little money, as my first year wage per week was twenty-five shillings and nine pence, we had to be very canny in the purchase of a vehicle. Our accommodation in Perth was in various boarding houses or with long suffering families. Home was in Bunbury, which was 115 miles south, and with little to keep us in Perth on weekends we naturally wanted to make the trip as often as we could. Extra money had been accumulated by picking grapes in season for the purpose of buying some mode of transport. The government provided us with generous sick leave entitlements, which everyone used for the second casual job. We counted the kitty one day and came up with the sum of fifteen pounds.

We surveyed the rusty frame and flat tyres of a 1927 model Harley Davidson motorbike under a pile of timber and wire netting in a chook house in West Midland and made our offer to the owner. Undeterred by the layers of rust and dust and chook manure, which coated our prize, we both agreed that the old machine had potential and dragged it home to begin restoration.

Our new acquisition had two folding running boards for the rider's feet, a single foot brake for the back wheel only and a spark control twist grip and throttle on the wide sweeping handlebars. A rocker foot pedal operated the clutch and the big knob on the left side of the tank was the gear change lever. A brass plunger pump sticking out of the middle of the petrol tank was there to lubricate the engine. We discovered later that the surplus oil ran out onto the rear chain and thence was flung off to paint a big black oily stripe up your back. The engine had external valve operating gear, which pounded up and down just by your knee as you rode along. There was no provision for lubrication here but we found from experience that the random use of an oilcan on the moving parts reduced the valve noise to a dull clatter.

With two of us to share the bike, we had a problem. There was no provision for carrying a passenger, though it looked big enough to handle the extra load and more.

Calling on all of our trade skills we cut a piece of plywood and bolted it to the top of the rear mudguard to act as a pillion seat. It needed some padding however to reduce the punishment on the spine of the passenger due to the rigid frame and total absence of springing. We improvised by using materials close at hand. Some strong green canvas was stitched up to form the seat cover, which was then nailed to the board. Stuffing the whole thing with chook feathers provided the padding. I remember watching Joe blast off up the road in a haze of blue smoke and noise one day followed by a swirling trail of white down. The seat had leaked apparently, which was not a major setback as we always kept a good reserve supply of feathers and the landlady's hens were on the moult at the time.

Joe astride our 1927 Harley - it had been restored to pristine condition.

One of the problems faced by drivers and riders of cars and motorbikes both during and after the war years was the savage cut in fuel supplies, most of which was diverted to fight the war. Our ration entitlement in the form of a small coupon, which had to be presented at the service station, was for one gallon of petrol per month. This was not going to take us very far on a big beast like the Harley, so we had to diversify. Some cars fitted gas producers which worked well for them, but what to substitute for petrol in our machine? After some experimenting with the likes of white spirit, Shellite and other volatile substances we found that the old girl would run on power kerosene, though it needed some coaxing. Kerosene was not rationed, and we discovered a ready supply from friendly farmers in the district. A motorcycling friend of ours showed great initiative as he rode around the suburbs. To keep his machine going he pulled into every service station en route – this usually after hours, and drained every one of the delivery hoses on the petrol bowsers into his tank. That's enterprise!

It must be all right in these later years to divulge the secret procedure we devised which saw us eating up the miles and leaving less fortunate motorbike riders behind with empty petrol tanks.

We very carefully siphoned our one gallon of precious petrol into the small reserve fuel tank on the Harley. Kerosene went into the main tank. Let's mount up and make a trip to Bunbury.

The old bike responded to a hefty boot on the kick-starter and chugged away happily on the diet of petrol. Experience taught us the precise moment when the engine was hot enough to continue to run on pure kerosene. By opening and closing the fuel taps in the right sequence, the kerosene began flowing into the carburettor. Riding on a dark night was a good time to observe the changes in behaviour and was also the best time to enjoy the special effects.

When we switched over the fuel taps and the motor began to gulp the kerosene, the rhythmic Harley engine beat changed to a kind of throaty bellow. In the darkness the exhaust pipes could be seen to glow red-hot up near the engine and then the incandescence of the metal gradually spread down the pipes towards the back. A tailpipe flame would appear rearwards like an aircraft jet engine on afterburner. It began in ragged red colours to change to an angry blue spearing out in the darkness upwards of half a meter behind.

There was another effect, which was described to me by a friend who saw us blast through the town of Waroona one night on the Bunbury road. We were apparently leaving in our wake an acrid cloud of kerosene fumes, which he described as like the passing by of a thousand hurricane lamps.

There was a neat ploy I devised while riding the bike that was guaranteed to startle any neighbourhood. Accelerating down a street at full throttle you suddenly retard the spark on the handgrip and switch off the ignition. This builds up a big charge of petrol - or kerosene fumes in the exhaust pipe and at the crucial moment the ignition is switched on again.

The resultant loud bang and bright flash out of the back would seriously challenge the performance of any one of Lord Nelson's ships' cannon.

In a fit of bravado intended to impress my girlfriend, who lived nearby in Bunbury, I repeated this performance rather late one night. After the customary deafening explosion, a shower of sparks along the road and a loud clatter suggested that the exhaust pipes had been blown clean off the cylinders.

The ride back home afterwards was a delicate affair of balancing the hot amputated exhaust plumbing across my knees, while trying to tone down the racket of the two big cylinders banging away and flaming from the exhaust ports.

A trip on our machine from Perth to Bunbury was one of high adventure, especially with the hazards of the dark and stormy nights of mid-winter.

Rider and passenger would both kit up to provide insulation against the cold and the rain. This usually consisted of old surplus military clothing bought cheaply at disposals stores. Army greatcoats, flying boots, overalls and a flying helmet or balaclava would complete the ensemble. Wearing the coat or jacket backwards, buttoning it up the back and stuffing newspapers down the front helped to delay the onset of hypothermia. A large pair of flying goggles of the type favoured by the Red Baron in his aerial dogfights during the First World War completed the image. A piece of tarpaulin was handy too. Used as a lap cover, it kept the lower parts relatively dry and trapped the heat from the engine too.

There was a yarn doing the rounds about that time of a motorbike rider who used to wear his jacket on backwards to defeat the cold just as we did. It seems our hero crashed one day and was lying in a groggy state by the side of the road. The attending policeman questioned a witness later as to the circumstances of the accident, as by this time the crash victim had tragically succumbed and was now deceased. The witness is reported to have said, -

"He seemed OK at first until I went over to help and turned his head back the right way around."

That same aforesaid girlfriend - an avid motorbike rider - was to become another of the victims of the Harley one other dark night. Returning from a spin out to Dardanup we approached the Picton road bridge. This was an ancient timber affair with crosswise decking. A vehicle going across the bridge ran its wheels along two parallel narrow rattling planks, which was all right for cars but decidedly hazardous for motorbikes even under normal conditions.

The shower of rain, which had wet the timber runners and made them dangerously slick, brought about our undoing. I don't ever remember our tyres boasting anything resembling a proper tread and as we approached the bridge around a curve, natural forces took over. It is an amazing sensation when a two hundred-pound motorbike disappears from beneath you, but it did and we crashed at some considerable speed in the middle of the bridge.

My flying body hit the bridge railing while the bike careered onwards protesting loudly with a lot of sparks. It seemed likely that I had suffered no serious injury, but was suddenly concerned for my pillion passenger who had separated from the bike also. She was following a rather higher trajectory than me and so was apparently airborne for longer. She was a well-built lass and actually thanked me later for providing my yielding body for her to land on. This secondary impact did knock the wind out of me however and I stayed down for the count.

On the runs down along the Southwest Highway on the bike, Joe and I took turn about to ride sitting in the big comfy sprung seat. The other perched in misery with gritted teeth, feeling every bump in the road through the custom-made pillion seat stuffed with chook feathers.

It was a dreadful night when we were hammering down the highway with Joe at the controls. With my head bowed to avoid the stinging rain, I began to sense after a time that all was not as it should be. The bike was bucking over rough ground and shrubbery and the white marker posts seemed to flashing by on the wrong side. Joe had gone to nigh-nighs and still at considerable speed we appeared to be cutting a new bush route to Bunbury.

To his credit, Joe rallied when I yelled in his ear, and taking immediate stock of the situation and with no reduction in pace, manoeuvred around a number of solid objects, like trees, and soon had us safely back on the road again. It occurred to me that in the confusion of weaving through the bush we could have reversed direction to find that we were travelling back the other way. This was not a worry however - Joe's sense of direction, even under those conditions of darkness and rain and confusion, was infallible.

The old bike almost met its end on one return trip and we along with it. Lulled into a sense of complacency by the steady thump of the engine, we were both startled by an ominous explosion from the motor. There was a loud bang, the bike began to cough and run on one cylinder and then there was an overpowering stink of hot kerosene fumes. Still trying to maintain

An original example of our J model Harley.

speed we looked for the cause of the trouble. The motor had suffered a catastrophic failure and blown a big chunk off the top of one cylinder head. The flying piece of engine had ripped a gash in the bottom of the fuel tank. The hot exhaust, our legs and the whole side of the bike were awash with fuming fuel and our instant immolation seemed certain. Like the fighter pilots of old who put their planes into a high speed dive to extinguish a fire in their aircraft, we kept going as fast as possible, leaving a trail of fuel and motorbike parts behind us.

Wheels

We must have been little different to the majority of boys approaching their seventeenth birthday. We shared the major goal for that time in life, which was to have our own transport. For me, a conveyance of some kind no matter how humble, spelled freedom and independence and the ability to go on the merest whim, to wherever I chose. Another passion manifesting itself about that time also was success in the pursuit of girls. These two desires were linked in that if you had the first, then the second would surely follow. This was not always true in practice, as we were to discover.

Apprentices in our day were synonymous with pushbikes. With little or no money in our pockets, the trusty treadly served us well, albeit with a lot of effort and acute discomfort. I rode the five miles to work each day for a number of years in an easterly direction, returning on a westerly course home. The summer weather pattern, apart from having very hot days, also provided trying conditions that when riding in the morning, the howling easterlies coming down off the hills were enough to almost stop you in your tracks. The sea breeze in the afternoon on the other hand, while pleasant for some, was another headwind that had to be fought all the way home. There was a time when it seemed certain that a headwind was the only kind you could get – when you are riding a bike that is. To further add to my misery, I was forced to labour in the wake of many of my older and more affluent workmates who zoomed past on motorbikes and in cars oblivious to my strain and pain.

While brother Joe and I shared a kind of love-hate relationship with our 1927 model Harley Davidson motorbike, which was considered pretty outdated even in those years around 1945, it did serve us well. Running happily on a diet of kerosene during the war years and for some time after, it carried us   for many miles.

Time moves on however and we sought to enjoy a more comfortable and less risky mode of transport. I looked forward to owning a car. Lack of finances precluded the chances of buying a new one however as a stroll through the city peering into showrooms confirmed my worst fears that a new vehicle was out of our reach. Some new cars were being offered after the war and we gazed longingly at the shiny models, realising that we could not come within a bull's roar of owning one. The asking prices will give you an idea. The imported English Morris tourer was a natty little car at 480 pounds ($760) cash. The Italian baby Fiat,

The Citroen and the Harley - typical transport

89

priced 'to suit everyone's budget' was a steal at 299 pounds ($598).

Our new stepmother had four brothers, Lloyd, Allan, Don and Bill. When Dad married her we found ourselves with four instant uncles - and good blokes to boot. We had much to do with Don and Bill who though older than us were the youngest in their own family. Both were married, but more importantly were comparatively affluent and owned cars. These they most generously offered us to use for that special occasion - like taking the girlfriend to the pictures or out on a picnic in style.

Don's only vehicle was a Baby Austin tourer. It was a real caricature of a car with a tiny body and large, skinny wire-spoked wheels. While not a startling performer it would always get you there and back. This was not the car's fault of course, but while I was driving the Austin one day it ran out of petrol and stopped. Not usually a big problem, but this time it was straddling a railway level crossing in Cannington. There were no red flashing lights or bells or boom gates in these days to warn motorists of approaching trains, you had to depend on your faculties of sight and hearing. Yes, there was a train coming towards me at a good speed, which suggested that I get the car off the tracks without delay. Being alone, I had to get out and push which was easy enough for a strong lad. The steep slope in the roadway however caused my vehicle to gather speed at a great rate and take off without a pilot. I thought later that the scenario was reminiscent of a Mack Sennet film. Pounding alongside the car striving valiantly to get in behind the wheel, I was at the same time aware of the roar of the train as it passed close behind me. The driver gave a loud contemptuous blast on his whistle and then I heard some passengers cheering.

Bill, the other instant uncle had a spare car, which he offered for our use. This one bore the impressive name of Austin Meteor. More in the style of a vintage sports car, it was a two-seater with the body style of one of the old Brooklands racers. It had a rounded tapering back end, which gave the impression of some race breeding, but with its little four-cylinder engine, a meteor it was not.

We lads had been collaborating with Uncle Bill, modifying this car to make it longer with the aim of fitting a rear seat. This was in the hope that we could fit in more girls – or some girls – in fact any girls who deigned to ride with us.

The body of the car we cut clean across behind the front seat and inserts were welded into panels and chassis to give it about two feet extra length. This work altered the original proportions of course and we had to bear with some small inconveniences as a result. The two side doors were welded shut to improve the strength of the body. As there was no top on the car this was not a bother. To get into the car you just grabbed the windscreen pillar and vaulted over the side into the seat. Also while in this state of partial modification it lacked any kind of a floor in the front section. There was little shine on the body panels as it was finished in a nice grey primer undercoat.

The vehicle was still in this state when I attained the magical age of seventeen and so was eager to take my driving test. There were no driving schools in existence around those times, so all aspiring motorists would have had instruction from fathers, brothers, uncles or neighbours willing to help. I had been driving my dad's car from the age of twelve, so I felt reasonably confident as I pulled up outside the Midland police station in the Meteor.

You may have picked up the comment elsewhere on the friendly attitude of the police towards the public in that era. More likely to understand and assist than persecute and prosecute, they were usually gruff but friendly. I was about to test the patience and forbearance of the big bloke in blue with the clipboard who was about to take me for the driving test – or was it me taking him? There were in those days no special categories for learners and probationers or any of that rubbish. You either passed the test or you didn't.

The fact that I had driven alone and unlicensed that day to the police station and arrived in one piece would have given me some credit points already I reckoned.

The big copper stood by the car while I explained some of the minor differences with this car that perhaps some other cars didn't have. The welded-up doors, the pine plank for a seat and the missing floor in the front was noted I am sure. Apparently unfazed up to this point, he rolled his eyes a bit as he stepped over the side of the car from the kerb and found himself inside standing on the bitumen roadway. As he eased his bulk back onto the plank, I indicated that he might be more comfortable with his feet resting on the piece of pipe running across the car under the dash. We took off up the road in a haze of blue smoke, in the direction of Guildford Grammar boys' school.

There was a hazard I thought it would be advisable to point out to my examiner before we had gone too far. It was a big flexible drive coupling exposed at the back of the gearbox. With bolts and split-pins whizzing around next to his leg it was difficult to see, but was capable of inflicting severe injury if contacted by the unwary. It was a relief to see him move his feet and cram his legs against the side of the body where the door used to be.

The particular driving skills for which I was tested seemed to be pretty much as they are today. The indication that you intend to change direction is however different now. We no longer stick an arm out at right angles for a right turn or have the forearm vertical with palm flat for a left turn or stop. He seemed bemused when I remarked that my failure to check the rear vision mirror was because it was still in a box under the bench at home. With no top on the car we had a 360-degree view anyway. This line of reasoning appeared to be acceptable.

There is a standard perennial test that has seen the undoing of many learners over the years – the hill start. Executed skilfully, the car should move off smoothly from a stationary position up a hill without running backwards too far. This was executed with panache on a very steep pinch on Morrison Road. The policeman seemed to have something on his mind. He asked – quite civilly of course, why I had not used the handbrake in the last manoeuvre as most other drivers did. Being truthful I had to admit that the car did not have a hand brake – this also as a result of the unfinished modifications. It was at home in the box under the bench - with the rear vision mirror. The hill start had been accomplished by heel-and-toeing. It takes practice but it works. With your toe on the brake and heel on the accelerator and transferring foot pressure from the first to the second, this has the desired effect in giving a smooth takeoff. We used to claim that it made a handbrake unnecessary.

A good trick my Dad taught me also was a technique that made gear changing quieter and smoother in those old crash boxes. Probably seldom resorted to in the cars of today, I did a bit of this fancy footwork to impress the policeman. The procedure is called the 'double declutch' - not double de clutch, and while he didn't remark on it specifically when I executed this classy procedure, I am sure that it did not go unnoticed.

Recent rain had formed a puddle in a dip in the road, which had to be forded on the way back to the station. For most drivers in most cars this would not pose a problem, but momentarily forgetting about the missing floor I just powered through the water hazard. It was disturbing to see all that muddy water shoot up and splatter the examiner's nice clean navy trousers. I wished for the earth to swallow me up. We didn't talk a lot after that.

You could be excused for predicting that my driver's licence was refused then and there. On the contrary, while this tolerant and friendly bloke was wiping the mud off his pants he explained that his job was to test the driver not the car. He declared that if I could drive **that** car I should be able to drive anything.

"But get that bloody thing off the road!"

Harry was our firm friend and a likeable larrikin with a great sense of humour. Just a few years our senior, he was financially viable and possessed a most desirable set of wheels – an English MG TF sports car. Joe and I were still indentured apprentices at the Midland Workshops and wanted to return as often as possible to our home in Bunbury. We seldom missed a weekend travelling the 120 miles either riding or driving one of a weird assortment of cars and motorbikes. We even rode pushbikes one Christmas, so strong was the homing instinct - but only once and never again. In fact I would rather not talk about it.

Driving in the MG with Harry however was pure class. Low slung, with a long bonnet and flared mudguards; this car was a real head-turner. There were many country runs I can recall, but there is one, the memory of which always causes me to smile.

Returning to Perth one dark night on the Bunbury road, the MG was purring along at a good speed. We saw a pair of wide-spaced taillights ahead that we were overhauling and would soon overtake. Harry always kept stashed in the car a man-sized army bugle, the blast from which would wake the dead. Not that the car didn't have the regular hooter, but this was more fun. He pulled it free as we drew level with the big powerful car and took a deep breath. Above us I could see the driver of the Jaguar relaxed at the wheel with scant concern for the impertinent MG sports car that had ranged alongside him. His disdainful look turned to one of alarm as Harry blasted his ears through the open window – TA DA – TA DA! Harry then put his foot down hard and we took off. The MG was straining and vibrating at its maximum speed of seventy-two miles an hour as we showed the Jag a clean pair of heels – for a while.

The Jaguar's headlights were enormous and we were caught in their twin beams like a bomber in a searchlight as we drew ahead. Still straining flat out we noticed the two beams lift perceptibly. Someone said "he's accelerating" and indeed he was. The two glaring disks on high beam grew rapidly closer and larger in the mirrors and the big car flashed past at a speed I estimated at MG+60 mph. Harry had to fight with the steering wheel in the turbulence behind the projectile that had just burned us off so contemptuously. The two taillights ahead contracted to pinpoints and then disappeared. We cackled like fools all the rest of the way home. That memory is just one of the many that rest easy on my mind

So involved were the young men of the day in their cars that they often gave them names. You get to know your vehicle pretty well when you have had it all in pieces with its entrails all over the yard, repaired the worn and broken parts and reassembled it into a viable running machine. Like any member of the family it deserved a title of some sort with which it could be identified. Dad's old Fiat was given the name 'Tipo.' This had something to do with an Italian inscription he found on the engine and the name stuck. 'Tin Lizzie' was the generic name for the model T Ford while Chevrolet had their 'Peacocks' and 'Grasshoppers' too. Possibly due to the response I got from one of my cars after an overhaul, it earned the title 'The Coughin' Coffin.' There are few who would not have heard of the famous old car of the films named 'Genevieve.' I have mentioned the teenager's dream of attracting girls if he had a car, which was probably the origin of the highly optimistic name given his car by one lad - 'The Mayflower.' When asked why he chose this title he would explain.-

"Because - many are the puritans that have come across in it!"

The first cars we could really call our own had to be cheap for obvious reasons. This also meant they were second hand and pretty ancient. They were also small, so they didn't use too much fuel. This wasn't all bad, because they were very basic and uncomplicated in design, which also meant they were easy to fix when they stopped.

Most drivers in the early times had to be their own mechanics, so with sufficient knowledge and a set of tools they were almost always assured of being able to get a car going and keep it going. There were some events and situations with our cars that I am sure will seldom be repeated in modern times. My first and last attributable-to-me motor accident occurred just after I obtained my driver's licence. We all know this is a dangerous time of your life when you have the piece of paper but you don't have a sufficient amount of the most important ingredient - which is experience.

My Dad's Fiat saloon car was built solid. Being solid, it was also heavy as most cars of the twenties were.

While driving the Fiat on some questionable errand into town in Bunbury, I was horrified to suddenly see another car appear in front, coming apparently from nowhere. The Fiat was aimed at the dead centre of the intruder and it seemed that a collision was inevitable. It was a big Essex tourer, possibly a bit larger and heavier than my vehicle and he had come flying down a hill out of the sun and across my intersection. The Fiat hardly faltered in her stride as she ploughed into the side of the other one, and breaking the grip of its tyres propelled it sideways across the road, pinning it to a power pole. We disentangled the two combatants and after an exchange of details, both parties just got back in and drove home. Dad was not amused at the sight of his bent baby and insisted that I do the right thing and straighten the front bumper, which I did. Chaining the middle of the bumper to a strainer post in the corner of the fence, it was then a matter of just backing her up. I assume the other driver would have enlisted the services of the local blacksmith to straighten some brackets on the side, which were bent. We heard no more about it - an accident is an accident - or it was then anyway.

While on the subject of motors, it would be remiss of me not to give my 'A' model Ford some space here. Bought in Derby in the Kimberley in 1960, I drove it around the country there for a number of years. This old bus had loads of character and provided many hours of enjoyment and reliable transport out in the roughest conditions. The old Ford proved to be just as roadworthy and dependable under these extremes as a new modern vehicle, if not more so.

We saw the old girl for the first time one day, chugging into the town of Derby to park outside the pub. The dusty character who disembarked walked stiff-legged into the bar, where I accosted him to ask a few questions.

"Come very far?" - It looked very much as though he had been on the track for a while.

"Just over from Queensland across the top. Me bum is pretty sore though 'cause the car doesn't have any seats and I've been sitting on a bloody plank all the way from bloody Cairns."

The car was a 1928 model 'A' Ford Roadster, which back then had appeared all shiny out of the factory with a black canvas hood and a dickie seat at the back. So it was thirty-two years old, but it was still licensed and running and had just breezed in from Queensland. It needed to have some stamina to do that. This specimen had been converted to a crude utility. So much interest did I show in his car that he asked me to look after it for him as he was working away for a spell. I said yes and bought him a beer. A month later he rang and offered me the Ford for twenty-five quid. Why would I not accept?

It seemed to have a good potential for me to use as a knock-about vehicle. It has been said that the rough and rudimentary roads and tracks of the early days were specifically made for the model Ts and the model As of the 1920s to drive on. Straight out of the factory it must have looked a treat and was capable of attaining a speed of 65 miles per hour – or 105 kilometres per hour. It was reputed to be able to burn off most of the sports cars of its day.

It proved to be the ideal car in the bush with its big wheels and lots of grunt.

My four kids loved the old bus and we had many trips and picnics out in the bush. The dry salt marshes near town are hard and quite smooth for most of the year. We were driving around and around in a big circle in the Ford with a capacity load of squealing kids. I told them to look at something to distract them and when they looked back I was gone. With the car still idling around happily they finally spotted me sitting on the ground waving to them from the middle of the big circuit.

Panic took over when they realised the car was driverless – it was a comical sight. Letting them complete another shrieking lap I then strolled back and caught the car and swung into the seat again as it went past.

93

You see, the steering gear on the old Ford would hold its angle of turn if you put it on a hard lock. This feature together with the setting of the very useful hand throttle provided more fun than you could ever have in a modern car. These features were also very useful when I was driving alone and came up to a cocky's gate that had to be opened and then closed again. The procedure was to stop, select first gear, and set the hand throttle to idle, aim the car at the gate and drop the clutch. The car would tonk along at a slow walking pace which gave you time to jog to the gate and open it. Waiting for the car to idle up to, and pass through the gate, you then closed the gate behind it, jogged again to catch the car, jumped in and just took off. I used this lurk hundreds of times and it always worked - without the car ever ramming a gatepost or running down its driver.

While I was sitting in the car in the main street of Derby one day, a big tourist bus pulled up opposite. With cameras at the ready, a group of passengers crossed the road. One said, -

"Hey this is a great old car. I've taken some pictures. Will you start the motor; I'd like to hear how it runs."

With a feigned and exaggerated sigh of resignation I replied, -

"Mate, it *is* running."

It was a very quiet engine that one.

One day I experienced an example of one of her most frightening antisocial stunts. After parking at the front gate after a long run she would give me a hell of a fright. Just as I was mounting the veranda steps there was a flash and a loud echoing report from the direction of the car. This shock tactic recurred regularly after that when the engine was hot, and about three minutes after the key was turned off. The deafening explosion from the exhaust tailpipe could be heard for blocks.

There was a city visitor with me on a trip one day in the Ford and I was telling him about this strange phenomenon which we had begun to call Henry Ford's Revenge. While we were unloading the car at the gate after the long trip, he disappeared around the back. It was a bit disturbing to see him squatting and peering up the exhaust pipe. Too late she cried, as the explosion bowled him over backwards to land flat on his bum in the dirt. Any further efforts on my part to convince him that the story was true would have been unnecessary and futile.

We had been on a trip to Oobagooma Station way up north of Derby and were returning at night, and took a shortcut across a plain just north of the Meda River. The recent wet season had accelerated the growth of spear grass, which was thick and matted and nine feet tall. Our track had been unused for years and was difficult to follow, and twisted and turned through the thick growth, which whipped the car as we drove at a pretty fast clip. The headlights failed to pick up the fallen tree trunk across the track, hidden in the grass and making a solid obstacle eighteen inches thick. The impact was unexpected and bone jarring. The old Ford just tucked up her front wheels, reared up and over the log and we were airborne for a spell, after which we crashed down and bounced a couple of times and finally slithered to a stop. A quick inspection of the car showed no apparent damage – we were still mobile, which was a blessing as we were fifty miles from home in a remote area of the Kimberley in the middle of a dark night.

My mate and I paused in our laughing and congratulatory backslapping long enough to hear a faint groan from the back of the car. My dear wife Bev had been saying earlier and for quite some time that she was a bit envious of our enjoyment on these frequent blokey safaris, so to placate her I offered that she could come with us next time. This was the time - and the groan from the back of the truck was hers. Feeling sleepy, she had curled up in the tray where we had cleared a space in the conglomeration of shovels and axes and ropes and other gear, which was all now pinning her down to the floor. A check on dates later suggested she might have been with child at that time and which was the possible cause of her feeling dozy.

Her shock at being awakened from a nice nap can only be imagined. While her numerous bruises were yet to become visible, her request to move back to the front seat was quite understandable. Bev's technicolour bruises did fade with time and no, she did not jump at our next offer to go on safari and yes, she did have a lovely baby in due course.

My financial investment in the old car was doubled when we left Derby. I sold her to the local storekeeper for a one hundred percent profit - fifty pounds. Well, I had done a lot of work on it.

The need to get out on the road and go places was very strong when we were young and we were not easily deterred. Shortages of many commodities were still being felt long after the war and car tyres were hard to get even if we had been in a financial position to buy them. We were carrying out a pre-trip inspection on one of our cars to ensure it was roadworthy for the hundred-mile run to Bunbury one weekend. We knew that the left front tyre was worn; in fact the tread had disappeared months previously. Now the tyre, like an old boot, showed signs that it was wearing right through. The oval patch revealed a series of concentric rings, which were the canvas plies of the tire exposed to view. Normally a cause for urgent replacement, this was impossible at that time, as we had no spare. The forthcoming suggestion showed a touch of brilliance. We looked at the wall of the tyre, which stated it was a six-ply type. We counted the number of layers that had been worn through by contact with the road – it was only two, which suggested we still had four more plies to wear through before suffering a blow out.

Yes, so we could make the trip to Bunbury after all.

Glancing at that tyre when we got back to Perth showed it had worn through the last four plies and just begun to run on the inner tube – that's timing!

One of my earliest cars was a little Morris tourer, which I had lovingly restored to pristine condition and painted a lovely green with a brush and good old Dulux enamel. This car had poor brakes when I bought it, so along with all the other jobs, I fixed the brakes so they could stop me on a threepenny bit. As many mechanics have found, when you fix one thing, it causes another fault to pop up. The really good shuddering stops I was so pleased with, now caused my two doors to unlatch and fly forward and with a double bang and a cloud of dust and rust, hitting the front mudguards on both sides. My attempts to remedy this situation met with failure, as the cause was in the flexing of the wood-framed body, which unlatched the doors when the brakes were applied. As long as you know about these small things you can live with them I reasoned, so with an awareness of this possibility, the problem was put behind me.

Feeling rather proud of my green wheels, I drove down the main city street in Bunbury one Saturday morning. The relatively high volume of traffic through town had prompted the authorities to install a policeman on point duty at the crossroads, and it was obvious that he was very busy as I approached. I had to admire this representative of law and order in his immaculate blue uniform standing imposingly on his little duckboard island. The white gauntlets he wore served to make him more visible to the motorists he was directing in different directions with the waving of his arms and shrill peeps on his whistle. A slight distraction must have claimed my attention as I was preparing to drive on past him. Looking up at the sudden blast from his whistle I guessed it to be the command to stop. Stop I did and suddenly, forgetting the thing with the doors. The squeal of my brakes was punctuated by the double impact of these two unlatching and hitting the mudguards with the usual cloud of dust. It sounded like a shootout in Chicago. All traffic flow ceased as the policeman held up both gauntlets and sternly wagged a finger at me. He recovered his dignity and composure as I sheepishly reached out and closed both my doors again. He waved me around him, as after putting the car back in order my hand was stuck out for a right turn. He might remember me giving him a sheepish grin as he just glared and shook his head sadly. The glare turned to fright, as cutting too close to him in my confusion I drove with two wheels over the corner of his duckboard, narrowly missing his foot. A car is heavier than a policeman is – even a big one, and the effect was to flip the duckboard up and dislodge him. The rear view mirror showed the poor bloke, stripped of all his dignity, with arms and legs windmilling, trying to keep his balance and his self-control.

It was decided then that I should try to avoid that particular intersection in Bunbury for a period of time in the future and certainly on Saturday mornings.

Harry rode a Harley with a sidecar too – a great hit with the young family.

Grease and Grime

Completing the Year Ten Certificate at the Bunbury High School was a milestone in my life. This was 1943 – I was sixteen and at a point where a young man should make some decisions regarding his future. The second world war still raged and the whole show was much closer to home now with the Japanese, having clobbered the American navy at Pearl Harbour, were spreading in a fearsome tide southwards through South-east Asia towards our shores. Our country was being bled of able-bodied and skilled men, most of which were drawn into the fighting services.

My future direction was discussed with my dad. Although a few fellow students were opting to complete high school and go on to university, this was vetoed in my case. Our isolation from Perth and the expense of further education were negating factors, also in my heart of hearts I was itching to get up and go out into the world.

Dad offered two choices for me – the first a career in the bank, which was a job he himself had held since he left school and which he hated, or an apprenticeship of some kind. I felt that I probably would have died in the confines of a bank, having lived such an active and free life up till this time. Given my apparent natural aptitude in the field of mechanical work, both in the understanding of mechanical theory and principles, together with good hand skills, this was my choice.

The town of Midland Junction, twelve miles East of Perth, owed much of its prosperity and importance to the Midland workshops. A force of three thousand worked there in support of the State railway system, which was further swelled by the requirement to manufacture munitions and take on other wartime work. Most of the workers were brought by rail from surrounding suburbs to converge on the sprawling complex of cavernous red brick buildings interlaced with shining railroad tracks. Every morning hordes of men, like a colony of jungle ants, poured from the trains, across the bridges and were engulfed by the workshops' red brick buildings. At precisely five in the afternoon, the seething ant colony poured out from the buildings and back into the waiting trains.

Midland Workshops in present times. 1944 would have seen more rail lines and everyone there in work clobber

There I was standing with my hand-me-down battered suitcase at my feet on the Midland station platform. Fresh from the green fields and blue of the ocean of my home in Bunbury, I was dispirited and depressed by the contrasts. The trees, the buildings and even the people were clothed in a universal dull coat of sooty grey. Coal smoke drifted across the maze of railway lines from the chimneys of busy locomotives clanking past. The general hubbub of activity was split by the shriek of engine whistles while tall jets of white steam rose like waving plumes from chimneys and safety valves.

My accommodation had been booked at a lodging house in the town, towards which I turned my feet to begin a new and uncertain future.

Sleeping in the family home in Bunbury had been in close proximity to the sea with the cool fresh breeze and the sound of distant surf as a lullaby. Here in the lodging house I felt like a caged animal in a bare room with no window and with the prevailing smell of boiled cabbage. Tossing and turning in that sweaty summer night feeling very homesick and sorry for myself, I had vivid dreams of being pursued by fire-breathing engines and fighting for breath in the choking smoke from their chimneys.

As an insignificant ant in the surging tide of workers next morning I was swept in through the gates to the timekeeper's office. I was determined to make a good impression as my dad had advised me to do, so with my Gladstone bag in hand and the pithy taste of the cold baked beans breakfast in my

The big steam hammer - the one Rod used to drive to do his trick with the pocket watch

mouth I reported for duty. It seemed they were expecting me along with a hundred or so other boys that had trickled in from around the country to swell the intake of apprentices for that year.

From the man in the shirtsleeves, the waistcoat and the greasy tie came the edict, -

"You can't be an apprentice right away. You'll be on probation as a junior worker for a few months and if you prove yourself reliable and willing to work, we'll sign you up."

Feeling very self-conscious in the new blue overalls and steel-shod army boots, I sensed every pair of eyes was upon me as I walked the length of the blacksmiths' shop looking for the foreman. The floor was black cinder. Tall steel columns reared up to the grimy saw-tooth roof and a weak sunlight filtered down through the gloom. The flickering red eyes of forges and furnaces glowed, with sweating men in leather aprons attending them. My insides shook at the impact of the noise and the smell and the heat of this place where I was destined to serve the first part of my sentence.

The blacksmith foreman seemed a decent bloke and kindly, which is what I could do with at that time I can tell you. He began to describe my duties with a hoarse voice raised against the roar of furnaces and the bang, bang, bang of steam hammers at the end of the shop.

"See those hammers?" said the foreman, pointing to several tall, black structures, "Your job will be to drive one of them."

These monsters came in various sizes. There was a row of them ranging in height from ten feet to about twenty-five feet looming above the men working around them. The largest was pounding with stupendous force on a long glowing billet of steel with splashes of bright sparks and the impression was awesome. The slim figures of young lads could be seen at the rear of the machines pushing long operating levers, which controlled the stroke of the hammer. A fat steam cylinder at the top was thrusting the shiny piston rod up and down. This was connected to a block of steel, which acted like the head of a hammer, pounding the red-hot metal into the desired shape.

This was going to be exciting stuff I could see.

The 'best boy' taught me the rudiments of steam hammer driving and after some practice, I picked up the technique and was given some real work. The roar of the steam exhaust, and the sheer brute force unleashed by the slightest movement of my hand on the lever gave me an enormous sense of power.

It was traditional in the workshops that apprentices and junior workers suffered as the butt of many practical jokes and embarrassing pranks played upon them by the men, especially during their early initiation period. It was a kind of baptism of fire and we all copped it. This apparently was a fact of life and the manner in which you took this rough treatment had a lot to do with whether you survived or whether you went under in shame and defeat. It was to be my first lesson in getting along in the rough world of industry while still maintaining my self-control and a measure of humour through some of the ordeals we had to suffer.

The 'best boy,' who taught me the tricks of hammer driving had been doing it for years and his skill with the biggest and most powerful hammer was the stuff of legends throughout the shop.

"It's not just a matter of belting away at the hot billet" he explained. "The trick is to use a really light touch on the lever to get the hammer swinging so you are still able to bring it down under full power but just kiss the anvil."

I practised hour after hour but never quite matched his performances.

Every one of the boys in the shop was assigned to a tradesman blacksmith. Depending on your luck in the draw these men could either be your mentor or your tormentor. The 'best boy' was a gangling, pimply redhead named Rod. He had drawn the short straw when it came to teaming up with his blacksmith partner who was a bullying and callous bastard. Most of the other blacksmiths were proud of their boys and boasted of their achievements. Though young Rod was certainly eligible for well-earned praise, woe-betide him if he made the slightest misjudgement in his work. He was naturally terrified of this man and his foul explosive temper.

One of Rod's showpiece tricks with the big hammer demonstrated his amazing skill. A matchbox was placed on its end on the anvil under the hammer with its match drawer pulled part way out. Rod would turn on full steam pressure and get the hammer swishing up and down rhythmically, gradually bringing it down lower and lower with each stroke. Showing extreme concentration he guided the massive fast-moving hammer towards its fragile target until the tray of the matchbox was tapped shut. The final well-timed stroke would cause the matchbox to bounce on the anvil without damage. There was a story around that he had done the same trick, but with a pocket watch as the target, tapping it ever so delicately on its crown glass, causing it to bounce on the anvil without a scratch.

Soon after starting work in the blacksmith shop I saw Rod in the grasp of his blacksmith partner. He had Rod's shirt tightly bunched in his big fist and was swinging a pocket watch on a chain in the other.

"You are going to bounce this watch of mine under the big hammer like you done before." He looked around for approval at the ring of grinning workmates. "I've got money on this and if you bust my watch I'll bloody well kill ya!"

We could see that poor Rod was in no fit state to perform such a difficult feat, especially under the threats of his aggressive blacksmith partner. He was showing his fear and his hand trembled as he hesitantly grasped the operating handle of the steam hammer and turned on the steam valve. There was the watch ticking away, sitting in the middle of the steel striking surface under the hammer, with a sweating lad trying to summon enough of his skill and courage to go ahead with the difficult stunt.

The last seen of the two main characters in this drama was Rod, running at full tilt for the door and safety, with his blacksmith mate roaring obscenities and swinging a seven-pound flogging hammer in his wake. We all prayed for Rod that day and hoped he was fleet enough to stay ahead of the murderous blacksmith and his hammer, or otherwise we would be reading his name in the obituaries column of the West Australian newspaper in a few days.

The effect on any pocket watch, no matter how robust, of the impact of one and a half tons of steel travelling at high speed is unbelievable. With an instant gain in diameter of about six times the original and a reduction in thickness to about that of a Gillette razor blade, it was seen to have engraved upon its surface, the faintest impressions of hands and numbers and numerous springs, gears and wheels.

It would most assuredly tick no more.

A bizarre thought occurred to me as the relic was dragged off the anvil by its flattened chain - it would make a fine medal for Rod to show off in commemoration of his achievement, though I doubted he would be game enough to wear it to work on Monday.

Overalls 'n Boots

Six months had passed since my admission into the Midland Junction Railway Workshops as a junior worker in 1944. The effect this new experience had upon me was a kind of culture shock. There was however very little in the way of the finer forms of culture in the duties of a railways junior worker. Having arrived alone, aged seventeen, with bag in hand, freshly separated from the country air and close family surroundings and finding myself in a strange world of grime and sweat and strange faces, was pretty unnerving for this young man.

When the day came for the new apprentices to move into their respective trade areas according to their job assignments, I turned my back on the blacksmith shop. That dingy world of glowing furnaces and thudding steam hammers was behind me. I presented myself to the foreman of the fitting shop. He was a stern faced man with greying hair and wore the obligatory navy pants and waistcoat, complete with enormous boots. A pocket watch chain looped across his belly completed the picture. This was the uniform of the traditional British tradesman of the time it seemed, though why one would want to wear a collar and tie in those circumstances was beyond me.

The total workforce in the Midland Workshops in 1944 numbered over three thousand. This had been swelled by the addition of a large annexe engaged in wartime armaments production. There was a large factory facility containing rows of special lathes making artillery shells for our diggers to fire at the Japanese in the Pacific war. Work was also going on in the repair of American submarines which were based in Fremantle. All the lathe operators were women and girls. Like thousands of their sisters, they had joined the war effort to take the places of the men of military age who were on active service.

Our Master of Apprentices was George Groves. He was a large, portly, congenial character who took us under his wing and was generally liked and respected by all in his care. George gathered the lads around him for a lecture. He explained the roles and responsibilities of a government apprentice and some of the hazards we might encounter along the way in the five-year term ahead. He stressed the importance of maintaining the workshops' tradition for good workmanship, encouraged us to be diligent and hard working and to respect our elders and betters at all times.

"You will each be assigned to a skilled tradesman who will teach you your craft and in return will expect hard work and respect," he said.

Respect for our elders was but one of the rules of the family upbringing for most youngsters, so we didn't have a problem with that one. Polite behaviour, coupled with consideration for others were qualities expected of us. I remember at home the stern reminders to stay in my place if ever I should attempt to interrupt adult conversation. 'Speak only when you are spoken to' was a rule, which eventually caused me some embarrassment in later life. If I wished to give one of the men in the workshops a message while he was engaged in conversation, I would hang around waiting for a lull to say my piece, not daring to butt in. I must have been standing like a timid fool one day for about ten minutes, until finally one of the men turned and roared, -

"For chrissakes waddya want?"

It was some time before I learned that I too had rights and should learn to exercise them. This was all part of the man-making process I suppose.

It was fortunate that I was one of the lucky ones in the draw for a tradesman partner. Jim Baker was a solid, powerful, curly-headed fellow and a top man on the tools. We got along famously from the start and I owe much of the expertise that I acquired to him. Jim saved my life in fact, the circumstances about which will be covered later. Some of the boys did not fare so well, with their senior partner turning out to be more an enemy than an ally. A word in the ear of the master of apprentices however, would have you transferred to another tradesman so that your training would not be disrupted. The management apparently saw the apprentice as a resource important to the survival of all the trade skills and in fact insurance for the future of the industry, so we were well looked after. The sheer number of craftsmen in such a wide range of trades' areas made the workshops an ideal training ground. I have had the opportunity since those days to compare the level of competence learned there against the best from other industries and even other countries and it was never found wanting.

From the boy's point of view there could be a down side to this master/apprentice

A group from the three thousand workers at the Midland Workshops

arrangement, as some will testify. A new apprentice, especially if he showed signs of being naive or immature, was vulnerable and subjected to the local sport of apprentice baiting. Any boy who had not received wise counselling and warnings in the early stages of his apprenticeship, as fortunately I had, could find himself the butt of many practical jokes. These came in infinite variety and most were quite elaborate and well rehearsed, with the prime object of embarrassment for the victim. It has been interesting to observe in later years that similar practices are labelled with terms like bastardisation, persecution and discrimination. We accepted all this as our lot however and I believe that on the whole it was a toughening process, experienced in a rough school where you learned to take it all with a sense of humour and come out wiser and a more mature as a consequence. Being on the bottom of the totem pole wasn't really so bad when you consider that the only direction you can go from there is up.

Tool stores were located in all the major work areas. Special equipment and materials were issued from these, as were required necessary for the work of that section.

A gangling first-year apprentice fronted up to the counter of a tool store one morning. The old shellback storeman, after an appropriate show of indifference for a considerable time, finally asks what it is that he wants.

"I want a bucket of rivet holes," pipes the lad, an innocent expression on his face, which branded him as a prime candidate for this scam.

"Ah now and what size would they be?" asks the storeman with a bland expression on his dial. "'Cause if you don't know, then you had better scarper back to the job and find out."

"They want three quarter inch," wheezed the messenger, now breathing hard after his run to the far end of the building and back.

"Yeah OK" says the helpful storeman "What thickness though? We have to know what thickness they want."

Another trip back to the job and the victim was back again at the store counter. He passed on the required information.

"Now we are getting somewhere, but we need to know the kind of steel "– etc. And so it went on until the storeman appeared satisfied with the requirements. It was important to be able to judge how far you could carry on a performance such as this before the victim woke up that he was being conned. The old hands were very good at knowing where to draw the line before going on to the next stage or winding it up.

Pointing to a row of big steel buckets lined up against a wall, the storeman directed the lad to take the third one from the right.

This fairly simple lurk was but one of many pulled on the inexperienced and the unwary. These particular buckets contained hundreds of steel discs that had been punched out of heavy plate to form rivet holes. They were nothing but scrap. The weight of what was literally solid steel was considerable, and much too heavy for one man to lift, even the most powerful. The storeman could be seen trying to maintain a straight face as a gaggle of boys sweated and strained to carry the mass of completely useless 'rivet holes' back to their leering workmates. No doubt they would have been ordered to return the load to the store – perhaps a little wiser now, to be ready for the trap to be set again for the next victim. It was obvious to me in time that the row of buckets of 'rivet holes' was placed there for the express purpose of playing out this little bit of industrial theatre.

Another well-worn apprentice trap involved the 'struggle bar.' When asked for this particular piece of equipment, the storeman pointed out a very large steel bar of great weight.

"You look like a strong lad," coos the storeman to the eager young man, "Latch on to that and take it back to your job."

Struggle he did, as he half-carried, half-dragged the useless billet of metal back to the grinning crowd at the work site.

One bright young fellow working at the East Perth power station was dispatched to the stores building to fetch a 'packet of high tension volts.' He was warned as to the dire consequences if should he drop or even bump the package. Red warning stickers were plastered all over it to this effect.

Tiptoeing the two miles back to the power station with the dangerous load grasped firmly in his sweating hands took more than an hour.

The allegedly lethal package was in fact a regular standard everyday house brick wrapped in brown paper. The revelation when it was disclosed as only a brick would hopefully have had the boy putting it all down to experience.

Another amusing incident I witnessed one day unfolded when a young hopeful poked his head over the store counter and asked if he could have a 'long weight.' The storeman, wise to the procedure for this one, nodded and went about his business. With all the signs if feeling he was being ignored the youngster piped up asking for the long weight again.

"OK son, you are getting it.!"

As he was a polite kind of a fellow, the lad bided his time through another period of being ignored but finally burst out, -

"What about my long weight?" The response that came from the storeman was, -

"You have been standing there for half an hour sonny - wasn't that wait long enough for you?"

Then we all felt sorry for the pasty looking lad in the machine shop who inadvertently revealed his Achilles heel to us. He showed all the signs of being a bit of a hypochondriac. This particular orchestrated stunt would begin with one of us stopping at his worksite and asking if he felt all right, as he didn't look too flash. After a further appropriate time lapse, another would put a hand on his shoulder and looking into his face would remark on his peaky appearance. All this attention soon had our man drooping visibly to the point where these sympathetic comments from 'trusted' workmates would see him heading for the nurse's station seeking medical advice. Bets were laid on the odds of him booking off on sick leave. In many instances he did. Providing that you didn't push it, this one worked every time - with him at any rate.

The inventory of practical jokes ready to be played on new boys was long and varied. Remembering that some of the senior workers had served in the workshops for up to fifty years, they had accumulated a lot of practice in skulduggery in that time. Work in industry can be monotonous and boring at times and a break from routine was always welcome. In the fresh crop of new apprentices starting each year, the likely ones were soon identified and tested for their levels of vulnerability. The system flourished in the five years I was there. It was considered very bad form to wise up a likely target and in doing so spoil the fun. Also, anyone who was the butt of a joke and turned nasty and belligerent as a result was in for a hard time. A sense of humour was a valuable asset in that place – nay a necessity.

It was not entirely a one-way street however as some of the practical jokes that were played on apprentices had surprising outcomes. A new apprentice had turned up in my section and we were keen to see how he fitted in to his new and strange work environment. We 'experienced' people had learned that a sense of humour and the ability to take a few knocks was a desirable trait. This fellow was a big quiet country boy and was a wild card, for as yet we had not seen his reaction to this kind of work-related stress.

New apprentices were allotted the worst and the dirtiest jobs as a matter of tradition – a kind of testing process it must be assumed. Ben seemed to be coping with all this but his mettle had not yet been properly assessed.

Aging steam locomotives are dirty beasts. They were rolled into the repair shop first to be dealt with in a section called the stripping pit. I spent the mandatory three months working there, as did all the other fitters' apprentices. It appeared to me to be all a case of brute strength and ignorance with few of the finer points to be learned – but just grease and oil and dirt day after day. I did learn a lot about locomotives however while pulling them apart, which was an essential part of the training process.

Having mentioned that a steam locomotive is dirty – nay filthy, there is a part of it that is the filthiest of all. This is inside the smoke-box at the front of the boiler. As the engines burn vast quantities of brown coal to generate steam, all the smoke and soot from the firebox is drawn through the boiler tubes, through the smoke-box and goes up the chimney. There is a wire mesh spark arrester in there which is intended to trap burning sparks and embers so that the dry farmland and crops are not ignited along the railway lines in passing, especially in summer.

Ben's allotted task one morning was to take a shovel, crawl into the confines of said smoke-box and start digging out the accumulation of soot and cinders so that repair work could proceed. It was clear what was likely to happen next as one of the men climbed up on the engine frame and closed and locked the heavy door confining Ben inside. Taking a seven-pond flogging hammer he swung it hard and clouted the side of the steel shell of the smoke-box. The victim probably gets the worst fright of his life as the concussion causes your ears to ring long afterwards. The second effect of the hammer blow is that of dislodging the thick layer of black soot lining the inside and dumping it on the unfortunate captive.

Quite a large crowd of onlookers had gathered to watch the fun and witness the emergence of the martyr. After the allotted space of five minutes during which time there was not a sound from within, someone climbed up and unlocked and opened the door. We all waited expectantly.

The unrecognisable figure that we could only assume was Ben, leapt out of his prison with his shovel poised apparently as if to decapitate the nearest man. In describing his appearance, the word dirty acquired a new meaning. The avalanche of soot had coated him from his head to his boots. Rivulets of the stuff trickled down his face and off his shoulders while two white eyeballs were swivelling in the black mask that had been his face. Then the most unexpected development took place.

With little puffs and eddies of black dust he went into a kind of soft-shoe dance routine on his platform above the crowd. His shovel, which we feared he was going to use as a lethal weapon, was held across his body. The left hand was doing some complicated fingering on the handle while the right hand was strumming imaginary strings at the other end. With a wonderful imitation of Al Jolson the singer he threw back his head and broke into a fine rendition of 'Mammy' to the cheers of his audience. You know the song – 'The sun shines East, the sun shines West, but I know where the sun shines best'...

We were all falling about with laughter as he finished his item, complete with the little soft shoe dance steps. He then spread his arms wide as on a theatre stage and made a low bow to his incredulous audience. The original Nigger Minstrels would have been turning in their respective graves that day I felt sure.

Ben's quick thinking and his ability to cope and come up smiling were established while his popularity and acceptance were assured after this little bit of workshop theatre. Just in passing, the colloquial term for a shovel among the old timers who used them is actually a 'banjo' – so it all fits.

Jack was a crabby old coot. He was the workshop's crane chaser and had been treading the floor directing the overhead cranes for longer than anyone could remember. Jack's major irritation, of which he made no secret, was apprentices. He had suffered many indignities at the hands of the boys over the years and would never make peace with them at any cost. The regular flare-ups between him and the lads were a source of great amusement so nobody tried to intervene. It might have been appropriate for the foreman to step in on safety grounds, having seen what was going on, but why spoil a good thing on such a fragile issue as safety.

Keeping watch from the height of a locomotive cab at the end of the line, boy number one signals ahead that Jack the crane chaser is coming. As Jack shuffles past the next locomotive on his way through the shop, boy number two leans over above him and delicately places a wad of kerosene-soaked cotton waste in the crown of his greasy hat. It is skilfully executed and Jack feels nothing. As he passes under the third boy in the line, the waste is ignited with a match.

Still oblivious to his hazardous condition, Jack now has a flame to rival the Olympic torch sprouting from his head. His progress down the line can be traced by the ruddy glow shed by the hat fire and the trail of smoke in his wake.

There was great conjecture among the men on just how far the victim could continue before the fire ate through his headgear and down to his bald head.

There was actually no mistaking the point at which the heat had penetrated. He bellowed and tore the blazing hat from his smoking scalp. He danced and kicked and stamped on it to quell the flames – and the language! As a responsibility to my readers I shall not repeat verbatim what

Jack actually said, or rather babbled in his rage. In his furious outburst he brought down terrible curses upon all apprentices, regardless of race, colour or creed and upon their forebears as well.

There was another rather spectacular joke pulled on a machinist with all the action being provided by the man himself after the trap was sprung.

Almost everyone carried a brown imitation crocodile skin Gladstone bag. It was a part of the railwayman's working uniform. The Phantom Runner was a star athlete – well, he must have been, to be able to sprint the considerable distance from his work area within the workshops to the main gate and stay ahead of most of the other three thousand-odd hopefuls at knock-off time. The Phantom Runner always placed his Gladstone bag in a predetermined spot on a workbench near his starting point for the dash every afternoon. His speed, while phenomenal, was overshadowed by his agility and grace as he wove through all the people and other obstacles - especially on Tuesdays. This was the day of the week he was looking forward to meeting his ladylove after work. This fact was known to all his workmates.

This particular Tuesday, after placing the Gladstone bag on the bench precisely for a pick up on the run, he moved back for his dash, which was due to begin at whistle time.

All the men around his Gladstone bag were tense and watching. As the notes of the knockoff whistle were still hanging in the air, he came at high speed and snatched his bag from the bench. Carried by considerable momentum, he crashed into a cupboard with the bag, now bottomless in his fist, wearing an incredulous look on his face. The previous contents of the Gladstone were scattered far and wide while its ragged bottom remained securely fastened as part of the timber bench top. It took him a little time to assess the situation, gather his scattered wits and possessions and find the cause for his ignominious downfall. It appeared that while someone was distracting the Phantom Runner's attention earlier, his bag was carefully opened and had two large railway dog spikes driven through its bottom into the timber beneath. This made it effectively part of the bench. Without moving its position, the doctored bag was then carefully closed to appear as innocent and undisturbed as before. So many laughing faces surrounded the victim after the spectacular crash that he found it impossible to guess the identity of the culprit. As it was a team effort though, he would have had to blame half of the workforce, so he had to grin and bear it. It was never a good idea to stand out in a crowd in that place as they would plot and scheme until they found the Achilles heel and the ways and means to bring you down.

As special purpose work safety boots had not been invented yet, we all wore surplus army boots, which were adequate protection. The standard army boot was fitted with a steel horseshoe plate on the heel to increase the life of the boot. One particularly vacant-looking lad was lolloping through the boiler shop one day. One of the welders spotted this likely victim and called him over.

"See this steel plate on the floor here," he pointed, "I want you to stand on it to hold it flat while I weld it to this other bit." The boy stood dutifully with his big clodhopper boots on the plate as directed and watched.

"Don't look down you dopey bugger - you'll damage your eyes with the flash from the welder!"

I can still see the skinny figure in his big boots with his hand clamped over his eyes while the welder carefully fastened his steel horseshoes to the steel plate on the floor.

On completion of the two welds, the tradesman packed his gear, rolled up his cables and left. Sensing that nothing much was happening and still with his hand clamped over his eyes the boy asked if he could go.

Heat from the welding on the heel of an army boot apparently takes a little time to penetrate to the wearer's foot. The temperature increase caused the boy to yelp and looking wildly around at the circle of grinning faces, he attempted to walk away. Swaying like a reed in the wind, he pivoted from his anchored feet, first forward, then back and from side to side, but was getting nowhere. With the inside of the boots getting hotter, he tore at the laces and ejected from the smouldering footwear in pain and panic.

The crowd roared with mirth.

The last I saw of the poor kid was his thin gangly figure in very holey socks chipping away with a borrowed hammer and a blunt chisel specially selected for him for the attempt to free his boots.

It was cold comfort for the lad that the fellow who did the welding job was a certified tradesman who could lay down a fillet as strong and as permanent as any engineering standards could possibly require.

The most highly organised and well-publicised scam aimed squarely at gullible apprentices was the annual Peanut Meeting. In the early part of each year a number of the young lads were invited to volunteer to be Peanut Kings and were given a glowing account of the importance of and the benefits to be gained from accepting this honour. The prestigious role of a peanut king they were told, involved collecting funds over a period of months from the men throughout the workshops. These donations were actually in the form of pledges, with the cash to be redeemed at a later date. The money was reputedly to be used as big donations to charity. The great attraction for these collectors, which usually convinced them to take on the job, was the hint that they would be allowed to retain a percentage of the money to keep for themselves at the conclusion of the funding drive. This could amount to a considerable sum they were told and obviously the more they collected, the bigger the bonus in it for each of them. The boys could later be seen with clipboards in hand, accosting men throughout the works, badgering them for promises of big donations for this worthy cause. The totals mounted up to considerable sums on their lists of prospective donors.

On the much publicised day of the peanut meeting a thousand men or more were gathered around a large makeshift raised platform in the yard. The master of ceremonies strode to centre stage and called for order. The six peanut kings, still with their clipboards, lined up in front of the cheering crowd. The youngsters looked embarrassed by all the attention but also seemed proud to have played an active part in this important charity event. They showed up in their best clothes too as befitted the occasion and the crowd was hushed.

"We are going to show you lot just how much effort these lads have put into collecting money for the peanut fund this year," shouted the speaker. "Each of the boys will read out the sum total of his collection – also the names of those men who promised the biggest donations" - cheers, "And the names of those miserable buggers who gave the least."- Boos.

With the formalities done the MC now comes to the highly anticipated part.

"Last year we decided on the amount of thirty percent of the money collected which to be retained by each lad. They have done a magnificent job this year and I reckon they should get more than that."

Someone shouted that they should get fifty percent of the take, which was met with more cheering. The boys meanwhile were carrying out furious mental calculations as to the amount of wealth involved for each of them as the percentage was raised by shouts from the crowd. They still did not twig when their share of the money reached ninety percent and they were literally millionaires on paper. Millionaires one minute and destroyed the next. Someone in the front of the crowd yelled.

"Give 'em the lot!"

This was the signal that spelled the doom of the peanut kings and with it, destruction of all their illusions of great riches.

A number of figures with buckets appeared suddenly above the stage and delivered a deluge of slime and muck down all over the peanut kings. There was no escape even then, as jets from fire hoses bowled them over in a tangle of arms and legs and clipboards. This dastardly act was also the signal for the initiation to begin for all of the new apprentices throughout the workshops. Young terrified lads could be seen being pursued in every direction. The usual treatment was to be stripped, coated with muck or painted and then thrown into a water tank. One very large and very strong apprentice was giving a good account of himself, but went down under the weight of numbers. After being lashed to a wheelbarrow he was wheeled under a spouting fire hydrant to cool off. He was forgotten and I heard that he later required emergency resuscitation.

Few were spared and I was finally to meet my fate too. Recalling some advice I was given earlier, which suggested that if you fought you couldn't win, and you may as well get it over with, I surrendered. With this in mind I made only a token resistance and then felt myself being lifted bodily and hurled into a tank of freezing black water in the blacksmiths' shop. Still shivering later and with a lump on the back of my head from hitting the edge of the tank, I considered myself relatively lucky compared with some.

This is the locomotive I helped to build, together with other apprentices, as my fifth-year final practical assessment.

There was however, a delightful twist to the peanut fund activities one particular year. As the peanut kings began their rounds collecting pledges, one of them devised a lovely scam of his own design. He discovered a group of new tradesmen who had recently been brought into the workforce and were still unaware of the details of this great tradition. Setting himself up as their dedicated Peanut Fund Representative, he convinced them of their obligations towards the fund, but requested the donations all to be in cash. He made a killing and there was nothing anyone could do about it. This young fellow would surely have earned the vote for the one most likely to succeed in business.

An institution within the main institution was that generally known as the foreign order department. It was a highly organised and structured defacto system of mutual assistance given to anyone who wanted a job done for him personally, as opposed to the authorised government work which all were paid to do. There were very many different crafts and skills being practised in the workshops complex. Fitting, machining, coach building, copper smithing and lock smithing were all to be found and many more. Once you had convinced a craftsman that the article you had asked him to make was for yourself and not for the government, he would get straight on to it. Tucked away in the loco drivers' shed there was even a fully equipped barber's shop with all the gear and a big chair. Payment for a haircut would be in the form of some little job you could do for the barber, sourced in your particular trade area – an exchange of services if you will, and it worked a treat.

Having that lovely table lamp made for your sister's wedding present was one thing, but spiriting it away back home was another. Many tales were told of the ingenious methods employed to this effect. Articles smuggled out included pieces of furniture, boats, bikes and racing cars, all made in the workshops.

One young fellow complained to the gatekeeper that his motorbike was being interfered with where it was parked outside the gates and could he bring it in so he could keep an eye on it. The gatekeeper relented and the bike was ridden inside. At knock-off time this same machine was seen leaving through the gate. No particular attention was paid to it, or to the shiny new sidecar that was attached to it. This new addition had been built and painted and fixed to the bike and then ridden out as bold as brass, right under the gatekeeper's nose.

A famous story was widely circulated about the worker who was seen wearing an overcoat on a warmish day as he left the works and was headed for the train home. His workmates noted he was a little unsteady on his feet and making heavy weather of it in the climb across the footbridge. He collapsed on the station platform with apparent breathing difficulties. It wasn't his heart that was causing his distress, but the fifty feet of steel chain wrapped in coils around his body under the coat. It is reported that once the overcoat was removed it was relatively easy to stand on the free end of the chain and then bowl him along the ground until all the chain was wound off. A stiff-legged walk on a man going home might suggest a piece of wood or metal up the trouser leg required for a construction project at home. Gladstone bags often suffered rupture from all the extra weight on the homeward journey – and this was after the owner had taken out his lunch!

There were many well-entrenched rituals and traditions to be observed in the workshops community. Events like birthdays, marriage, and the birth of a child or departure from the service, or admission to the service were recognised and celebrated in different ways. After witnessing some of these I resolved never to divulge any of my personal life's details to my work colleagues for fear they would make something of it.

These events as practised there helped to provide a social framework in a similar way to those seen in clubs and lodges and other associations and they played an important part in life at the workshops. Much of the real work carried out by the men was downright dull, boring and repetitive and the fun generated by miscellaneous pranks and rituals was a welcome diversion. In general I have found the British working man - and there were many there at that time, to be a bit of a wit with a good sense of humour and a liking for fun.

Word got around that the wife of a young tradesman in our section had presented him with his first offspring. It was rumoured that 'The Stork' might be coming to visit him to mark this important event.

The new dad was busy at his workbench unaware of the movement around him as men filtered over to surround him. The Stork was seen flying down the length of the long building headed directly for the unsuspecting worker. This very large and very realistic bird was constructed from wood and wire and papier-mache and measured three metres from wingtip to wingtip and was snowy white. It made a spectacular arrival gliding in with a baby pram in its bill with a doll in it. The 'flight' of the stork was accomplished by having it suspended from the hook of the overhead crane, which travelled the length of the shop.

The swelling crowd that was now packed around the red-faced target cheered as the stork descended serenely and landed at his feet. One of the wags delivered an eloquent speech praising the young father's virility and other matters pertaining to his manhood. He called for a response. The guest of honour's attempt to speak was drowned in the roar of cheering and chiacking and heckling and then the crowd melted back to their workstations. The last seen of the subject of this tomfoolery was his retreating figure bent under the weight of the stork as he returned it to its traditional resting place at the far end of the building. This too was part of the ritual it seemed, as he had to run the gauntlet of his applauding mates. It was required that he did this, almost like Jesus bearing the cross, as tradition demanded.

Colin Campbell was one of the finest tradesmen I have ever had the privilege of knowing or working with. It was under his guidance as an apprentice that I served in the creme-de-la-creme of the fitters' craft, which was the maintenance of all the precision machine tools in the workshops. Colin had a handicap common among many senior tradesmen in heavy industry – he had poor hearing. His deafness however was relative - as I found that in a quiet environment away from the workshops he was profoundly affected. Place him in the constant clatter and noise of machinery however and he could understand every word spoken. Walking with him in through the main gates one day we were chatting. I used a loud voice so he could understand me, which he was apparently happy with until we entered the din inside the building. Continuing our conversation he suddenly turned on me and yelled, -

"You don't have to shout Ivan - I'm not bloody deaf you know!"

Other tricks and practical jokes were rife and you had to be on your guard to avoid becoming a victim. While appearing somewhat childish in nature, the horsing around did tend to ease the tedium and it should be remembered also that a large proportion of the workforce comprised apprentices in their teen years.

The pay packet trick worked well and had the effect of revealing much about a man's character. Planted on the floor somewhere between the machines was the bait. The trap having been set, it was watched by those working in close proximity, for each passer-bye's reactions. The bait was one of the universally familiar brown pay envelopes. From the open flap a banknote protruded. This was actually the corner torn off a note suggesting there was quite a sum inside.

A young fresh-faced apprentice spotted it first. Scooping it up without even looking inside, he headed straight to the foreman's office to hand it in. Another picked it up without a trace of furtiveness, looked inside and with a grin replaced it to reset the trap. The next to see it there was the one we had hoped to catch. He was a mean-looking weasel of a man and his manner suggested he was a bit shifty. On spotting the envelope he walked past, made a casual circuit, returned and put his boot on it. In a feigned innocent show of tying his laces he bent down, palmed the envelope and stuck it in his sock. This bloke was our star turn. The furtive one was followed as he scurried away by a chorus of boos and catcalls and ribald comments. He appeared to wither under the storm of derision

There was an idle practice followed by one worker, which he had evolved to lighten his day. His regular job was to operate an enormous lathe used to machine the end plates of boiler shells. It was a deadly slow and hypnotic task requiring the patience of Job.

Rex had to climb a ladder at the start of each day and sit perched high on the machine staring at the cutting tool for a solid eight hours. Beneath his seat was an electrical switchbox containing the gear that controlled a two hundred horsepower motor. A small hole in the top of this cabinet had caught his eye and to pass the time he began dropping short pieces of metal welding stubs from his great height aimed at the hole. He had been amusing himself for years in this pastime without scoring a hit. We all knew of his little game and could hear the regular ping, ping as the missiles bounced off the top of the cabinet all day.

I was about twenty yards away on the day Rex scored a bullseye. He finally got one of his metal missiles through the small hole and into the high voltage cabinet. The resultant electrical short-circuit was resounding with a brilliant flash and a cloud of smoke. The cover blew off the cabinet as the equipment fused and all machinery in the area ground to a halt.

My last recollection of the scene was an image of the figure of Rex up on his perch. Like a demented dervish, he danced up and down with glee, waving his arms and shouting.

"I did it – I got it in the hole – you bloody beauty!"

The electrical repairs arising from the incident took a considerable time to complete before that section of the machine shop would run again. The 'eyedropper' himself was unrepentant and considered that he had achieved the oracle. After all, he had been attempting this difficult feat for years without success – until that lucky day.

I'm sure you will agree that it is good to have friends. One of mine in particular called Harry Murray can take the credit for this yarn. Perhaps the workmates in his workplace had been suffering the bouts of boredom as we did at times and had decided to liven things up a bit.

The victim in this case had been testing household vacuum cleaners on a bench in his repair shop but had gone to lunch. On his return to the job he flipped the power switch to start a machine - with a most unexpected result. The exhaust end of the Hoover, which normally expels air, erupted with a roar and a belch of flame and took off across the shop like a Scud missile. It seems that this unusual behaviour, which is most uncharacteristic of this humble appliance, was the result of a bit of doctoring during the lunch hour. A petrol-soaked rag mysteriously became inserted in the sucking end and when the machine started, the volatile fumes were drawn through the machine, past the sparking electric motor and ignited with the aforementioned spectacular results.

As an apprentice fitter in Midland Workshops - with two mates.

The Star Boarder

Country apprentices from the workshops were billeted around the suburbs with families who were willing to take them. It was a very congenial arrangement especially if you scored the right people and you all got along well together. Having grown up in a large family where we had to do our bit and pull our weight with the work around the place, it was not such a big step to do the same for the families we were boarding with. While the amount of money I had to pay for the all-found accommodation swallowed up ninety five percent of my wages, most homes provided the normalcy and stability for a boy on this, his first time away from home. Many were the accounts both good and bad from my workmates about the experiences that they encountered in their respective foster homes.

My parents did some groundwork on my behalf and had found a family in Bayswater that was willing to take on an apprentice. My stepmother had turned out to be a committed follower of the faith. She had connections through the church, which turned up what she described as a fine and most suitable Christian family in the suburb of Bayswater.

They were migrants from the old country. He was a foreman at the Midland Workshops, which was a bonus for me as he was able to wise me up on a lot of the tribal customs that flourished there. In particular I was indebted to him for forewarning me of the stunts and the tricks played on new apprentices by the old hands. His wife was a dour woman with a poker face and the sense of humour of giraffe with a sore throat. Over time I think she delighted in lording it over me and issuing orders to do all the odd jobs around the place. It was advisable, for the sake of peace to go along with it. They were a pious and holier-than-thou lot who dragged me along to church every Sunday and pushed me into involvement in all the bible study groups and youth club dos too. As I was by nature a bit of an agnostic and could see little purpose in the religious palaver, I felt that it was better to conform and not spoil my parents' good intentions. They must have decided that I would be in little danger in the big city with guardians of my moral character such as these. It was a matter of accepting - with feigned enthusiasm, the labour of chopping firewood, mowing lawns, repairing things, doing my own laundry and even washing the dishes.

Their teenage daughter was attractive and a bright spark and was in fact the only relief in that grim and rigid household atmosphere. We two scrubbed along pretty well and she managed to talk her parents around to the idea of having me take her out to various functions. This was good for me as the company was pleasant, but of even greater significance was the offer from the parents to pay for everything too. So we went to church picnics and to the beach and gradually gaining the parents' trust and confidence sallied out at night as well. Ballroom dancing was very popular and it was a requirement at most social functions that you were reasonably good at it so you didn't make a fool of yourself. The actual dance steps were either Old Time or Modern and had to be learned. The intricacies of the Waltz Oxford where you do the number gliding around a lovely polished dance floor literally glued to an attractive partner held a particular attraction.

With some reservations expressed and a set of strict rules on behaviour to follow, I was given the bus fare and the entry money to take the daughter to a dance studio in the city one night. Purely at my companion's instigation of course we learned all the steps in record time but kept extending the number of trips to the city to fulfil other obligations and desires. Most Thursday nights would find us down on the Esplanade enjoying a mild snogging session. This was better than dancing I reckoned, and as it was all paid for - well I ask you...! After one of these pleasant interludes it would be a mad dash up the Terrace to catch the last bus home. Then one night we nearly came unstuck.

My dancing partner was making the trip report to her parents at home after we had attended our 'classes' and was extolling the benefit to her future social life in being a good dancer. Standing back in the doorway I suddenly felt the urge to demonstrate one of the steps as an example to the parents. We executed a lovely pax de deux during which I steered her backwards out of the room. To her credit she did see the reason for my impulsive behaviour when it was pointed out that the back of her woolly blue coat was matted with dry grass seeds. You don't pick up that kind of incriminating evidence at dancing classes. It is my belief that I would have lost more than my social privileges that night had the old man spotted them.

Brother Joe and I were looking for a place that would take two boarders and where we could share the accommodation. The opportunity presented itself with a home in Maylands. This couple had no kids and while they were obviously very devoted to each other they were the oddest pair I have ever met. Very small and slight, the husband was a rooster of a man with a shining bald head. He trotted off to work every day at the local barber's shop across the railway lines. His wife was a kindly, jolly whale of a woman who waddled around in old carpet slippers and doted on her little husband, smothering him with fat and affection. When we shared a joke and she laughed, which was often, she laughed all over, with ripples of mirth travelling up and down her body. Her heaving bosom at these times threatened to sweep all before it and topple the furniture. The dear soul was a chronic asthmatic however and this caused her to pant with any exertion and wheeze with each breath like a locomotive at the level crossing across the road.

To the uninitiated and unsuspecting, Friday night was unforgettable in that house. Just as Joe and I would have our heads down about to be claimed by sleep after a hard day's work, the husband would trip in through the front door. He was always in great spirits on Friday nights having just left a session with his mates at the pub nearby. His bride, in anticipation of his homecoming would be languorously reposing on the big brass bed in her sexiest diaphanous nightie. I saw this remarkable picture once through their open doorway.

The walls were thin and when the lights went out on Friday nights all hell broke loose. It was plain that the little stallion was feeling his oats as he claimed his conjugal prize. Even with a pillow over my head I could still hear and even feel the frenzied mating ritual through the wall in the room next door.

Their big brass bed protested loudly with the squeaking of the mattress springs while the bed head beat a tattoo on the wall. Copulating elephants could not have made more noise as she flung her bulk around in the throes of ecstasy. Then there was the asthma. In cadence with the increasing passion, her breathing went from initial deep sighs to rapid shuddering gasps and then she began to whistle. While not quite as loud as that of a football umpire's, it was however a good imitation. It was impossible to sleep in that house until the lovers' tryst was over. As I learned more about this kind of thing I reflected that from the sound effects Dr. Kinsey could have accurately plotted the rise, the plateau, the climax and then the afterglow for one of his charts on human behaviour of this kind. Then I thought - good luck to them anyway.

The next unsuspecting family to take us on was in Guildford. It was no accident that all of the homes where we stayed were close to a railway station. As this was our main form of transport to and from work it made sense to find a place in the close proximity where we could jump on a train every day. At the conclusion of this topic you may be thinking that we did not reside for very long periods at each home. In mitigation, it should be remembered that this was over a stretch of five years, so it was not as if we wore out our welcome in a short space of time and were thrown out to go and find another place. On reflection however, after one landlady said she needed our room urgently for a long lost cousin or whatever, I never did see any suitcases on the front verandah after we had left.

While our stay at this particular family home was generally uneventful there was one time when the quiet pool of domesticity had a few ripples on it. Joe and I bought an old and decrepit Harley Davidson motorbike and dragged it home to restore it. We began to take it apart in their garage with a view to having it reborn like new. This was a time-consuming job during which there was a steady stream of filthy clothes to the laundry and the sounds of labour continuing far into the night. She was without doubt a kind and understanding lady who accepted our reassurances that all would be finished soon with peace restored. Her forbearance could not have lasted forever - and it didn't.

It was ideal conditions for us as we could come and go and work on the bike as we pleased. It was good too to have the chook house close handy. When we made the pillion seat for the bike we were able to collect enough feathers with which to stuff it. I swear that we had collected a sufficient quantity with what we found lying around and never resorted to having to pluck a live leghorn.

An event which occurred coincidentally just before we were relocated from that house had to do with some prize-winning orchids. These cosseted and treasured blooms flourished in their prime in a bed near the garage.

We painted the motorbike a nice royal blue after we had fixed it all up. A tin of turps we had used to clean the Dulux enamel from the paintbrushes was on the floor of the shed near the door. Being naturally tidy by nature I pelted the contents of the tin out into the darkness one night in the final cleanup.

Our landlady, a caring soul, had the stricken look on her face next morning of one who has just lost a relative near and dear to her. Her accusing finger was pointing in the direction of the aforesaid display of prized orchids. Having only a passing interest in things botanical I could not recall what vibrant colours the orchids were displaying yesterday. Today to a plant, all flowers and stems and leaves were a royal blue, the same colour as our motorbike. Additionally it seems that mineral turpentine is not a good substitute for Thrive so the plants were not only blue, but also appeared to be in the advanced stages of an untimely death.

In my recollections of those years, if ever there was a home from home it was at a house in East Guildford. The wife and mother was an aging, chirpy, birdlike woman with a fine sense of humour. We started off well and before long she was treating me like her own son. Her husband was a bluff and kindly bloke and we seemed to hit it off from the first day. There was a nineteen-year-old daughter too with red hair and I noticed her straight away. That relationship developed to be close, but simple and uncomplicated, like brother and sister.

The old couple had done it really tough in the years of the Great Depression in the thirties. During those years he was one of thousands of men on a kind of working dole arrangement called Sustenance. The couple had spent years in the outback in primitive conditions where he worked on the construction of the Transcontinental Railway across the Nullarbor Plain. They could relate stories of hardships encountered while living in a tin and hessian hut while battling the heat and the dust and the flies and the snakes. Born of necessity in those hard times she had developed a collection of fine bush recipes for kangaroo cuisine.

One day I challenged her to produce one of her famous dishes. She rightfully argued that she would need a kangaroo and reminded me that you couldn't buy one of those at the local shop. I set off to get one for her. They were plentiful just sitting around in the shade in the bush up at Red Hill on the Darling Scarp. Just as you might choose a prime cut in the butcher's shop, I examined a number of potential dinners over the sights of my old .303 rifle. OK I mused – not too big or too small, not too young or too old, just right – bang! With scarcely a glance from passers-by I rode back down the hill on the Harley with the rifle slung on my back and the roo draped over the tank.

"We've got to hang the meat first," said my landlady, "so it will be nice and tender and won't taste gamy."

Slipping a haunch into a clean pillowslip I suspended it with a piece of rope from the branch of the old Cape Lilac in the back yard and then sat down to wait.

"You dopey bugger" giggled the chef, "it takes three or four days. We'll have it for dinner Friday night."

My taste buds still twitch and the mouth waters when I recall that sumptuous meal. The meat, when we cut it down from the tree was black and scaly and crusty on the outside, which is just as it should be apparently.

The name given to her triumph from the oven was pocket steak - a la marsupial. Baked to a succulent juicy turn with bacon and herbs and spices in the pocket, surrounded by crisp brown roasted vegetables and swamped with rich gravy, it was unforgettable. The dear soul preened under the unstinting praise heaped on her for her culinary skills. Her giggles dissolved into her usual hearty laughter and then the inevitable happened – in the middle of her merriment the top plate of her false teeth came unstuck and clattered down onto the bottom ones. Completely unabashed, she hoicked them back up with her thumb and through her chuckles said again that she would have them fixed one day. She was guilty from time to time of letting go with her favourite expletive which was what we heard when the teeth came down. I have never heard a bullock driver even let rip with this one - even though I have never known a bullock driver. In her consternation she exploded, -

"Ooohhh – FOOT!" I had never before heard that term or have I since. She was a grand lady.

It was very late and very cold when I rode into the yard and climbed stiffly off the motorbike. It had been a busy night tomcatting down Fremantle way. This was also my last night in Perth for a while, having booked a seat on a plane to go up North, which was due to leave early next morning.

Usually early to bed, the family was still up at this late hour as I could see that all the lights were still on. Fearing a crisis I met my dear old landlady at the front door in her gown and slippers with a look of concern on her face.

"We were worried about you on that motor bike and land sakes, you are frozen stiff." She was right - I was damn cold.

Leading me down the passageway she clucked her concerns.

"You can't go out to the sleep-out on the veranda now, you will freeze. Why don't you hop into bed with our little girl, she will warm you up." Little girl - I mused?

Not one to knock back an attractive offer, I could see through her bedroom door that she was in on the plan too. She was sitting up in her double bed with a wide smile, patting the pillow beside her. This was a family conspiracy it seemed – blimey!

So knackered was I – and this is the honest truth – I was in Noddy Land soon after hitting the pillow, but with the chill rapidly disappearing from my bones, thanks to my obliging and considerate and warm bed mate.

I still don't know if the dad was a party to this unexpected development or not, which was why I panicked when he stood over the bed and shook me awake in the dawn light. With the first seconds of awakening usually a bit confusing at the best of times, I felt a sense of dread as I realised where I was. The feminine aspects of the room and the sensation of the warm female body against my back told me I wasn't in my own lonely bed. The old chap put a restraining hand on my shoulder as I scrambled to escape.

"Thought I had caught you eh?"

Then his face broke into a grin and he roared with laughter. The friendly daughter woke up and joined in too, while Mum stood at the bedroom door and just smiled and nodded approvingly.

115

How Far is Possible?

Some new machine tools had been installed in what was known as the Shell Annexe at the Midland workshops during the war. It was 1944. This complex was dedicated to manufacturing large quantities of ammunition to supply our armed forces that were fighting the Japanese in the Pacific theatres.

Jim Baker and I were assigned to install and commission a number of shell lathes. These were special purpose machines for shaping the projectiles for twenty-five-pounder guns, the field artillery piece favoured at that time by the Australian army.

High in the steelwork of the building, we were busy installing a long drive shaft to power the lathes. It was situated about eight metres above the concrete floor. Jim and I were on ladders at this considerable height making some adjustments. With a metal spirit level gripped in both hands I turned and to maintain my balance, steadied against one of the steel support columns. Running down this column on insulators was a set of high voltage electrical power cables. The old insulation on the wires was perished and shredded and I just caught the gleam of bare stranded copper - but it was too late. As the next few seconds of time were a blur to me, I will relate my partner Jim's account of the events that followed.

He said there were blue flashes and a shower of sparks as my metal spirit level bridged the bare wires and shorted out the 500-volt power running through them. I apparently then did a good rendition of a tribal dance, dropped the heavy spirit level with no small risk to those below and proceeded to fall backwards off the ladder. It appears that Jim's lightning reflexes and considerable strength prevented me taking the long and possibly fatal drop to the concrete and machinery below. He lashed out with one big hand and grabbed the only part of my anatomy he could reach, which happened to be my throat.

Regaining consciousness fairly soon after the electric shock I wondered in my confused state why all the little figures of men were dashing around below. I had been dragged bodily and by the neck in Jim's vice-like grip, to be draped like a sack unceremoniously but safely across the steel shaft high up there under the roof.

The crabby old railway nurse, whom I fear had a great dislike for apprentices anyway, was very matter-of-fact about my physical condition when I staggered into the clinic. Most I am sure regarded her in her starched apron and with the scowling face as not being out of place if she were seen patching up soldiers in the Crimea or in the mud of Flanders. The name given her by the men revealed the general esteem in which she was held. She was referred to as Iodine Annie.

"I've got your burns dressed, but what happened to your throat?"

I could only manage to croak the reply that my partner had done a good job on my larynx, for which I had to be profoundly grateful really.

"OK you can go back to work now son" she grated, and pushed me out through the swinging flyscreen door.

Still rather dazed and uncoordinated, I was following one of the rail lines back to my work area feeling decidedly sorry for myself. Apparently about to take another dive I lost all interest in the world once more and began to fold. A big hairy arm intervened before I could hit the railway line. It was Jim Baker again who was walking to the clinic to check on my progress there. I heard later, but was not a witness to the blast given the horsy nursing sister by Jim, for trying to send me back to work in such a shabby state.

My next clear recollection was associated with the rustle of starched nurses' uniforms and pungent hospital smells.

The pretty young thing in the white cap was clicking her disapproval and scolding me for being too grubby to put in one of her clean hospital beds where it was required I should stay for a period under observation. Strangely I felt too lethargic and tired to be embarrassed or even mildly excited as the nurse stripped me stark naked and steered me under a hot shower. Wielding a big foaming sponge, she set about the task of scrubbing my body to the required shade of pink as all evidence of my previous grotty state sluiced in rivulets down the drain hole. It was obvious that she had regard for my finer feelings and I will always remember her words as she set about the task of purification.

"Ivan," she said between giggles, "what do you say I start at the top and work down as far as possible - then I'll start at the bottom and work up as far as possible" and then, with another giggle she said, -

"And you can wash possible."

In the end after ablutions were done I recall in my dopey state that it was she who actually washed 'possible' after all and believe it or not, without any unseemly or unsightly reaction on the part of 'possible.' There was this new word for my vocabulary too with regard to body parts that I learned that day.

Later in the hospital bed as the lights were dimmed and the sleeping pill began to take hold, I was trying to recall the events of the day. The effects of electric shock and the burns I suffered from my contact with the five hundred volt cables were bad enough, but the effect – or lack of,

Workshops brass time docket was your ID. It was moved from the board on the gate to the one at the place where you worked. This told the foreman whether you were there or not.

on 'possible' was the most unnerving of all. I never want to have another experience like that.

Seriously though there is an interesting comparison to be drawn between the working conditions and attitudes towards accident prevention in 1944 as compared to the present day. In fact the potentially fatal accident in this account was the result of such negligence as to be almost inconceivable in the present enlightened age.

Such is the clarity of hindsight.

Payday

Every Friday afternoon at the Midland Workshops I would join a long queue of fellow-workers, which snaked across the yard and up to the pay window. Shuffling forward in this line would finally bring me face-to-face with the old paymaster. He had a shiny pink bald head and side-whiskers and rimless spectacles perched on the end of his nose. He looked for all the world like a Dickensian clerk.

After being checked off his list, you were handed a small brown envelope with the week's pay in cash inside and your name and number and pay amount on the front. The pay packets I collected in the first year of my apprenticeship were pitifully thin – almost anorexic in fact, though I believe that that particular human disorder in those days had not yet been invented.

It seems that the amount of cash received for a first year apprentice was a hangover from the days of the industrial revolution in Britain. An apprentice's contribution to his employer's business then was considered very low in value and his remuneration reflected his status. Some even had to pay their employer up front to begin to learn their craft I believe. It was good to know that we had progressed to the point where we were at least rewarded, even though it was in small change, for a hard week's work.

I counted the money into my hand. There was one crisp pound note, which I fondled, as I knew I was not destined to have it for long. In the small change was counted two florins, a shilling, a sixpence and a threepenny bit. In apprentices' parlance that was one quid, five bob, a zac and a trey bit. Oh, and the taxation department had shaved off a shilling which I never saw. The equivalent of this in today's currency is two dollars and sixty cents. While you may counter with the argument that money was worth a lot more in terms of buying power than it is today, you must concede that it was still a meagre sum. Not that the money was mine to keep however, as the landlady would have her hand out when I got home for twenty-five shillings from the envelope for my board and lodging. All this financial haemorrhaging left me with ninepence to splurge as I pleased until payday the next week.

Everyone wrote letters in those days, as mail was the major form of communication between friends and families. A postage stamp cost twopence. Dad insisted I write home regularly to the family in Bunbury while away. I would put the bite on him now and then for a handout to buy essentials, like a pair of socks to replace the holey ones, some postage stamps, a tube of toothpaste or a ticket to the local picture theatre. If I was lucky enough to be going with a girl, she would have to pay for herself and we would meet at a prearranged seat inside the theatre. It was a kind of test for the relationship I suppose and pride would not have allowed me to have her pay for my ticket even if she offered. Dad had a great scam designed to keep up my writing efforts and it worked once or twice. He would send me a letter with a footnote that he had enclosed a pound note 'to keep you going.' Turning the envelope inside out I could find no pound note in it. Thinking he had forgotten to enclose the money, I would write back immediately pointing out his oversight. It became clear to me after a while that the cunning old devil had done it on purpose so I would write back more often.

As I was nearly always in financial straits with not even the cash to buy a newspaper, it felt pretty isolated in the workers' train on the way home where almost everyone else had their noses buried in the Daily News. It was a habit I was trying to break, which was to surreptitiously read another man's newspaper over his shoulder as we rattled along. One old fellow seemed totally unaware of my attentions to his paper for a considerable number of miles, but then without turning his head, asked,-

"Is it OK if I turn over to the next page now?"

Most of the tradesmen in the workshops showed a kindly, almost paternal attitude towards apprentices. While the boys found themselves frequently the target for scams and practical jokes, particularly those boys who were found to be naive and gullible, we all had at least one mentor. Jim Baker was mine.

Jim was a fine skilled tradesman mechanical fitter. I was paired with him for several months in the repair and maintenance of machine tools in the vast machine shop. This was the creme-de-la-creme of the fitters' craft. The work involved a high level of precision in the care of lathes, milling machines and a huge range of other equipment employed in heavy industry. This work was also a welcome change from the heavier and less precise tasks involved in the repair and building of locomotives, which was the major focus of work in my skill area. There were many technical challenges involved in this work and being by nature a keen student, I learned a lot of the finer points under Jim's guidance. My level of job satisfaction was so high in those learning years that I looked forward to every day for its variety and interest. Jim was a bull of a man with tight curly hair and had great physical strength, to which I was at another time able to testify. He also would listen carefully as I poured out my woes, was always ready to offer sage advice and was known to tuck the odd ten-shilling note into my shirt pocket.

Misfortune struck when I lost my pay envelope one payday. It must have contained accumulated holiday money, as there was more cash in it than usual. Most unusual however was the fact that it had not been returned to me. In those days it was rare to lose something and not have it come back, so I assumed it had disappeared down a drain or something. The loss was a serious blow to my already precarious financial state, but unbeknown to me a relief initiative was soon swinging into action.

Montgomery or 'Monty' as he was known was a good friend and also a wit and a good cartoonist. He made a name for himself I believe in later years in Australian Rules football as one of the top umpires.

There soon appeared half a dozen lads threading their way through the works rattling tins and pushing them under noses. I wondered at this strange activity, but was soon brought into the picture. Later in the day a delegation of my apprentice workmates backed me up against a bench. Monty stepped forward and made a lovely speech describing my predicament and then presented me with one of the collection tins full of cash. They had staged a whip-around and collected the full amount that I had lost, plus a bit over 'so we can all have a drink.'

It was a very gratifying experience for me, made even more moving when I examined one of the collection tins. Monty had prepared white labels, one of which was pasted around each of them. On these his clever pen showed a very good cartoon likeness of me, portrayed as a dejected, sad and sorry character dressed in rags. There were two headings in big letters on the labels too, which I will always remember. They were, –

'FOOD FOR MACMILLAN APPEAL' and,

'ALMS FOR THE LOVE OF IKE.'

This made me feel so grateful and realise what a great spirit of mateship existed in that community of men.

Eight Minutes

It must be pretty obvious that in a large industrial facility like the Midland Workshops, which employed three thousand men, that some of the support facilities and services would need to be on a grand scale.

I am speaking at this time of the works lavatories - toilets, loos, or whatsoever.

During my first days in the government service at the workshops I became aware of a term used frequently by almost everyone.

It was 'Eight Minutes.'

This was the time allotted to every man once per day to go -'to answer the call of nature.' There was no time limit apparently on just going for a splash.

Seriously now, can you imagine the chaos that could reign if this allowance was not tightly monitored and the three thousand men were free to go as often as they liked and sit for as long as they chose? You must see clearly that this lack of control might lead to dire consequences with disruption to work schedules and targets and God knows what other problems in the organisation – perhaps a state of anarchy yet.

On one of my first days in the locomotive shop I felt the strongest urge to go. The foreman whom I approached for permission for my mission was a large British gentleman with a thick North Country accent. His greasy necktie sloped down over his expansive belly and he was standing with his thumbs hooked in the pockets of his equally greasy waistcoat. It occurred to me that perhaps I should scrunch my cap in my fist and touch my forelock in the presence of such a symbol of power and authority. This was not necessary however as he readily agreed to my request to go for eight minutes. I was new at the time and didn't really know the drill.

The works lavatories were large and impressive brick buildings. Architecturally ornate in the style of the turn of the century, they were a fine tribute in the service of one of man's basic bodily functions. Just inside the entrance there was a kind of foyer, like a picture theatre entrance, but with no big posters portraying the actors in their performances – for reasons of delicacy I suppose. It was complete with a small grimy window with a keen-eyed clerk peering through it.

"Name and number?" was the demand from the window.

"Ivan Macmillan number 996" I responded.

You see I had my brother Joe in the workshops as well at the time so I needed to state my full name and number to allay any possible confusion of identities in this very important and closely monitored procedure.

A quick check by the clerk in the big ledger on his desk verified that I had not passed through these portals at any previous time on that day, so he nodded to affirm that I could enter. He then carefully spooled off two perforations of toilet paper squares from a large roll by his desk, pushed them through the window as I initialled the book, then shouted, -

"Next!"

Deep in the bowels of the building was a long corridor with rows of cubicles on two sides that seemed to go on forever. Sometimes you had to walk a very long way before finding one that didn't have two white knees and a pair of boots showing in the murk.

It was always a good idea to note the time of being seated, from the big railway clock on the wall above the entrance, so you could take full advantage of your eight minutes but without going into overtime. It was a mystery to me just how management monitored that eight minutes, considering that there were approximately three thousand straining customers passing through the system each day. There was always the lurking suspicion that Big Brother was keeping tabs and if you breached the time limit there might be a tap on the shoulder at any time. To my knowledge this never came. You would need to use your full eight minutes each time too as there appeared to be no way of accumulating time credits you may have been entitled to, which arose from a number of short visits which were under the legal time allowance

As I write, I have the feeling that we are missing something in our present toilet routines and experiences. I'm referring to the flushing arrangements we used then. These latest in dunny engineering achievements had a forbidding look and an intimidating flush. Disappearing up the wall at the back of the pedestal was a two-inch pipe, which connected with the water reservoir up near the roof. There must have been twenty gallons of water poised above your head in a rough concrete cistern with a long chain hanging down. When the main event was concluded and the paperwork done, you grasped the handy loop on the end of the chain and gave it an almighty yank. The sound effects were awesome as the stored water literally thundered down the pipe and into the pedestal. Just as this plumbing arrangement has slipped into dimmest toilet history, so did the question 'did you pull the chain?' disappear the same way?

Quite apart from its prime purpose there was an entertainment element to be enjoyed in the eight minutes ritual. The wall graffiti was a rich canvas of the expressions of the industrial worker of fifty years past. It seems that when you sit quietly at peace with nothing more on your hands than two squares of plain Government Issue toilet paper, the mind wanders and the imagination flourishes. The kaleidoscope of graffiti art and ribald humour was so intriguing that I would make a point of noting which cubicle I had occupied and on the next visit take the one adjacent further down the line. Also, with new material being added by the other 2,999 workers on a regular basis and having completed one cycle, it was feasible to start back at number one cubicle and go through sequentially again. Reading the walls beat the feeling of isolation and boredom and eight minutes seemed to pass in no time at all.

In later years living in the remote Kimberley region I used to travel extensively in the outback, where I was privileged to see pictorial evidence of another very different race of people. In the caves and the canyons I would wonder at the picture stories left by the ancient hunter-gatherers many thousands of years ago. It would be great to think that the unique inscriptions left on the walls of the workshop toilets by the white man could be preserved as a kind of time capsule like these for the benefit of our future generations. The problem I could see was that they were only displayed on bricks and mortar. Their chances of surviving down through the ages could only have been guaranteed if they were etched on more permanent surfaces as the aborigines' drawings had been – like Eyre's Rock for instance.

May I offer a sample as one of the many writings and witticisms I observed there in 1945? There is a suggestion of a breach of copyright in this, but it is nevertheless still one of my favourites. It appears to have been borrowed from the walls of one of the underground toilets which could be found right in the middle of St. George's Terrace in Perth city where you had to pay a penny to unlock each door to go to the bog. It goes something like this, –

"Here I sit broken-hearted. -
Signed the book but only farted."

There was a bit of advertising there too. One was promoting 'Knacker Lacquer' which claimed to give "Glamour to your hammer and - lustre to your cluster." I couldn't locate the supplier and so never had the opportunity to try this product.

All in all, the institution seemed to work well and was a credit to the brains behind it. It provided relief to the needy and entertainment too. It also created employment for those privileged and respected few that gave such a dedicated service at the window in the foyer day by day. My hope is that its architectural and cultural contributions to our society are not completely forgotten, perhaps to disappear in the mists of the passing years.

The Young Salt.

With the end of the war coming at last in 1945, followed by all the wild celebrations, our city of Perth came to life again. Everyone heaved a sigh of relief and the all the lights came on and the blackout blinds were removed from the windows and normality resumed its place as before. There were smiles on the faces of the people you passed on the street.

My apprentice colleagues and I at the age of nineteen realised then that we had missed the opportunity to volunteer for military service and so after the war's end we were free to lead a normal and carefree existence. I played lacrosse for Bayswater B Grade and while not by any means the star player, made a lot of friends and enjoyed the sport immensely. Then there were the dances we went to at the Embassy Ballroom, the Pagoda or the Drill Hall, or for something a bit more intimate a good film with an attractive companion at one of the picture theatres in Perth. One of the most enjoyable and possibly the healthiest sport I became involved in was through a close friend and contemporary I knew as Brendan.

This friend had a rich dad and the rich dad had a lot of expensive toys and among these were two yachts. One was an eighteen-foot skiff with a lot of sail and was capable of great racing capability, while the other was an H28 class yacht called Nova, an ocean-going cabin cruiser – a deep keeler. Brendan who was the fortunate son of the rich boat owner asked if I would like to join his crew and sail with them. The first weekend out on the water was enough to get me hooked and I signed up as a member of the Royal Perth Yacht Club.

There was no doubt in my mind that these were serious sailors. Not one of their boats had an engine, not even an auxiliary engine and to these purists any boat that did was looked down upon, reviled and called a stink boat. We all wore clean crisp whites on board, a windcheater with the boat's name on the back and canvas shoes that would not mark the varnished teak decks. The skipper wore an impressive navy peaked cap with braid on it. He smoked a pipe as well.

My job on Nova was that of forrard hand and once I had memorised which was the pointy end of the craft, it was just a matter of time before I mastered the strenuous business of wrestling jibs and spinnakers. I learned the ropes as it were under Brendan's patient tutelage and while I tried hard to master the job, I wouldn't say I actually went overboard with it. It turned out to be a wonderful weekend pastime with excitement to spare and with the added bonus that you didn't have to kick in for the petrol.

We had finished competing in a race on Perth Water in Nova one warm summer Saturday and had returned to the moorings. Brendan asked his dad could we young blokes sail the boat downriver to Fremantle after dark for a moonlight cruise. The olds agreed and headed for home and the four of us hoisted sail and tacked downriver in the fading sea breeze. We had made a prior arrangement to pick up four supernumerary crewmembers on the way.

The girls were waiting on the little jetty on the shore below the university and clambered eagerly on board. They were all students and said they looked forward with great anticipation to these trips – it was a welcome break from studies I suppose.

These very pleasant images of the good life return to me in great clarity even now as if they were formed only yesterday. The surface of the river that night was rumpled just enough to blur the edges of the shining track laid down by the yellow summer moon lifting above the Darling Ranges. We eight young people, not in the mood to talk, gazed at the scene while listening to smoochy music on the cabin radio and the sound of the water lapping along the hull. We had all climbed up on the boom of the mains'l and reclined back, shoulder touching shoulder in the curve of the canvas. A bottle of wine was pulled from the icebox and passed hand-to-hand down the line followed by another. It was a chilled sweet sauterne, which seemed to perfectly compliment the atmosphere of the quiet companionship we shared.

We had to put the boat about at Fremantle, which interrupted the idyllic atmosphere and broke the spell as we pointed the bows back upriver. With the moon now on the other beam we sailed on sedately until the wind died altogether and our vessel came to rest, rocking gently in a flat calm. Drifting for some time beneath the high banks and the twinkling lights of Bicton and Mosman Park, we waited for the land breeze to pick up to get us under way again. I don't think at that stage that anyone really cared. My partner was certainly relaxed as she rested her head on my shoulder and went to sleep. So much for the great Latin lover - but in mitigation I remember she did say she had had a very busy day. Anyway it was soon back to work at the pointy end of the boat again attending to the jib as the warm summer airs stirred and the sails began to fill. We found some more sauterne in the icebox and toasted the lightening sky in the East and welcomed the land breeze coming across to us over the hills. With the boat now making way and answering the skipper's hand on the helm we headed back.

It was still early morning but in full sun already when the girls disembarked and straggled up the path to the university buildings. We found our berth at the yacht club and did all the things yachties have to do to stow gear and secure the boat, and then headed off to a late breakfast and a soft bed. While I doubt if I fully appreciated it then, it felt so good to be young and fancy free.

One of the highlights on the yachting calendar was the blue water Fremantle to Albany ocean race, which we were going to contest in the H28. As an established member of the crew now I was caught up in the planning and the training and the preparation of the boat in the weeks before the big day. It appeared first as I was filling in the forms for leave of absence from work - there was a small red pimple on the back of my left hand. It felt different to the usual eruptions occurring around puberty time and it soon asserted itself. My whole hand swelled rapidly and red streaks appeared up my arm. It throbbed like hell. It was just two days before the big race when I found myself in a hospital being told to breathe deeply as I was knocked out for some surgery on the hand. It was a case of blood poisoning they said. There is still the memory of the bitter disappointment I felt and the operation scar is still visible to remind me of it today. The crew came to visit me in hospital before they left for the big adventure, leaving a bag of peppermints and the reassurance that there would be other times for me to look forward to.

There were about a dozen yachts one sparkling morning all jockeying for position behind the starting line in Fremantle harbour. This was to be another entry on the yachties' calendar – the race to Rottnest Island and the Winterbottom Trophy held over the forthcoming holiday long weekend. It was reputed to be three days of ocean racing and sheer decadence to look forward to – for some anyway.

A sleek fragile-looking sailing canoe of great length and narrow beam ranged up alongside us. This boat was one man down in its three-man crew and their skipper shouted, asking for a volunteer to go with them so that they would be permitted to compete in the race. Our skipper replied that he could spare a man and that Ivan was the volunteer. With bag in hand and a camera around my neck I dropped down into the other craft. After the smooth and stable handling of the cabin cruiser this was akin to dropping onto the back of a wild Brahman bull in a rodeo ring – and we were still in the harbour.

The starter's gun boomed from the mole and we were away. Clear of the heads, the boat heeled to the sou-wester, which was very stiff and getting stiffer. In gage roads I noticed a peculiarity with the way this boat handled the big swells. She did not lift her bows and crest the wave like most well-mannered yachts do, but punched straight through like a torpedo, with solid green water foaming over the coaming every time.

My required function on that boat became crystal clear immediately and can be summed up in one word – bailing. This craft carried a lot of sail too – we were in a race remember? My two companions were as busy as one-armed wallpaper hangers with an itch, wrestling with the sheets and the tiller while I bent with a will to my laborious and crucial job with the bucket.

Halfway to our destination, saturated and chilled to the marrow in the keen wind I thought I might be able to offset my advancing hypothermia by bailing faster. Thinking that there might be another advantage in an increase in gallons per minute over the side, I might also be able to delay the point when I was sure our boat would inevitably fill up and founder.

Now at the age of nineteen no one could classify me as a seasoned drinker – anyway we weren't supposed to be involved with spirituous liquors - according to the laws of the land, until we became of age at twenty one. My Bacchanalian experiences had been limited to a few small beers or a drop of wine on rare occasions. The big black-labelled bottle of Johnny Walker was retrieved from a locker by the skipper, uncapped and passed around with the promise that it would ease the pain and discomfort that I was suffering. So here I was skolling neat scotch with a salt-water chaser over an almost virgin palate. The whisky soon lit a fire in my belly and infused my limbs with a wonderful glow. There was this indomitable feeling that I could bail and sail forever. I mused later that ours was probably the first boat in history to be saved by a bottle of whisky, though there was probably a number that had perished as a consequence of it.

While I was still head down and bailing with a will, we slid serenely into the sheltered waters of Thompson Bay. My efforts only ceased when the bucket was scraping the ribs on the bottom of the boat and the skipper with a restraining hand, patted me on the back with words of praise for a job well done.

My own crew in Nova, moored some distance away, sculled the dinghy across to pick me up as I waited for them, swinging from a stay and clutching my bag and camera. There could be no doubt in anyone's mind there that Ivan was thoroughly inebriated as a result of his involvement with the big black bottle in the boat. My otherwise sturdy constitution had proved to be no match for the ingestion of a considerable quantity of neat grain spirit and I will allow you to choose from a list of adjectives describing my condition. Mrs. Macmillan's little boy Ivan was - crocked, wasted, legless, elephant's trunk, snakes hissed and magotted. On an inebriation scale of ten I would have scored about 9 _. I was still standing - in a fashion, which accounts for the remaining half a point though this wasn't to last it seemed.

The picture I formed of the approaching dinghy was rather vague and it seemed to be moving on a strangely erratic course towards me and that gave me cause for concern. Releasing my grasp on the stay I stepped out in the direction of the dinghy – into the clear waters of Thompson Bay. This is not surprising as it was still some distance away and I fell and disappeared a few yards short of my target as it were. When asked later to describe this event I could only offer that which could be remembered. Without any of the normal sensations encountered when being suddenly submerged in the ocean, I do recall that my eyes were still functioning. Visual impressions at that depth were firstly the lovely turquoise green colour underwater. The stream of bright bubbles held my attention for a while as they zigzagged upwards and then my camera floated past my face heading for the surface too. It was a surprise to see that the camera could float and I remember thinking it to be a good feature in a camera for people like me in a situation like this.

From that point on I must have reached the ten on the inebriation scale, so the details of further developments are pure hearsay and I cannot claim to actually verify what became of me after that. Suffice to say that I was dragged up the beach unconscious by the crew, wrapped in towels and parked in a hole in a gully behind a sand hill to sleep it off.

The time at which it all ceased to become just hearsay was when I returned to the world and crawled up to the crest of the sand hill in the hot sun next morning. I swept the bay with my eyes searching for Nova. It was seemingly difficult for me to focus and I didn't spot her straight away, as there seemed to be a problem of identification that I had to contend with. With apparently some of the demon drink still in my system it looked to me as though there were two Novas – or was it three? Not being too sure of which of them I should go and hail I thought it best to give that a miss for the time being. Sliding back down the dune and back into the gully, I re-excavated my hole in the sand, wrapped myself in the towels and surrendered myself again to oblivion.

It was a relief to learn later that the owner of the sailing canoe had said that I would not be needed, as they would be able to cope unaided on the voyage back to the mainland with the weather fine and with relatively calm sea conditions. The realisation of what I had missed by sleeping for those two days came home to me as my shipmates told their tales of the weekend they had with the holiday crowd. There were wondrous tales of their experiences at the big dance, at the pub and with sheilas galore. What they were telling me wasn't very clear as I was suffering blockages with all the sand in my ears and eyes and was still burping stale whisky fumes.

With this unfortunate drinking episode literally under my belt it was just a case of learning about moderation in the light of experience. Like most young people I enjoyed some wild and wonderful times, but based on that episode I seldom disgraced myself in the same way after that. None of my friends have ever had to report that I was at any time crocked, wasted, legless, elephant's trunk, snakes hissed or magotted ever again. It was my decision to try and stay clear of the line between being happily inebriated and being downright disgusting. There's really no point in crossing that line looking for a good time I found, as you probably wouldn't remember any of it anyway.

My more experienced friends advised me that when attending the big dance where nubile maidens were in great numbers, I should carry a big stick. There was a reason for this. Our standard outfit when ashore included a sporty white windcheater with the name of the boat emblazoned on the back and a pair of natty little white shorts. It appears that the tanned beauties would check out the boats in the bay in the afternoon to pick out the most expensive and best-looking craft. Nova was a picture at the moorings, which must have been the reason why our crew attracted so much attention. Our dinghy, when returning to the Nova in the early hours after the dance, was loaded to the gunwales with very garrulous and happy people – usually twice as many as on the trip to shore at least. We played hosts to the visitors of course with the best of maritime hospitality. Very soon after climbing on board, you could see recumbent bodies scattered everywhere. One morning-after I woke after a night in the sail locker in the bow. My young and attractive companion woke too and we introduced ourselves one to the other. The skipper on these occasions, true to seafaring tradition, treated us all to a fine breakfast of bacon and eggs and toast and strong coffee and the aromas wafting out of the cabin brought the rest of the company to life. Sitting around in the cockpit in the morning sun was a memorable experience. Most of the crew did not know the names of their respective companions, or did not remember them, much less the names of any of the other visitors. While munching on a sumptuous breakfast everyone introduced himself or herself, and we were just one big riotous happy family.

This account is related to you with due and humble respect for all the genuine *old* salts out there in the world.

The Running Sheds

For those readers unfamiliar with railways and railway jargon, the title of this story may be puzzling. In 1944 when I was serving my apprenticeship in the WAGR Midland Railway Workshops, steam was the prime mover for haulage on the State's railways. The clank and rattle and chuffing of steam locomotives could be heard on all suburban rail tracks and also along the lines stretching to the major regional and agricultural centres.

When a loco has completed its run from Perth to a country town like Bunbury for instance, it is uncoupled from its train and steams over to the running sheds close to the rail station. Here maintenance crews swarm all over it, replenishing water and coal supplies, checking the state of the fire in the boiler and carrying out any running repairs as required for the return journey. To clear up any mystery regarding the title of this story then – a 'running shed' is a depot where running repairs and maintenance are carried out on locomotives, which have been sidetracked from a country run before continuing on their scheduled trip.

All major running sheds had a turntable. Steam locomotives are all capable of running in either direction, forwards or backwards with equal aplomb. The majority of the locos on suburban service, mainly tank engines, as opposed to country service were seldom turned around but would run say, boiler-first to Fremantle and then tender-first back to Perth on the return trip. The more powerful and faster mainline locos however, preferred to run in the forward direction – hence the presence of the turntables in the running sheds, which were used to rotate them on the lines and point them back the other way. It could be imagined as a matter of class distinction and dignity, which forbad the pride of the fleet, the long-legged powerful stars of the track to be seen travelling backwards. In my naivety I reasoned that if a loco is travelling backwards, the chimney smoke doesn't get in the driver's eyes. This train of logic did not seem to be sufficient justification to change anything however.

One of the perks of my particular trade designation lay in the requirement to be transferred to and work for a term in a country running shed. This opportunity to get out of the main workshops in Midland was most welcome. Four years incarcerated there pulling engines to bits and rebuilding them began to pall, and I yearned for some experience in the field on 'live' locos where I could observe my powerful charges in action. Most lads in the other trades were workshops bound for the full term of their apprenticeship, which was a minimum of five years. I felt like a kid released from school on a Friday afternoon as I left the grim buildings of the workshops behind. With great anticipation I looked forward to a change of pace and a change of scene and some new experiences too.

For my relocation I chose the Bunbury running sheds. This was my hometown and my family lived there still. The prospect of being at home again after four years in the city was a great treat with the opportunity to pick up on the circle of friends from my schooldays there. It was also a good start to the major break I would soon be making when my railway service came to an end and I could venture out into the world as a qualified tradesman and seek my fortune.

Running sheds are not the most pleasant or comfortable places in which to work. Summer time was trying in the extreme under the iron roof. Steam locomotives are by nature very hot with the big coal fire in their boiler and scalding steam escaping wherever it can. Most running repairs have to be done with the boiler hot and up to pressure so all in all it was a very sweaty and rather grimy existence. Winter could be equally trying too with freezing wind and rain swirling in through the open sides of the buildings. There were no walls to speak of. It was common to experience the greatest extremes of hot and cold when you were exposed to heat on one side, while your back was wet and chilled on the other. I suffered even more discomfort during a stint in the running sheds at Collie, the big coal town up in the hills Northeast of Bunbury. Nightshift in a running shed can be an ordeal.

Collie is renowned for its inclement weather and miserably cold winters. As it was a temporary assignment, my accommodation provided turned out to be less than salubrious. An old gangers' van parked in the yards was our new home. It was not insulated, had no heating and the roof leaked. When I turned in on the first night I found myself in cold sodden blankets due to a leaky roof, but with no alternative I had to tough it out. As I was working in open sheds, exposed to the rotten weather, I was miserable in the extreme. It was almost inevitable that my health suffered and sure enough, probably as a result of the conditions I went down with the 'flu.

The lads returning to the van after their shift were most sympathetic to my deteriorated state of health and came up with the remedy, which they declared would sort out my problems and cure my ills. It was customary for us to walk the two miles into town and spend some of the evenings at the local pub - telling lies and drinking beer. This they had decided to do. I was dragged from my sick bed, steered to the pub, propped up at the bar and doctored with copious amounts of Bundaberg Rum and cloves. This treatment they declared would burn out my infection and put me straight again. I was reassured too that even if I was ill I wouldn't care. My recollections from that point become confused. It was true that at that juncture I was feeling little pain, but decided to head back to the van in the railway yards to get my head down again. There are no recollections of anything at all after leaving the pub. Having pointed me in the right direction and watched me drag myself off into the rainy night they retired again to the bar until closing time.

Accounts of the rest of my movements are based on second-hand information.

Apparently when my workmates eventually arrived back at the van later that night I wasn't there, so they sent a search party back along the most likely route I would have taken from the pub. Being a dark and stormy night it appears that I was difficult to find, but was eventually located in a paddock by the side of the road like a bundle of sodden rags completely out to it. I can only relate the events as my rescue party told them to me and I treat this one part with extreme scepticism. They had deduced that I must have been seeking some shelter and warmth before collapsing and passing out, as they found me curled up and cosy with a dozing dairy cow in a paddock. With no recollections at all about this phase, I was loath to call them liars. Insofar as giving them the benefit of the doubt, it seems likely that I owe that cow a debt of gratitude. Perhaps she saved me from an untimely death from exposure in that cold and soggy paddock. I had thought to stay in touch with her somehow, with the idea perhaps of finding an opportunity to express my appreciation, but trouble was, all cows look the same to me.

My first awareness of a return to the land of the living was with the feel of starched bed sheets and the face of a nurse appearing through the fog. It was two days later and I had been unconscious with pneumonia in the Bunbury District Hospital. There are still serious doubts in my mind as to the medicinal value of Bundaberg rum.

The nature of the work in the running sheds was exciting at times and never boring, but it was not without its risks and dangerous situations. In retrospect I valued the experience and would have changed nothing. To me a life behind a desk or sitting in a teller's window would have seen me die of boredom at the tedium.

Most people would have seen the little rail trolleys scooting along the lines, transporting crews to various destinations. These were a common mode of transport and in the running sheds we had our own, being used to visit the various points on the rail system on work assignments. There were various types in use ranging from the small trikes used for two men and a toolkit, up to the bigger 'Groper', which could carry a gang of eight men at a good speed along the rails.

It was a privilege to witness a very entertaining scenario played out on a country rail line near Donnybrook one day. The major players were eight fettlers travelling to a job on their Groper rail trolley, and the other was a large locomotive.

Firstly, as with all vehicles on rails, the problem arises how to turn them around when it is required to stop and go back in the opposite direction. Unlike the situation with a bus for instance, you can't effect a three-point turn on a rail line. The original designer of the Groper rail trolley overcame the problem of having it run in reverse by installing a two-stroke engine. These as we all should know, are capable of driving in either rotation. With the Groper, you just apply the brake to stop it, throw over the spark advance lever, give it a push start and you are away back to where you came from without fuss.

On the top of a hill on this lovely clear day I sat on a rock to rest and admire the green sweep of scenery below. A single rail line came from my right, curved around the base of the hill and disappeared to my left. The unmistakable sound of an approaching steam locomotive claimed my attention coming from this direction. Swaying to the thrust of its pistons and with a tall plume of smoke it came at a fast pace drawing a long line of loaded freight wagons.

Then from my left came the staccato bark of the groper engine at full throttle. It signalled the approach of a rail trolley carrying a gang of fettlers. It was approaching the curve around my hill from the opposite direction and it seemed that there would be some interesting developments soon. Firstly the goods train and the rail trolley were travelling at speed towards each other, neither could see the other, and of greatest interest, they were both on the same line. With no way of signalling or intervening in any way I decided to wait and watch as events unfolded.

The fettlers on the trolley were a happy lot by the sounds of laughter and chyacking coming faintly to my ears. This merriment stopped abruptly when they became aware of the goods train bearing down upon them from around the curve ahead.

The driver of the train for his part reacted quickly by pulling the whistle cord, shutting his throttle and applying his brakes – hard. As you would suspect, you cannot bring a five hundred-ton speeding train to a stop in anything like its own length. Giving credit where it's due, with showers of sparks streaming from his brakes and the whistle shrieking, he was doing all that was humanly possible to avert a head-on collision with the Groper.

Back on the Groper, the engine cut its racket and the brakes squealed as the crew manfully hauled back on the long brake handle. As far as I could determine, there appeared to be little frivolity among them at this point.

At a range of about two hundred yards and closing, despite frantic efforts to prevent the seemingly inevitable collision, the men on the trolley must have reassessed their situation and began one by one to abandon ship.

I winced as the hurtling bodies landed on the rough rock ballast and sleepers at the side of the track. They skidded and rolled and were left in tangled heaps of arms and legs and dust in the trail of the Groper. None stood up again by the side of the track that I could see.

Perhaps partly due to the reduction in the weight of its human cargo, the Groper was slowing its pace appreciably. There was however this figure of one man remaining like Horatio at the gate, wrestling with the brake handle trying to bring the trolley to a complete stop.

The goods train bore down relentlessly and the intervening gap was shrinking as the Groper finally slithered to a halt. The lone remaining fettler knocked the spark lever over, then leapt off about fifty yards in front of the train. His heroic straining to get the trolley started – normally the task of two men was rewarded as the groper engine fired. The trolley gathered speed with our hero clinging on for dear life and scrambling back on board, heading away from impending and almost certain disaster.

At the point when the increasing speed of the Groper away from the train exceeded the decreasing speed of the train, I could scarcely see a space between the two.

The closing scene in this drama was the trolley accelerating away from the big black locomotive, with its pilot hunched over the controls. This was until he began to go back past the rest of his companions. They were a sorry sight, scattered along the sides of the track in untidy heaps. He leapt to his feet and returned their feeble waves with a gallant sweep of his hat and a deep bow of acknowledgment.

A feature I enjoyed in the running sheds was that the unexpected became the norm and every day was different with new situations appearing without notice almost every day. The work had its dangers too and I felt the need to keep my wits about me at all times.

During the war years a new crop of steam locomotives appeared on the lines. These had tall driving wheels and were reputed to be capable of high speeds. This was the U class and we knew them as U-boats in the railway jargon. The one distinguishing feature of these engines was that they were fired with oil instead of coal. The tender at the back had a large tank and the oil was sprayed into the firebox under steam pressure. They were the loco fireman's dream, as instead of having to shovel many tons of coal in through the firebox door, an extremely arduous job at the best of times on a long haul, he just twiddled a few valves.

A U-boat rolled into the sheds one day for some repairs. Much of our work was concerned with the greasy bits underneath, which were accessed from a pit between the rails. With spanner in hand I dropped into the pit and stooped to pass underneath the engine to find the fault and fix it. This particular loco still had a full head of steam up, as it was required to turn around and head back to Perth as soon as the minor repairs were done. The fire in the boiler had been shut down which was a great consolation to me, as I had to work in the pit underneath it. The heat would have been intense and as it was also a very hot summer day I was grateful for such small mercies.

While busy at the task under the loco, I failed to notice the activity in the driver's cab. The fireman had decided that due to the falling steam pressure, he would need to light up the boiler again. The fuel oil he sprayed into the hot firebox began to vaporise, at which point he was supposed to light it immediately. Dense white oil vapour soon poured down and rolled along the pit. I smelled it first and then looked down to find myself up to my waist in the stuff. A quick exit was called for, which I executed in haste, but not without sustaining a crack on the head from one of the hard bits. The fireman at last managed to light the fire, which then began to consume all the fumes, including those in the pit, which I had just vacated. It didn't take long. The big red ball of flame raced from one end of the pit to the other leaving everything combustible behind it as charred, smouldering remains. The radiated heat was intense as I stood back and watched. Had I been a little slower in my evacuation, I fear I may have been reduced to something resembling one of those smoking remnants.

Needless to say I had a few words with the inconsiderate fireman and as I said earlier – the unexpected becomes the norm when working in the running sheds.

We were notified one day of a serious pile-up, which had occurred on the main line South of Bunbury near a siding called Yornup. A goods train travelling at speed had left the rails and someone had to go down there and put it back on again. All rail traffic had been halted and the line had to be cleared and the tracks repaired as soon as possible.

The breakdown van was made ready, tools and salvage equipment loaded, the team climbed aboard and we were on our way within the hour. Our string of vans and a couple of wagons headed southwards rolling along behind a fast locomotive.

The derailment fortunately was that of a goods train that carried no passengers and so had incurred no serious injuries or loss of life, but the scene we witnessed as we neared the site was one of chaos. Darkness was approaching and a light rain had started to fall. As the details of the situation became clearer I could make out the two hundred yards of the train scattered and piled like the toys of a petulant toddler. There was a jumble of twisted wreckage of the wagons with their loads spilt in a tangle of timber and goods and mountains of bulk wheat. The train's locomotive had ploughed off the tracks too and was listing at a steep angle with its nose buried in the mud. Wisps of steam drifted forlornly from its boiler fittings. To my inexperienced eyes the task of clearing this lot was well nigh impossible. To compound the difficulty of the job which faced us was the fact that the line at the smash scene ran along the top of a high embankment with steep drops on both sides into deep gullies.

All fifty tons of the derailed locomotive had to be stabilised so it wouldn't topple over, then lifted inch by inch on big mechanical jacks and replaced on the rails. I thought I knew the meaning of hard work, but the Herculean effort required, using crowbars, twenty-eight-pound sledgehammers, chain blocks and other gear, was to tax every man's strength and stamina to the limit.

Two locomotives, one from either end of the derailment site began to drag wreckage out of the piled conglomeration of smashed wagons. This was simply jettisoned down the steep embankments on both sides of the track. It was then set on fire to burn all the timber panelling and flooring and reduce the wagons to steel skeletons.

All through that night the team continued the bullocking work until we felt ready to drop in our tracks. A quick mug of scalding black tea and a doorstep cheese sandwich around midnight provided a short break and then we turned to again.

The bizarre images on that black winter night made an indelible impression on me. A ring of kerosene flares hissed and spluttered in the rain, shedding a yellow flickering glow as we worked. The piles of shattered wagons, which had been torched, made their own funeral pyres with leaping flames and showers of sparks. Men sweated, stripped bare to the waist in the light of the flames and still the misty rain fell on shining backs and bunched muscles and wreckage.

The foreman said he had a special task for me that was an important part of the efforts required in clearing away the mess. I was to assume responsibility for all oxy cutting of steel as was required by the team of men who were clearing the piles of wreckage. Actually, being in charge of the cutting meant I had to actually do it all myself, so pulling on a leather welder's jacket and donning long gloves and goggles, I set to. I began to really enjoy myself and was feeling pretty important in this key role. This was certainly better than the stresses and strains of the manual labour.

It was soon time to earn my stripes however. Almost in the centre of the crash scene there was a situation that looked decidedly hazardous to resolve. Two fully loaded timber wagons had slewed at right angles across the track but were still linked together by a safety chain. One wagon was hanging out over the drop on one side of the embankment while the other counterbalanced it on the other side. Apparently the sum total of their weight was one hundred tons and they would have to be separated to clear the line by cutting through the chain that remained connecting the two.

Several men manhandled my oxyacetylene gear to the top of the bank where the two wagons were still linked and then retired to a safe distance to watch developments. The one chain remaining was stretched like a rigid steel bar under the combined weights of the loads suspended upon it. All I had to do was cut it. I must admit to having a feeling of vulnerability at that stage, but the barracking from the others told me it was no time to shirk and there was no way of avoiding the job even if I had wanted to.

The result of that chain being cut was pretty spectacular. The two freed wagons plunged down the bank and crashed into the gullies. Both parts of the chain lashed back viciously, the last link having parted before I had completely severed it.

When the grey dawn broke it was still raining, with curtains of drizzle saturating everything. We were all dog weary but there was little more we fitters had to do. A team of fettlers was replacing the buckled rails. The little galley in the breakdown wagon turned out some greasy fried eggs clinging to thick slabs of bread. Washed down with scalding black tea from a chipped enamel mug I swear it was the finest breakfast I had ever eaten.

I looked back through bleary eyes along the shining rails as we pulled away from the scene of our night's work. The images of the twisted and blackened remains still wreathed in blue smoke disappeared around a bend and were hidden by the trees. Turning to a nearby bunk I can just recall the gentle swaying of the van and the click of the wheels for scant seconds before sleeping like a dead man.

We were building locomotives during my apprenticeship in the Midland Workshops - this is one of my babies.

Episode Six
Wittenoom Gorge

The five years spent in the railways as an indentured apprentice had a very claustrophobic effect upon me.

Walking for the last time out through the main gate and past the time office of the Midland Workshops, I was conscious of the newly won trade certificate in my shirt pocket. I was stepping out into the wide world. Work was plentiful with many opportunities for a qualified tradesman. It was in these times not so much a matter of whether a job could be secured, but just which job to take.

This was 1949, the year of my 22nd birthday.

The apprentice workmates with whom I had served and had 'finished their time' along with me, scattered in all directions immediately they were released from bondage. There was a comment I recall rather sardonically, made to us by the Master of Apprentices not long before I left the service, -

"We have had our eye on you and a couple of the other lads. You have a lot of potential and we would like to see you stay on working here."

Any thoughts I may have entertained about forging a career in the railways were quickly dashed when querying my future prospects had I stayed on.

"You could be the workshop foreman here in less than ten years," he said.

The prospect of the continued grind building and repairing locomotives had no further appeal for me - no more than the promised pay incentive of an extra three pounds a week if I decided to wait and actually won the promotion. These prospects paled in comparison with what I considered the big wide world had to offer. The offer was declined with thanks.

Wittenoom Gorge was the location for a mine, the product of which was blue asbestos. The Australian Blue Asbestos Company at that time in 1949 was booming and expanding its mining operations in the Pilbara. Qualified tradesmen were urgently required. It held the prospects for me of good pay, wider technical experience and the chance to see the mysterious North, which I had heard so much about. My job application to the Australian Blue Asbestos Company as a maintenance fitter was accepted.

Flying was a previously unknown experience. At the Perth Domestic Airport I found myself huddled and freezing in the pre-dawn together with others of my travelling companions in a cavernous hangar that passed as a passenger terminal. The flight departure was scheduled for 6 am. The boarding call came and we straggled, cold and stiff across the tarmac to the waiting aircraft just as the horizon's grey mists were clearing to give way to sunrise.

The Douglas DC3 aircraft was the workhorse of our Airlines WA, the major carrier in the State, as indeed it had been for many years in war and civilian service in many parts of the world. Our plane squatted, nose pointing to the sky on two fat tyres and a single tail wheel. Two Pratt and Whitney radial engines jutted forward from the wings sporting large three-bladed propellers. Once inside the plane I found that the floor between the two rows of seats was on such a steep upward slope that I had to literally pull myself along the aisle to find my allotted seat and then fall back into it. The sun rose just after takeoff and the morning sky was clear and blue.

It was unfortunate that my maiden flight northwards was undertaken at the height of the Pilbara summer. The DC3 did not have the benefit of pressurization or air-conditioning. This meant that we flew at relatively low altitudes where the atmosphere was dense enough to provide the oxygen for breathing, but which found us battling all manner of extreme air turbulence. This was especially noticeable over the superheated landscape of the northern areas as the day progressed. Trying to doze with the seat laid back, my head rolled to the side and contacted the hard window frame through which the numbing vibration of the big piston engines was transmitted. My skull resonated with the beat and sleep was impossible. Combined with these discomforts we were forced to sit in the airless aircraft baking in the sun on the dirt airstrips of the many whistle stop ports of call on the journey. The further north we travelled the hotter it became, as evidenced by the amount of sweat produced at each stopover. The memories of this first flight will always remain with me as a bad dream and then there was the nausea. There was a chance to stretch our legs at the Meekatharra 'airport.' Along with the other passengers I headed over to the passenger lounge which was a pipe-framed humpy with a tin roof, a weatherworn wooden bench and a water bag swinging in the hot wind.

Our in-flight breakfast on the plane comprising two hard-boiled eggs with a slice of bread and butter washed down with luke-warm coffee was destined for only an uncertain stay in my stomach. For some of the other passengers the sick bag was their constant flight accessory.

A chatty airhostess revealed to me one of the airline's secrets. It appears that the strange taste of the coffee served with breakfast was due to the demands on the limited catering facilities on the plane. It was necessary she explained, to use the hot water to make the coffee, which had first been used in which to boil the eggs. This also accounted for the fact that the coffee was always served after the eggs and not before. Then it all made sense somehow.

From the air on our banking approach there was the small town of Wittenoom below straddling the road, which curved from the river flats and wound up the gorge to the mine. The scene was set against the awesome backdrop of the Hammersley Ranges. Their walls of red rock mottled with spinifex clumps enclosed the group of tiny buildings like a great amphitheatre. The asbestos mine was hidden five miles up inside the fold of the ranges while the pub, the store and town houses were huddled together at the mouth of the gorge. I could see a couple of vehicles, like toys, each with the attendant plume of red dust in its wake, moving out along the road to meet our aircraft.

The big door in the fuselage was swung open and a rickety stair on wheels was pushed up allowing the passengers to disembark. First impressions were not reassuring. Having left the cooler, greener and softer climate of Perth just hours before, the contrasts struck me like a fist. Worst of all were the heat and the glare. The sun beat down from a copper sky and the landscape danced and eddied with the mirages. The spinifex clumps intrigued me though with their profusion of puffball shapes against the red rock.

This was to be my work place and home for a year, or as long as I could hack it - whichever came first. As Dad once said in a rare benevolent mood, -

"I don't think I raised a quitter."

So, trying to live up to this and appear nonchalant at the same time, I picked up my possessions from the baggage trolley, boarded the grimy bus and headed for town.

A large proportion of the workforce at Wittenoom proved to be single men, or at least men without partners. It was said that it was a retreat where those who wanted to dodge the arm of the law, escape from their wives and families, or were avoiding any other oppressive social obligations, could disappear. Many were migrants, some having left the snowy winter landscapes of Europe to be shunted directly into what they must have regarded as a furnace-like arid environment of rocks and dust and spinifex.

My first sight of the accommodation offered as my home for the ensuing stay created a feeling of mild despair. The determination that I would have to make the most of the situation had pushed me into this, though first impressions were far from encouraging.

My salubrious accommodation for one year at Wittenoom Gorge

The allocated housing was in the shape of an unlined, concrete-walled hut with fibro partitions and a flat tin roof, all perched on a rough concrete slab. The window frames were devoid of glass. Some previous occupant had attempted to make it more homely by planting a creeper against the wall. The plant had died in the dry dust. About ten feet square, the hut was furnished with two steel-framed shearers' beds comprising rusty steel pipe frames with cyclone wire mesh bottoms. The occasional chairs and tables were old empty beer crates while rude shelves and a piece of pipe across the corner provided the walk-in wardrobe. My bed had a thin horsehair-stuffed mattress and a grubby ticking pillow. To supplement these sleeping arrangements I was issued on arrival with one thin cotton blanket with red stripes, of the kind frequently seen in horse stables. Later in the year the nights turned bitterly cold with a piercing wind off the Great Sandy Desert. We had to become innovative or freeze, so we sewed some heavy hessian bags together to make a bedcover. They called them waggas – a kind of bushman's doona. One of the men came from a sheep station and was lamenting the fact that he could not do as they did back on the station on cold nights – "pull up a couple more dogs". A single bare light bulb hanging by its cord from a rafter completed the picture.

We ate at the communal mess hall loosely termed Ross's Restaurant. An example of the level of care and attention given to preparation of our meals by the sweating staff was seen in the daily cut lunches. These we watched being made up ready for the dusty journey to the mine at midday.

Rows of thick slices of bread, well past their eat-by date were laid out on a table in the mess kitchen. Butter was heated in a pan until it assumed the consistency of yellow oil. One of the kitchen hands with this pan in one hand and a large paintbrush in the other dipped the brush in the oil and perfunctorily swiped it over the surface of the bread. Special fillings used in these sandwiches included vegemite, limp lettuce, or a kind of petrified cheese, or polony slices resembling dry autumn leaves in both taste and texture. Given that the hungry miners might already be feeling disgruntled at these offerings, the situation was further worsened by the condition of the food on arrival at the mine. Wrapped individually in old newspaper, the packages more often than not spent an hour in the heat and the dust on the back of a truck, all of which rendered them most unappetizing if not completely inedible.

Later, in agreement with some of my newfound mates it was decided that we would have to seek a better deal on lunches. An approach to the kitchen staff at the pub produced results. For a small fee they supplied us daily with very tasty lunches including fresh bread, fancy fillings, fruit and cake. Due perhaps to the fact that we were young, single and male and the hotel staff were young, single and female, it could have been seen as a case of favoured treatment for us. Upon that situation however hangs another story.

A number of us young single men soon banded together and became good friends in what turned out in some instances to be a rather hostile environment. We were all qualified tradesmen of different disciplines and together we had some great times. We also could offer a united front if it was ever needed against some of the criminal element that existed in the workforce and as a result we were seldom bothered or preyed upon as many others were.

Well established in the men's camp at that time was a group of bullyboys who exploited some of the weaker members of the work force. Using sometimes-brutal standover tactics their aim was to fleece the poor unfortunates of their pay packets through the gambling rackets they were running. Betting, two-up, dice and card schools ran almost continuously with the richest pickings being on pay nights. Some of the victims of these scams, although earning good wages in underground mining, seldom had enough money to buy a ticket out of the place, had they wanted to leave. Between being scalped at the schools and passing over their hard-earned at the company pub, many were in a no-win situation and laboured for nothing, having found themselves caught up in a cycle of hopelessness. The threat of violence was never far below the surface and we witnessed many examples of it during our time there.

The small police force of two officers in the town was usually conspicuous by its absence when needed. While they were seen together at the pub occasionally, they always seemed to use the back streets and were never seen monitoring the real trouble spots centred on the gambling schools in the men's camp. The publican had his own unconventional but effective method of ejecting undesirables from his front bar – and an undesirable is a bloke who has spent all his pay. This licensee was a big fellow with the distinguishing feature of an enormous beer gut. This he would use as a weapon. Dragging some poor inebriated miner to the swing doors at the front of the pub he would slam the offender against his belly. This would propel him down the front steps to sprawl in the dusty street outside. There was no shortage of live entertainment for the customers there, especially on pay nights.

Inside the mill the primary crusher was the first step in processing the asbestos ore to extract the fibres

Not only did many of the miners lose heavily at the gambling, but some also ran up debts and owed the standover men considerable sums. A big ugly raw-boned street fighter known only as Fitzy ran the 'swy' school with an iron fist - or two. Almost anyone who could not pay his gambling debts to this villain was marked for a beating. One of the poor unfortunates who were deeply indebted to Fitzy absconded from Wittenoom without telling anyone. Fitzy sent one of his henchmen to the coastal town of Roebourne where he soon tracked down the man and beat him badly. This low life was seen after his return proudly showing everyone the split knuckles on his fists as evidence of the deed. Knife fights and the use of guns were not uncommon incidents either.

Our group of young men was engaged in the maintenance of machinery and equipment used in the mining and milling of blue asbestos at the mine site deep in the gorge. Large clouds of fine grey asbestos dust belched from the open sides of the mill building, drifted away on the wind and coated the landscape and everything and everybody in its path with a thick pall.

The asbestos ore emerged from the mineshaft in the wall of the gorge and was fed onto conveyor belts at the back of the mill. Crushing and separating reduced the ore to fines and the fluffy blue fibre was captured and accumulated in the bagging area. The bagging operator worked in a permanent thick fog of asbestos dust while filling the hessian sacks for export. At day's end, all who had worked in the mill area had the appearance of grey ghosts. The fibre permeated every part of the body and for days was blown out of noses, dug out of ears and coughed up in sputum. We had no idea of any health hazard associated with the fibres. Neither did our employers warn us that we could suffer future serious consequences. As the years go by the legacy of asbestosis and often-fatal mesothelioma diseases has claimed many of the lives of those who worked there - ignorance is bliss.

It seems that I was soon to witness the rough justice meted out in Northwest mining towns and would have found it hard to believe that it was happening.

It was Monday morning. Just of late there had been a growing anger and bitterness building up against the standover men who ran the gambling schools in the camp.

From a conversation with a group of miners I picked up that something sinister was about to happen. They were climbing onto the train of wagons hauled by a small locomotive heading down into the mine workings. It appeared that one of these hated characters had been 'fingered' by the miners as having lived out his term. The name was mentioned and it appeared that everyone knew of the plot against him. It was apparently the infamous Fitzy. He was a member of the underground miners' team and

The married quarters Wittenoom town site - after the town was abandoned

was seen boarding the little train with the others. The word was that he was going into the mine this morning but would not be coming out.

As I understood it, there was to be an 'accident' involving high voltage electrical equipment from which the safety cover had been carelessly removed. A sudden push at the crucial spot and the victim's body was to complete a fatal short circuit down there in the darkness of the mineshaft

With morbid interest I kept watch on the mine entrance for some time until a scene unfolded which confirmed the rumours. A single flat railcar emerged from the mine into the sunlight and lying prone on it was a still figure with a tarpaulin thrown over it. It seemed that I was a witness to the sequel of an alleged act that can only be described as a homicide. Justice had been rough and swift - and terminal.

Wittenoom showed its ugly face in the events following the demise of Fitzy. Any death and the subsequent funeral, which I was previously accustomed to witness, was usually accompanied by a degree of sadness or at least a quiet celebration of the life of the one recently departed. My version of a normal civilised ceremony was about to be severely mangled.

In the eyes of the workforce in general, and Fitzy's enemies in particular, it promised to be a day of celebration. There were two reasons for this. His debtors were now free of their obligations and threats - these died when he did. The second lay in the custom in the mining industry, of every man taking a day off work when one of his comrades went to the big mine up in the sky.

It was plain that Fitzy would not have made a good-looking corpse. As he was no oil painting in life, he could never be imagined as resting in repose with hands crossed on his chest surrounded by floral tributes and looking dignified in death. For this reason, and due also to the extremely high temperatures at that time of year, it was deemed advisable to commit him to the ground without delay.

The funeral cortege was ready to depart for the deceased's last resting-place, but there appeared to be some delay down at the graveyard. Two hard rock miners had won the lucrative contract to dig the grave in the unforgiving ground and were working together, their skills most appropriate to carry out this grave responsibility. The reason for the hold-up in the proceedings appeared to be the disagreement which arose between them, which had sunk to the level of a punch-up in the hole. Fitzy had been waiting for some time in his coffin for the situation to resolve itself, but was at long last seen to be setting off on his final journey.

The funeral procession consisted of a pilot vehicle to show the way, then the hearse, closely followed by two busloads of miners who were the mourners. This term is used loosely to describe the vehicle transporting the coffin as a sign of respect for tradition, but a conventional hearse it was not. A five-ton tip truck had been commandeered from the mine transport pool and pressed into service for the job. It was a dilapidated vehicle of ancient origins and happened to have a load of asbestos mine tailings still in the back. The coffin had been heaved up on top of this heap of mullock for the funeral procession. Hard on the heels of the hearse travelled the two buses packed with the miner mourners. They had been collected from the pub and were already well into their cups by this time.

Raising long trails of red dust, the sorry procession rattled along the gravel track to the cemetery. I feared for Fitzy in his long box that was balanced on top of the load in the back of the bouncing truck. It seemed only a matter of time before he teetered too far and plummeted off his precarious perch.

For reasons already given, the mourners were in great fettle. They cheered and sang drunken songs and showed a brave spirit even in these hours of solemn observance.

The tone of the graveside service was suddenly marred when during the stage of committing him to the earth; one of the inebriated pallbearers lost his grip on the rope and

The scenery along the Hammersley Ranges can be spectacular. The colours change as the day goes by

pitched Fitzy headfirst into the hole. This was seen purely as a humorous distraction for the mob. The task of retrieving the coffin, though difficult, was finally accomplished.

Standing around bareheaded in the scorching sun can raise a terrible thirst in a person, so with all due respect the service was short. Only the two gravediggers remained when the buses departed on their way back to the pub, where I imagined the staff was again bracing themselves for the celebration of the wake.

The two gravediggers completed their task of interment in record time with their previous animosity apparently buried too. It was hot, thirsty work and I felt sure that the lure of a foaming beer pot in the cool bar could have had a bearing on their speed in completing the job.

Digressing somewhat but still on the subject of dead people, to maintain your interest I recall vividly what was for me the stuff of nightmares for a long time after the event.

It was a Sunday night and I had just returned with a group of mates from a trip to one of the local scenic spots in the ranges. Feeling rather thirsty I remembered the can of drink that was hidden in the town freezer, and walked across in the darkness to retrieve it. This large cold store was used to hold a lot of the meat and other perishable foodstuffs for the shop and the workers' mess. It was necessary to reach high to unlock the heavy door, which was raised about five feet off the ground. Familiar with the position of the switch for the interior light located just inside I flicked it on. The small bulb in the ceiling cast a weak glow in the vapour from my breath as I fumbled around in the shadows for the drink.

Then I nearly died of fright.

Through the fog I could see there in front of me at eye level two bare human feet. They were a dark colour and swollen. These were connected to a pair of legs, which, as my eyes strained upwards appeared to belong to a corpse. The dim yellow light shone on the features of a man with plastered-down hair and with closed eyes set in a swollen face. The body was rimmed with ice crystals and it looked to be frozen solid. Standing there rigid like a grim sentry he scared the living daylights out of me. Shaking with the shock of the gruesome find, I swung the door shut to block out the sight of the apparition. Heading unsteadily back to my quarters, I remembered what I had come for, but was not about to go back for the drink. There was little sleep for me that night.

The background to the mysterious case of the body in the freezer was revealed to me next day by a close friend of the deceased.

A Sunday swimming party had been organised by a group of townspeople at a large and deep rock pool in the Gorge where a large quantity of beer was consumed. According to accounts, everyone was pretty smashed when the water sports began. One of the party dived from a high rock ledge down into the pool – and stayed there. He wasn't missed for quite a while, which gave him plenty of time to drown. Found eventually trapped under a ledge deep below the surface, his body was not pulled out until some time later.

The body of a person drowned in fresh water in that torrid climate definitely has to be rated perishable. It was decided then, while waiting for the formalities to be processed, to store him in the town freezer along with all the other perishables. It was said that at first he was seen to be taking up considerable space on the floor, but of course after a while became stiff and manageable. Some bright spark must have suggested they stand him up against the wall. This is just how he was when he surprised me that night.

One particular pay night, urged by the sounds of argument and abuse that was rising to a new pitch, my group rolled up our bedding and departed the camp. Rather than risk becoming embroiled in one of the drunken brawls, we decided to leave them to their own devices and sleep elsewhere, preferably quieter and safer. We subsequently spent the night in swags under the stars in a nearby dry riverbed. We felt reassured in the knowledge that perhaps the most dangerous neighbour likely to be encountered down there was a stray dingo.

138

My hut mate was a gem. Bill was a retired soldier from the British Army who had opted for a life in this remote place for God-knows-what reasons. Everyone must have had an explanation for being there but one didn't ask questions. Always immaculately clean and neat in his personal habits he epitomised the British soldier who had been shaped by the strict discipline of an army barracks. Never a stickler for neatness myself, as anyone will tell you, Bill kept our hut in perfect order - it was almost as if I had an army batman on call. Blankets on both the bunks were carefully folded and drum-tight presumably according to the army manuals. We were the best of friends and generally looked after each other.

For my part, it was Bill Cape who made Wittenoom Gorge a better place.

It may sound as though I depended on Bill for the high degree of neatness and order in our little shack. This is true but I did try to make some contributions to our home comforts.

There was this idea I hatched of employing some advanced technology to streamline the running of our little household in order to make the routine domestic tasks a little easier and more efficient. We shared an alarm clock, which was set every night to explode in a cacophony of noise next morning to awaken those who would otherwise have not. No electronic, LED, solid state and programmable timepiece this, it looked like the product of a Russian foundry. The state of the art alarm clock in that era was pure bog simple mechanical, powered by springs and gears, and it weighed around two pounds. Like all clocks and watches of the times it had to be wound up daily or it would stop. Our timepiece had two large windup keys on the back for tensioning the springs, one for the clock part and the other for the alarm mechanism. The audible department on top comprised two shiny bells side by side with a metal ball on a rod between, which at the allotted time in the morning would oscillate at a great rate. This bashed both the bells, the noise from which had the desired effect. I had noticed that when the alarm went off the big winding key on the back rotated as the spring wound down.

We had electricity in our hut, which was something. It powered the wireless on the shelf, we could see at night time by means of the bare bulb hanging from a rafter and most luxurious of all we could boil the electric jug to make a brew. All these appliances we had to plug in and switch on in sequence every morning. My dream was to be awakened by the alarm, hear the jug boiling merrily and have the light on as well so I could find my socks upon rolling out of the bunk.

It occurred to me that there was a source of power in the rotation of the big alarm winder on the back of the clock that was unleashed at the right moment in the morning. Now - how

The mine entrance in the gorge - Previously a hive of activity it is now abandoned and derelict.

to make the alarm clock turn on the electricity in the hut? Sensing a surge of creative enthusiasm, I removed the key from the alarm winder on the back and replaced it with a small pulley. Around this pulley were wound four turns of a piece of string and I took the free end of this up the wall and tied it to the toggle of the light switch. So far so good but this if it worked, was only going to turn on the light. We then connected the wireless power plug and the cord to the kettle together at the overhead lamp socket in a power distribution arrangement that would have turned a licensed electrician's hair grey overnight.

Lying sleepless in bed in the early hours I was running through in my mind the preparations we had made the previous night.

1. - Clock wound and time set. 2. - Light switch off. 3. - String tensioned and wound on pulley. 4. - Wireless and jug switched on.

The alarm went off on time, the pulley turned but it didn't operate the light switch as planned - instead the damn clock hoisted itself off the floor and was winching up the wall on the string towards the switch. Few of man's greatest discoveries I felt sure, perform as intended the very first time. Some extra weight was obviously required at the bottom of the cunning device to overcome the resistance of the light switch. We tied the clock to a brick on the floor and it all worked superbly.

There were some awful Joe McGee arrangements that I saw in use there. Compared with some of these, our rustic electrical system in the hut paled into insignificance. There was a continual problem when doing your laundry in the men's camp. Hot water was needed to restore the miners' work clothes to something approaching clean, but the damn wood-fired boiler would seldom work and when it did, you had to almost get in a fight to use it.

The camp and in fact the whole town was affected by an enormous drop in electrical line voltage from time to time. Our wireless groaned and the lights went dim. This remained a mystery until purely by accident I found the cause of the problem a few huts away from us. There was a dull rumbling sound coming from behind this hut accompanied by lots of steam, so I came in from the back to check the source. This bloke was boiling his clothes in a 44-gallon drum, which I thought was not too

Ross's restaurant. The door in the centre is the one where Bill stood when he was fired upon.

unusual, until I saw how he was doing it. There I could see two heavy cables hanging down from a nearby street power pole and these disappeared into the depths of his Joe McGee washing machine. It seemed that on the ends of these two cables in the drum were connected two steel pick heads. Electrical current passing between the pick heads in the suds was giving him all the heat and agitation he needed, but it must have given the bloke at the little power station a king-sized headache as he saw his meters virtually jumping off the switchboard and the diesels straining with the massive load.

Bill had been a bugler in the army. On special occasions and only in front of a select few he would proudly unroll a red velvet cloth containing his war medals and memorabilia from military campaigns long past. Hidden away from careless hands was his most prized possession - lovingly polished to a golden glow was his army bugle. There was some engraving on it - references to theatres of war in which he had served I think.

That bugle was not just a curiosity though - he could still play it with great skill. The blast from a big army bugle was however not appropriate in a mining camp such as ours, full of soreheads with enduring hangovers. To continue his practice without unnecessary risk, Bill devised a method of playing which would not disturb the neighbours. Removing the mouthpiece from the instrument, he would hunch up on his bunk, pull the blankets over himself and tootle away to his heart's content. I can still remember hearing the haunting strains of the Last Post, Retreat, Reveille and other calls filtering through the blankets as he ran through his repertoire.

Anzac Day was approaching.

"I'm going up on the hill tomorrow at sunup to sound Reveille," announced Bill on the night before the big day.

Checking the dates, I found Anzac Day would fall on the day after pay night and felt that Bill's life could be in jeopardy if he woke the camp at that early hour. Expressing my fears to him was not going to put him off. Bill seemed to think it was his duty as an old soldier to mark the occasion in the traditional manner and no amount of talking was going to convince him otherwise.

The first blast from Bill's bugle jarred me awake at dawn on Anzac day. Stumbling across to the window I looked out to see my friend up on the rise against the wall of the mess hall playing his heart out.

He really was a comical sight. The thin, white, bare body in all its glory was bathed in the first rays of the rising sun. With chest expanded to its maximum and with his stomach sucked in, he was having trouble keeping up his baggy cotton underpants. These he was trying to hold in place with his free hand. His two skinny white legs disappeared into a pair of army boots - hardly parade ground dress uniform.

The ringing notes of Reveille rolled out and then the whole camp started coming to life with curses and mutterings. My worst fears were

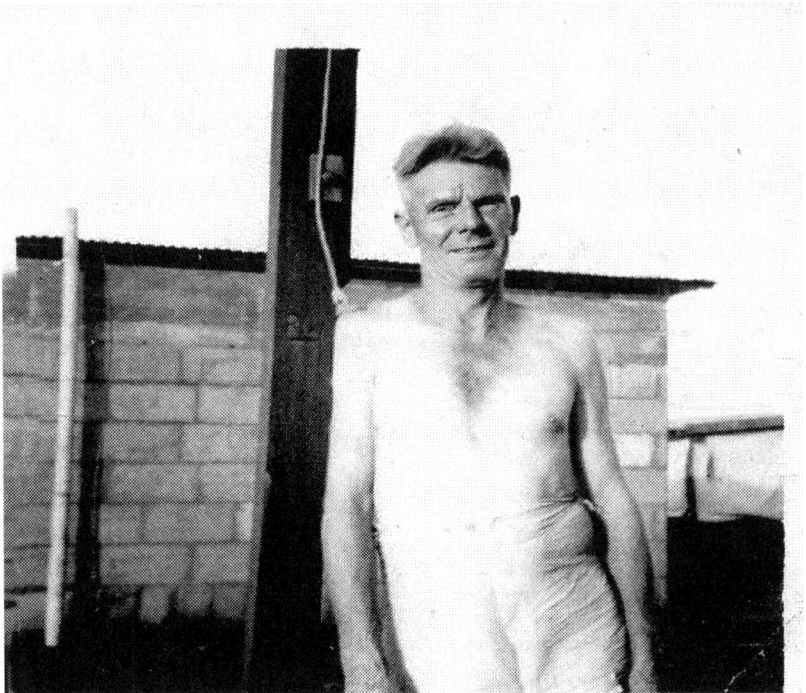

Bill Cape - my hut mate bugler

being realized.

One particularly nasty character in an adjoining hut, who was no doubt nursing his usual prodigious pay night hangover, started yelling obscenities at our bugler from his window. His tirade was of course drowned out by Bill's clarion calls from the hill and then more angry faces appeared at windows and doors down the line.

The bully's shout came through.

"Shut up that bloody racket Bill or I'll bloody well shoot ya!"

Hearing the threat, I was astounded to see the muzzle of a rifle being pushed out of the window in Bill's direction about forty yards away. It was with a feeling of helpless frustration that I watched, while being unable to warn my old friend of the mortal danger he appeared to be facing.

So Bill played on. I knew that a .303 rifle going off in the confines of the camp would certainly arouse the rest of the inmates from their sleep and stir up a veritable hornet's nest of angry men.

I saw the bullet strike on the wall of the mess building just above Bill's head as the echoes of the report bounced off the surrounding buildings. The combined shock of the shot and the shower of fragments had an immediate effect on Bill. The bugle call stopped in mid-note, he lost his grip on his baggy underpants, which slid down to his boots - and then he dropped his bugle with a clatter on the stony ground.

It wasn't all hard work at Wittenoom

The simultaneous discharge of the rifle and the abrupt interruption to Reveille seemed to mark the sudden demise of Bill Cape - not as it is believed with old soldiers who only fade away, but by a much more violent end.

The animal that fired the shot was beside himself with uncontrollable laughter.

"That will teach the silly old bastard," he shouted to the now fully awakened assembly of stunned onlookers, as he put the rifle away and rolled back into his bunk.

I ran up the hill and helped Bill collect his bugle and pull up his underpants and then led him unsteadily back to our hut. It was a much shaken old soldier who must have thought he had met his own personal Waterloo that Anzac Day, far from the battlefields where he had served so long before.

The Breakout

There was one memorable bright spot in the unforgiving twelve-month stretch that I worked in the asbestos mine at Wittenoom Gorge in 1950. The provision of leave back to Perth during the year was unfortunately not a feature of the work contract. We signed up for a full year and at the conclusion of this time on the job, the Company most generously offered to pay the plane fare back home. This was mind you, after they had already deducted from your pay the cost of the airfare to get you there in the first place.

As a tradesman working in the mine and in the mill located deep in the gorge, I found that the living and working conditions were harsh and far from congenial. The northern summer sun beat down on the workings, closely hemmed in by the high walls of the narrow defile, reflecting and concentrating its heat to generate searing temperatures. While the mill building itself was open-sided, there was seldom a breeze to penetrate and cool the men and machines confined there.

On my first day reporting for duty as a maintenance fitter, the mine manager showed me a list of repair work he wanted me to attend to. Hard rock from the mine passing through the mill crushers takes its toll in wearing away components of the plant, so the list was long and the work constant. On the work schedule allotted to me was the job of repairing the mill's dust collector. This contraption sucked the asbestos-laden dust from points in the mill where it was escaping into the atmosphere. It was then piped back to pass through a series of large cloth filters, looking like rows of Xmas stockings hanging from the mantelpiece. These were supposed to trap the dust and remove it, thus reducing the level of pollution in the air we all had to breathe. The unserviceable dust collector showing on my list of jobs to be done was marked as the lowest priority. This meant that it would receive no attention at all and was in effect non-existent.

"It's a dinosaur," complained the manager. "It's not working, but I don't want you to spend too much time trying to fix it, there's plenty of much more important work for you here."

On one particularly windless morning we approached the mill on the trip up the dusty road in the workers' bus from the townsite. As the mill came into view, we could see the whole building and its surroundings obscured by a thick billowing asbestos/rock fog that layered the ground and blanketed the trees and landscape around it.

We knew nothing of the life-threatening potential of the grey clouds. It infiltrated our ears, eyes, and hair and was sucked into our lungs with every breath. You could judge the amount you had carried by the grey stream to the shower drain each night. A clearing cough into a handkerchief left a dirty stain.

There was a small and exclusive club formed at Wittenoom. We were six keen and very optimistic young males with much in common. All were qualified tradesmen of one discipline or another and our ages were similar too - I was 23. We gravitated together to form a close-knit group, determined to make the most of what was shaping up to be a pretty hard though interesting life. We explored the country together, drank together at the local pub and generally enjoyed each other's company. We didn't much like the mindless behaviour of chronic drunkenness and gambling and fighting rife among the miners there. I was most upset when one of the criminal types - it was a haven from the law, took a rifle shot at my dear old mate Bill Cape on Anzac Day morning. This solid alliance gave every one of us a comforting feeling that should a member be picked on or persecuted by the troublemakers, there was a backup of five strong and capable mates to depend on. For this alone or for any other reasons, we experienced little trouble. Some others were not so fortunate.

There was an occasion when one member of our club was threatened with violence but he handled it very capably on his own. Hedley was an amateur wrestler. You wouldn't know it, though while powerfully built, he was a naturally quiet and unassuming member of the clan. Hedley found himself one night in the path of a vicious attack by one of the local standover men called Benson. Hedley had escorted this man's wife home from the pub one dark night as a gentlemanly gesture and Benson took exception to this. In a drunken rage he threatened to do Hedley some serious damage. I stood on the veranda of a house watching the street and observed Hedley returning. The figure of Benson came charging down the hill shouting threats and abuse. Whether Hedley saw him first or heard my shout I am not certain, but he turned to face his attacker. I was then privileged to witness the coolest act of self-defence I had ever seen. Hedley crouched, and moving off line, grabbed the fist that was aimed at his head. He rolled his attacker off balance and using the momentum of his charge, threw him. The ground there was hard and covered in flinty rock. Benson's own momentum carried him in an uncontrolled arc for about three metres, to land with a sickening crunch - on his face.

Hedley spoke a few quiet words to his would-be attacker sprawled on the rocky ground and then joined me on the veranda. Benson collected his wits after a time and with his face a bloody mask rounded up on Hedley again. I picked up a steel chair to make me feel better. After a few more words from Hedley about expecting a repeat performance if he didn't desist, the antagonist retreated back into the darkness. Having no more use for my intended weapon, I put down the steel chair.

One day in a Northwest town some years later I spotted Benson. The wounds had healed, but his face still bore the ugly black scars from his forehead to his jawbone. I wondered if his experience and the resultant scarring had put a dent in his reputation as the invincible street fighter, but thought then on the other hand that it may have served only to enhance his unsavoury image.

The time had come in the Northwest regions for the local population to stage what was known as the Race Round. It was a series of horseracing meetings held every year at the same time in sequence between the major Pilbara towns. It was a time for them to let their hair down and forget the heat and the dust and the flies and the isolation, to join together in the carnival atmosphere of racing and sporting contests. Later called rodeos, in following the American terminology, the games the station people played were then known as bushmen's carnivals. In addition to the main racing events, there was rough riding, calf roping, flag races and athletic contests. The station owners and managers, or squatters, as they were known, would trot out their best horseflesh, their wives all dressed in the latest fashions and their nubile daughters if they had any. The finale to the wild weekend was the race ball, where everyone turned up dressed to kill. They drank the pub dry, slept where they fell and all considered had a fine old time.

Four members of our group decided to relieve the monotony of our existence and seek a bit of fun by breaking out of Wittenoom to travel to one of these fabled race meetings.

The coastal town of Roebourne was the first to host the races in that year of 1950, so we laid our plans accordingly. We accosted the mine manager requesting leave of absence over the long weekend, to go to Roebourne for the races and return immediately after. He agreed with reservations and made the point that we were essential to the running of mine and were needed to restore production at the works should breakdowns occur. Hedley, Ken, Lock and I were all maintenance fitters and represented a fair percentage of essential personnel at the mine. We were only planning to go for the weekend however, so thought that there should be little disruption due to our absence from the job.

Motorbikes were a preferred form of transport among the young bloods of the time. Hedley owned a World War 2 side valve Harley Davidson of the type used by army dispatch riders. It was a rugged machine and well suited to the rough conditions on the Pilbara dirt and gravel roads. Ken, on the other hand, was the proud owner of a new ES2 Norton 500cc single. An English import, the bike had plenty of power and a fabulous bark from the exhaust, which was typical of the big singles of the day.

We rode two-up of course with our belongings in packs and other gear and spares strapped on everywhere. I rode pillion with Ken on the Norton, while Lock was mounted behind Hedley on the Harley. It was quite a risky mode of travel, as the so-called roads were either deeply corrugated, strewn with random potholes, or had stretches of loose sand or powdery bulldust. I carried a rifle slung across my back, which was a fairly common sight in those days.

We rode into the town of Roebourne after the run down the Big Hill without incident. The trip

Two up with Ken on the Norton. The bike is a 500cc ES2 with a rigid frame.

proved to be worth the effort as we all threw ourselves into the festivities and the carnival atmosphere and enjoyed the excitement and the spectacle. We also did our best to impress the squatters' daughters at the race ball, but most, though enticing and attractive seemed to have eyes only for the progeny of the rich families who were wearing the moleskin pants and the big hats. We all agreed that a return to Wittenoom and work would be premature and anticlimactic at this point and seeing we had come this far...

The next meeting in the race round was to be at Port Hedland, a further two hundred miles up the coast. We felt sure that we had left the mine in such good operating condition, that a few more days away would not cause any undue inconvenience back at Wittenoom. Over a few ales to stiffen our resolve, it was decided that we would travel further north and join the Port Hedland revellers at their race meeting next.

The run up the coast was not without incident. On one stretch of the track, the coastal sand dunes had shifted, due to the effect of prevailing winds and were piled across the road in deep drifts. This is not good if you are on a motorbike and Ken and I came a real cropper on a bend. Too late to stop, we ploughed into a dry creek crossing full of soft sand and found ourselves airborne at some considerable speed. When the dust cleared, we picked ourselves up

145

and began collecting scattered belongings. Apart from a few grazes everything including arms and legs seemed to working all right which was a relief. Not so with the Norton however, as I could hear Ken cursing as he tried to restart the motor. We were miles from anywhere in the heat of the day with a sick motorbike and the situation looked a bit dicey. A result of the crash into the soft dry sand was that the bike's motor had ingested some of the stuff and the engine valves were jammed open. I peered up the winding track ahead as it disappeared into the shimmering hazy distance and wondered if we would be continuing our trip that day, or any other day for that matter, the way things were. Ken swore and jumped on the kick-starter again and again, but the only response to his sweaty efforts was a weak wheezing and chuffing sound, which was very ominous.

"Definitely no compression" gasped Ken. "We have got to get the sand out of the motor for it to run. We'll have to try and bum-start it."

We set to with frantic determination and pushed that machine for about half a mile over loose corrugated gravel, dropping the clutch from time to time to crank it over to coax some life from the engine. The heat was starting to take its toll and we were close to knackered when the exhaust gave one loud bang and the engine fired up. With the sand blown out of the motor, the bike began to run as if nothing had happened. We mounted up and headed northwards again.

The Port Hedland race meeting turned out to be even better than the previous one. In a fit of exuberance we booked into the Pier Hotel. As I remember it, of the two hotels on the waterfront, the Pier Hotel is the one on the esplanade and the Esplanade Hotel is the one near the pier. This makes it easy to remember which is which. Apparently no one was keeping tabs on our expenditure and after all the frivolity was over and it was time to move on, we checked our financial resources. It was a bit of a shock to realise that there was not enough money in the kitty to pay the hotel bill, much less to buy petrol and supplies for the days ahead also. An emergency meeting considered the alternatives open to us. We could wash dishes and do yard duties at the pub for a while to defray the accommodation costs, or we could do the other thing - reluctantly we settled on the other thing.

We were quietly preparing to mount up and were feeling really guilty at the thought of absconding from the hotel when fate took a hand. By sheer chance we ran into a fair-weather friend we had spent some bar time with, who asked if we wanted some paid work. It seems that one of the State ships was to unload its cargo at the jetty the following day. As many of the usual workforce were either drunk or in jail, we could almost be guaranteed jobs as lumpers if we showed for pickup at the wharfinger's office the next morning.

My recollections of our time unloading the ship at the Port Hedland jetty are still with me though in somewhat fuzzy images. We worked slinging loads in the hold, to have them hoisted by the derrick up to the waiting crew on the jetty. As in most of these northern ports, a major item in the cargoes was beer. It came in the big brown bottles in wooden crates and also in kegs. There seemed to be an element of carelessness, which I thought strange, as it was evident even among the more experienced lumpers. Every so often a crate of beer would crash to the deck, dislodged from a carelessly slung load. The resultant accidental fall would fracture the crate and spill some of the bottles on the deck. Everyone said "Oh no!" and seemed really sorry - and then they pounced on the loose bottles and drank the lot.

According to union rules, the men must never be overworked, and a continuous effort on anyone's part was apparently considered to be harmful. Consequently the drill was that while half of the team worked, the other half lounged in the shade and had an extended rest period. 'Smoko' times occurred frequently. The foreman, who signalled the time for each break would stop the work and then send down the refreshments – some sandwiches and more beer.

It appeared to me that anyone reporting for work as a lumper could start stone cold sober and almost assuredly leave drunk. Those who reported for work drunk, which was not unusual, would weave off down the jetty even drunker.

There is a true story of the time in Darwin when a cyclone stopped the arrival of ships' cargoes for a considerable period. In order to resupply the town with essentials after the long break, the first ship to arrive would alleviate the most pressing needs of the townsfolk. This ship carried an entire cargo of beer. It must have been reasoned that if shortages of life-giving commodities had begun to bite, a cargo of beer arriving would ensure that nobody really cared!

The lumpers' pay was good and with our wages combined, we comfortably settled the hotel bill with some left over. Our party departed the pub with heads held high and a cheery wave – by the front door. The road ahead opened up and the two bikes were on the track again heading east the same day. We reasoned that returning to Wittenoom down the coast by the way we had come, so retracing our tracks seemed a really dull proposition and incidentally the next race meeting was to be held the following weekend – in the gold town of Marble Bar. Returning to Wittenoom via Marble Bar was considered therefore to be the best course of action.

Our map showed the mining town to be about 120 miles inland over a rough and stony gravel track, which passed for a major road in those parts. While it was unlikely we would encounter more sand drifts as on the coast road, there were still hazards for a motorbike travelling at speed. Luckily we had not collided with any kangaroos to this point, though they were extremely numerous, so a sharp lookout was certainly called for. The sheer mass and speed of a big plains buck or a flying doe coming in at right angles leaves the motorbike at a severe disadvantage.

Ken's Norton had survived the ingestion of sand, apparently with little long term harm. The punishing potholes and badly corrugated roads took their toll however, with the frame fracturing just under the steering head. Undeterred, we levered the break apart, hammered a penny between the ends to ensure alignment and twitched it securely with good old fencing wire. At the cost of one penny together with a free contribution from a cocky's fence, it was a cheap but effective temporary repair.

We eventually bounced across the stony crossing of a dry river bed and followed the rising track, which became the main street of our destination.

Marble Bar has the dubious distinction of being among the most famous towns in the whole of the country for its extremely high summer temperatures. Not only does the thermometer climb to well above 100 degrees F (38 C), in the summer but also maintains this level of daily heat for months at a time without respite.

The Ironclad Hotel is aptly named. All is corrugated iron, which is a common sight almost everywhere in the North. It has the quality however in summer conditions, of reflecting the sun's rays and also cools rapidly in the evenings. Some clever building designs with wide verandas and breezeways help to keep comfort at a tolerable level for those who live there.

Our party of four was going to be staying in Marble Bar for longer than we had anticipated, as the Norton needed a new frame, which would have to be brought from Perth by airfreight. We doubted that the temporary repair with the penny and the wire would last the distance back to Wittenoom so plans were made to settle in for a spell.

The town was filling up with visitors coming in for the forthcoming race meeting and they had booked all the available accommodation. Hotel management was sympathetic to our plight however and asked if we would like to doss down in the meat safe.

One of the ingenious contrivances built to beat the heat on many station homesteads and in the towns was the spinifex cooler. The meat safe at the hotel was one such innovation. There was little or no refrigerated air conditioning in those times, so by utilising the evaporative cooling properties of water, together with some cheap local materials, a comfort zone was established into which you could retreat on the hottest days.

Spinifex coolers ranged in size up to the area of several large rooms. A typical example was timber framed with a thatch or tin roof. The walls were constructed with a double layer of chicken wire mesh with the cavity between stuffed with wild dry spinifex. It was not so tightly compacted as to prevent the flow of air in a decent breeze. A perforated pipe ran around the top of the walls and served to maintain a trickle system of water to keep the spinifex damp. In the hot semi-desert regions of the country the extremely low air humidity ensured the maximum cooling effect and they worked a treat. On a good hot dry day a temperature drop of ten to fifteen degrees inside was not uncommon.

We accepted the offer on the spot and moved some shearers' beds into our new lodgings and unpacked our gear. It was not the most salubrious digs in town, but it promised to be the coolest.

There were delays in bringing the bike parts from Perth and as our funds were depleted again, there was only one thing for it – go to work.

Dave Rodgers was a good sort whom we met in Marble Bar, who had a freight transport business there. I first saw his lanky frame leaning against one of the veranda posts of the Ironclad Hotel and we struck up a conversation. Actually the whole outfit comprised but one truck, which he drove himself. He said he was in need of a relief driver to spell him out as he proposed to keep the truck on the road twenty-four hours a day for a week or so on a pretty lucrative contract. I got the job then and there.

So for me, it looked like turning my back on the bright lights and fleshpots of town and resigning myself to punching a truck for ten days.

Dave's vehicle was a forward cab Austin diesel of about eight tons load capacity. I remarked on the fact that his rear tandem drive axles were running big single wheels instead of the usual duals. This he explained was so he could traverse the narrow rutted camel tracks through the hills, which passed for roads where he would be operating. Things were starting to sound interesting. His loads would comprise lead concentrate in steel drums at about one ton weight each. The run was from the lead mine at a godforsaken place called Ragged Hills near Bamboo Springs, to the Coongan Siding situated on the railway line to the coast.

Reasonably easy travel on the main road soon gave way to country resembling a moonscape when we branched off and followed the wheel tracks towards Ragged Hills. After a few miles there was no doubt in my mind as to how the remote place earned its name.

Over the clatter of the diesel Dave yelled, -

"OK, you drive and I'll get my head down. I had a few jars last night see. I'll take over when you feel stuffed."

He came to about midday. We had a bite to eat on the run after which he collapsed in the corner of the bucking, lurching cab and surrendered to sleep again.

The road condition deteriorated further which I thought could not have been possible. The faint wheel tracks weaved between rocky outcrops bearded with spinifex, rose steeply over jagged breakaways and dropped into dry wash creek beds. I felt a growing respect for the truck when it just bellowed a bit louder as it handled the steepest and roughest terrain without complaint.

My eyes were sore and gritty from the dust and the glare and both arms ached from wrestling with the heavy manual steering. Parking the truck in the shade of some rivergum woke my recumbent partner.

"You're going well" he yawned. "That was a good kip I had, so I'll take over after we've had some grub."

I was feeling decidedly peckish and after a long swig from the waterbag swinging from the bumper, I sank down gratefully on a rock while the boss prepared dinner. It would prove to be the strangest meal of my life.

There was this big enamel mug thrust in my hand, with the suggestion, -

"Get your laughin' gear around that." It was one third full with whisky, topped up from the waterbag. Dave slopped a liberal portion for himself into a large pannikin.

With a good campfire going and an inner glow ignited by the whisky, I relaxed and watched the bush chef at work.

He lifted a wooden crate off the back of the truck and dumped it by the fire, then added an enormous frying pan and a black battered billy. Several food tins of all shapes and sizes were retrieved from the box and set on the ground around the pan. Not one of the tins bore a label to suggest what it might contain. Dave explained, -

"Bouncing around on the back of the truck on the corrugations wears off all the labels, but I don't give a bugger, it's all good grub."

He stabbed each tin and hacked it open with a big sheath knife, then dumped the contents into the frying pan. I watched unbelieving as each one, regardless of its contents, followed the previous lot in a slushy pile.

"I call it hashmagandy," Dave explained as he added beetroot to the mess of beans, Irish stew and steak and onions.

After the tropical fruit salad had joined the conglomeration, he poked a couple of frankfurters out of sight and put the pan over the flames.

Being naturally ravenous I wolfed down my share, but what a taste sensation it was. Steaming hot, the pineapple chunks lent a sweet and sour piquancy to the curried sausages while the baked beans and camp pie gave a nice chewy texture to the dish. The sliced peaches were scraped over to the side of the chipped enamel plate for desert.

"Why stuff around? It all finishes up in the same place and it cuts down on the washing up after. It tastes different every time too I reckon." His brand of logic was convincing.

He was now fresh and eager to get the truck rolling again as the sun was setting red behind the hills. We still had quite a way to go to reach the mine.

It seemed it would be impossible for me to get a decent sleep in the passenger seat of the cab, so I asked where the relief driver usually dossed down.

"We've got the sleeping arrangements out here behind the cab, so slip in there and get your head down."

I had failed to notice a bunk of any kind up to this point, but the 'sleeping arrangements' were now being explained to me. Three empty burlap wool bales were hanging down the back of the cab. Their ends were secured to the roof and turned back up to form a sort of continuous loop or sling – a bit like a crude hammock. Using the door handle as a step, I climbed into one end of my pouch and wormed my body along until I was lying out full length inside it. Some adjustment was necessary as my head protruded from the other end, so the feet then had to hang out a bit instead. The whole business was coated in a thick layer of red dust. Any question about my being able to sleep under these conditions was soon answered - there was no recollection of the truck starting up and moving off.

There were some cool and shady spots in the Pilbara. This was one oasis where you could find relief from the heat and dust.

The rising sun next morning began to heat the burlap bed and I became conscious of my body beating a tattoo against the back of the cab as we approached the mine over the rutted track. It was time to turn to and look lively again.

Having serviced and refuelled the truck, we hoisted the load on and secured it and then took the road which lead to the rail siding. Then I was behind the wheel again as Dave slept like a baby in the dusty cocoon I had vacated - out in the blazing sun and the choking dust.

Almost without a break we drove the shuttle back and forth over the rugged Ragged Hills track for a week. There was little casual conversation between us, as while one was driving the other was most times hanging out in the sack - out the back.

An occasional plunge into a station cattle trough or river pool washed off a lot of the dust and sweat with blessed relief from the heat, although I didn't shave and had no change of clothes which I suppose would have made me smell rather indelicate at close quarters. The old saying that it's good to see yourself as others see you could perhaps be modified to – to smell yourself as others...etc. It is a fact that no matter how much on the nose you are to others, it's unlikely to offend you - personally. Meals were taken stretched out in the shade of a tree if in fact some shade could be found. Otherwise it was a case of duck under the truck.

It was a punishing pace for me, this continuous driving, but Dave was happy as he was making good money as we reeled off the miles. It's natural to assume that if the boss is happy, everyone is happy. On the positive side there was always hashmagandy to look forward to with good old billy tea and a chunk of dry bread with oily butter - out of a tin as well. The traditional half mug of rough whisky with cool water bag water served to ease the aches and pains. After a while I could appreciate how these old timers had become very partial to it especially as beer,

150

with no means of it keeping cold, was not a proposition. I did meet some characters in the north who would drink their beer hot regardless. Now that is an acquired preference.

On several occasions Dave said he was in a hurry to hit the road early in the morning and would settle for a 'dingo's breakfast.' Not having heard the expression before I asked him to explain. Looking at me as if I was abysmally ignorant he said with a grin, -

"It's just a drink of water and a look around." This explanation seemed to fit the circumstances pretty well.

Behind the wheel again I had just topped a rise and was rolling down a hill with a sweeping view of the rocky spinifex plain shimmering out to the horizon. On the ribbon of road ahead there was a plume of dust, which when my vision focussed in the glare, was soon seen to be raised by an approaching truck. A dip in the road hid him for a while and when he reappeared the dust had gone. The other truck was now stopped and actually turned side-on blocking the entire width of the narrow road.

Dave seemed to have recognised the stumpy figure walking purposefully towards us.

"That's Diesel Dick" he exclaimed, "I didn't know he was in these parts. We are going to have to have a drink with the old bugger."

After introductions, we squatted in the shade of the truck and sure enough Diesel Dick produced and poured the drinks all round. He put down beside him a canvas water bag retrieved from hooks on his front bumper and which was plastered with red dust - just to match the colour of almost everything else. I noticed that there was a pocket sewn into the side of the bag and into this handy accessory was stuffed a full bottle of whisky. It was the truckies' favourite tipple, a cheap Australian spirit known as Corio. It was pretty rough stuff and the regular drinkers had given it the somewhat derogatory title of 'COR.-10.'

The whisky was pleasant enough with the cool canvas-tasting water mix, but as the party progressed while I swigged tentatively from the big mug, two things were becoming clear - relatively speaking. Firstly, old Dick was determined that we would empty the bottle – secondly, he was not about to go and move his truck off the road to let us pass until we had. Being relatively young and inexperienced in drinking the hard stuff, I began to feel the effects quite soon. This was even after surreptitiously decanting part of every drink into the dirt behind the wheel of the truck. What I did swallow however hit me fair between the eyes and though not exactly legless, I must have been pretty garrulous and the two truckies - great blokes, had soon assumed the roles of my instant special lifelong friends - blood brothers almost.

Either as a result of driver inebriation or pure misadventure, I lost part of the truck's load further along the track. Muffing a gear change in the crash gearbox on a steep rise the subsequent lurch caused one drum of lead to roll off the back and down the hill. It was as hot as Hades in the midday sun. The runaway drum weighing a ton had trundled back down the hill into a windrow channel alongside the road. We couldn't leave it there, although the likelihood of anyone pinching it was remote, but the freight just had to get through no matter what.

Dave's grumbling about my poor driving was countered by my claim that had he secured the load properly we wouldn't have lost it. We could both see there was no time for argument though.

We dragged a collection of planks, ropes and chain blocks off the truck and set to work. An hour of sweating and straining in the sun saw the task of retrieval of the one-ton drum completed and we set off again, though with some very careful gear changing now. I learned that day that there is no more rapid or effective way to rid one's system of alcohol than to pump it out through the sweat glands. We continued the trip pretty much stone cold sober – or should that be just stone sober, as cold we were not.

Dave dropped me off back at the Ironclad Hotel with my pay in my pocket and departed with a cheery ribald comment and a wave. I found my three travelling companions sitting around looking rather disconsolate and dejected in the meat safe. In my absence they had apparently picked up a playmate in town in the shape of a statuesque young Negress who answered to the name of Audrey. She did not appear to be the cause of their dejection though as she was there in the group sitting on my bed, obviously having a fine time. I could only surmise that the lads must have been missing me a lot - even to the point of inviting a total stranger into their midst like this. As the last one likely to spoil a good thing or ruin a fine party as they were all acting in a very friendly way towards each other, I had to put a dampener on things by saying I was really tired and needed my bed back.

The boys were strapped for cash when I returned from the trucking job. It seemed this was the case even though they had all taken part time jobs while I was away. According to our standard and long-standing mutual agreement, I threw my wages onto the blanket together with theirs and we split it four ways. They agreed that my financial contribution was most welcome to ease the drought and get us back into the black. It was their opinion that I had been very lucky in that I didn't have to face the enticement of spending any money while I was out in the bush and unlike them didn't have the problem of coping with all the temptations they faced in town, both financially and otherwise.

They were very understanding about that.

The race ball at The Bar was a great success with all the locals and visitors abandoning their inhibitions and having an excellent time. We four felt somewhat disadvantaged in our quest for female company at the ball however, as we were rank outsiders who faced the stiff opposition from all the young colts in the moleskin pants as before. All was not lost however as there was a lovely and charming young lady who seemed to enjoy our attention. Her great body as I recall was sheathed in a black strapless gown and she seemed intent on enjoying herself. We were very happy and not a little surprised when this belle of the ball accepted the invitation to party on with us afterwards. She hitched her dress to reveal two nice legs as we climbed on the bikes and whisked her out of town. There is a wonderful geological feature upriver, which is a favourite picnic area for the townspeople. It is a white marble-like natural outcrop of jasper from which the town derives its name. It was to this lovely and secluded spot that we headed in the warm and balmy wee small hours of the morning.

In retrospect I guessed that our attractive companion must have believed in the old contention that there is safety in numbers. If she was nervous or apprehensive she certainly didn't show it. In the company of three strange men who had taken her to a remote spot, miles from human habitation, she seemed unfazed by her situation. Our private party lasting the rest of the night was most enjoyable and of course and to our credit we behaved like gentlemen. It was bright moonlight, as I remember seeing the white marble of the bar reflected in the surface of the black river pool fringed with river gums.

We rolled back into town after sunup and weaved arm-in-arm down the main street. She still cut a stunning figure in her black strapless gown – even if a little dusty by now. She must have sensed right away that we were honourable chaps, who would treat her with respect, which was the normal expectation for a lone woman among strangers in those days anyway.

While the Norton was raring to go again after being rebuilt with a new frame, the Harley now had a problem. The generator had burnt out and so was failing to charge the battery on which the bike depends for its ignition spark. Following some complex calculations involving current draw, anticipated running time and battery capacity in amp hours, we decided to take a chance and make the dash back to Wittenoom. We felt that our jobs would be on the line if we were delayed further.

A large fully charged car battery was stuffed into one of the saddlebags. We ran a power cable with crocodile clips to the distributor. Our Heath Robinson rig worked well and soon after the two bikes with four hopefuls on board left Marble Bar in a cloud of dust on the last leg of our extended long weekend.

We travelled south to Roy Hill and then west along the Hamersley Ranges, as the heat intensified. It would be difficult to forget that day, as in the scorching wind, all exposed skin surfaces felt as if they were on fire. Making a stop to simmer down at Marilana sheep station on the Fortescue River flats, a cold drink was offered to us while we took a break in their spinifex cooler. A thermometer in the shade of the house stood at 118 F (48 C). We had to push on to Wittenoom without delay though as we had absented ourselves for four weeks instead of the three days we had promised when we left the job. The other reason was that the car battery, which was keeping the Harley alive, was showing signs of finally going flat.

This stony track in following the line of the ranges, intersected dozens of deep gullies scoured out by storm water flooding down from the hills to the river flats. This was concern enough, but ten miles short of our destination the Harley began to cough and misfire. Suspecting that the battery had at last given up the ghost, Hedley put his head down, opened the throttle wide and nursed the spluttering, stammering machine all the rest of the way home.

An inspection of the lifesaving battery in the saddlebag showed not only was it devoid of electrical energy but the wires had disconnected too. We surmised that with no power to the spark plugs the engine had been running purely on pre-ignition aided by the extreme temperatures. To reinforce this theory, I can say that even when the ignition key was switched off the motor just kept on running. The old girl had done such a good job to get us home she didn't want to give in.

All that remained now was to face the music at the manager's office and try to explain away our prolonged absence. While the boss did not smile a lot – it's hard to smile and grind your teeth at the same time, he picked on us one at a time to tell our story. When it came to my turn I could only open my mouth and croak unintelligibly and wave my arms around. I was suffering a bad case of laryngitis, which had struck me dumb, probably as a consequence of exposure to heat and dust.

Within the hour we were back at work again in the mill.

Pilbara Sensations

As soon as the formalities and procedures of my new job and lifestyle were completed at Wittenoom I walked out of town and up into the Hamersley Ranges. My eagerness to explore and to learn what this country had to offer found me scrambling up the rocky flanks of the hills to the plateau on top of the range and later entering into the cool and the quiet hush of the gorges. I felt that this was going to be an interesting experience for a lad from the city.

The contrasts and the colours intrigued me, the like of which I had never before seen in the course of one day. Sitting alone on the crest of the ranges as the landscape shimmered below, I followed the path of the cockeyed bobs as they twisted like whirling dervishes across the plain. The writhing and spinning shafts rapidly changed colour as they sucked up the dust and debris from the land over which they travelled. There was a transformation when the cockeyed bob left the dust of the plains and whirled across a patch of burnt spinifex ash. It was changed from red to black in the twinkling of an eye. Then the light faded over the hills and I had to reluctantly head back to the camp.

Never the same from one day to the next, I remember the time when with night falling over the hills and black clouds rolling in across the plain, the jagged blue flashes of numerous lightning strikes lit up the landscape in a summer storm. These could be heard as rolling cracks of thunder, leaving behind the red glow of spinifex fires, which they had ignited and then left to burn and spread in bright flickering circles.

One of my expeditions began at sunup on a hot Sunday morning. The objective was to hike as far as possible up into the Wittenoom Gorge. They told me that some of the ancient watercourses had carved the rock to great depths for many miles back into the spine of the ranges. This urged me to go and try to discover remote places where I could imagine that not many had gone before.

Leaving the glare and the heat of the plain, I shouldered the small pack and picked my way upwards following the dry riverbed. Soon the sloping rock sides and scree changed and rose on both sides to form sheer vertical walls, while the strip of sky far above narrowed to become just a blue ribbon ahead. In the quiet of the shadows an old rock python reared up from the track to confront me. His big weaving head rose to the height of my shoulder to issue a challenge like a guard on duty. Stopping and remaining as still as one of the rocks, I held my breath as he slowly sank to the canyon floor to rustle away from my unwelcome intrusion and disappear among the boulders.

The bottom of the gorge, previously the course of a foaming torrent in the rainy season, now held a chain of fresh clear pools that had been left behind. They lie shaded and cool throughout the hot summer.

It became more difficult to find a way around and across the pools that increased in size and number as I moved on. I had to negotiate by swimming those that were contained from wall to wall in the narrow defile, to where the rocky bottom could be felt again in the shallows at the far end.

Rounding a bend in the gorge now deep within the range, I stopped to enjoy an unforgettable sight. Sunlight filtered down weakly from the narrow skylight above and illuminated a gloomy, echoing grotto. Resting on the smooth cool surface of a river boulder gave me the chance to absorb the features of the place detail by detail.

The black mouth of a cave fringed with lichen appeared high above in the far wall and from its depths curved a graceful silver waterfall. It was all sparkle and lace as it dissolved into floating mist and drifted down towards the black pool at its foot.

Many years have now passed in my life. I am on an oil production platform surrounded by sea, with the mainland out of sight below the horizon to the East. There is a hundred and thirty kilometres of restless grey waves separating me from the coastline. All around is a desert of blue or grey or green depending on the mood of the ocean. There is no grass, no red earth, no wildlife and only the salt smell of the sea.

Just as dawn broke one day, while standing at the railing high on the upper deck of the platform, I watched the golden disc widen and grow on the horizon to become the rising summer sun. As if by some order from old Sol himself, the first eddies of the offshore morning wind arrived, coming to me across the water. Gusting at first over sparkling sea ripples it fanned my face as I stood quietly. As the land breeze gathered strength I picked up the familiar and unmistakable pungent aromas of the Pilbara landscape that was warming over there in the beginning of the new day. The scent of dust and spinifex and kangaroos was unmistakeable.

As if to ensure that my memories were stirred, the land had also sent me a messenger. Fluttering through the bright sun haze I saw a tiny scrap of life from the direction of distant coast winging unerringly towards me. I was reminded of the butterflies around the pools and the reed beds along the banks of the Fortescue River where I had walked all those years ago. Surely this could not be one of their numbers, which was carried so far on the East wind to meet me and remind me of those earlier days in the bush.

We penetrated as far as possible about six miles up into the gorge. Walking was impossible so we took a canoe

Alone and high on that steel pinnacle above the sea, I watched as the butterfly circled my head in search of a resting place. Tired after the long journey from the shore, it sank at last gratefully and delicately onto the back of my hand, folding its wings as butterflies do to rest.

The recollection of this experience will stay with me for all time I think.

Hit the Road

My twelve-month's work contract at Wittenoom was almost completed and Christmas 1950 was just a week away. Ten of us decided to travel home to Perth by road and took stock of the available transport. There were two motorbikes – a Triumph Tiger 100 and a Don R Harley. The Triumph owner rode it well, almost as if he was part of the machine. The one with the Harley rode it poorly and was frequently separated from his. We had two up on the bikes and the remaining six were crammed into an International utility truck. With no tears of sorrow we left Wittenoom behind and headed south with the road opening up before us for the next 2,300 kilometres.

We carried provisions for the troops, and there was a big water bag hanging on the back of the ute and a rifle clipped onto the handlebars of the Triumph. My experience in the north taught me that you couldn't travel for a distance of over 2,000 kilometres on the atrocious roads of gravel corrugations and potholes without suffering an incident or two.

Well down the long track a large flock of sheep scattered when we roared up to a cattle grid. The good Triumph rider headed out into the bush and rounded them all up again for the grateful drover. The not-so-good driver of the Harley hit a washout too fast and dislodged his pillion passenger who then hit the road at some considerable speed and skidded for an appreciable distance on his rear end. The International ute had a catastrophic clutch failure, which meant we would have to try and drive it for the next eight hundred miles or so without the benefit of that handy component. A whip around amongst the passengers disclosed the only one who had ever driven a truck with a crash gearbox without the use of a clutch. This volunteer was Ivan, so that was my job for the remainder of the trip. Having in the past piloted some old cars where a clutch was a luxury didn't pose too many problems with this one.

The beer was cold at the Chequers Hotel at Bullsbrook and as we were now on the bitumen we drank to the safe conclusion to our journey. The locals could not believe that we had just come down from Wittenoom and considered it a foolhardy act on that road, especially on a motorbike. This gave me an idea.

From the phone box on the pub veranda I rang the Daily News office in the city. Posing as a local resident, I described the situation of the ten ruffians who had apparently just lobbed in from the north. The reporter was keen, as he needed a good human-interest story for the Christmas edition and asked me to speak to these people.

"Tell those blokes to meet me in Perth. We will have a camera and a reporter waiting on the Esplanade in an hour."

"OK I'll tell them," I assured him.

The eventful trip was not over yet it seemed as the not-so-good Harley driver roared down the incline from the pub to the main road just a bit too fast. Failing to take the turn onto the bitumen he, the bike and his terrified passenger disappeared out of control into the scrub on the other side. To everyone's relief after an anxious time missing, they were seen to emerge from the bush further along and accelerate off in the appropriate direction.

Our group photo appeared on the front page of the paper next day. It showed a scruffy bunch standing around the two motorbikes and a dusty ute. The water bag still hangs on the back of the vehicle and the rifle is still clipped to the handlebars of the bike.

Medical evidence emerging later had proved that exposure to blue asbestos could be damaging to health and even fatal. It manifested itself in later years in the thousands of cases of asbestosis and mesothelioma amongst those who had lived and worked there, being cited as the cause of many fatalities. It is still claiming victims these fifty years later. A government edict eventually closed the town, ordered the termination of the mine workings and now there is little to be seen of what it was in its heyday. There are only a few pitiful shells of houses and the empty mess hall on the hill and of course the tunnel into the wall of the gorge, which was originally the location of the mine entrance.

When I returned many years later on a car trip with my sons, I stood reminiscing over those fading images in my mind. In trying to recall old acquaintances, my eye tracked up to the mess hall that had once been the hub of so much activity and drama. The scar made by the bullet was still there above the empty frame of the doorway - mute evidence of the day when an old soldier thought he had met his own personal Waterloo.

Wherever you are Bill, I hope it is a kindly place where you are welcome to practise playing your bugle in peace and safety.

Picture of our convoy from Wittenoom as it appeared in the newspaper the following day. That's me leaning on the shoulder of Lock, the Harley passenger

Episode Seven

The Trip Around Australia

We four young men were very close friends in 1951 and I am happy to say, we still are to this day. Joe, my brother, was an obvious choice of a travel companion as we were close in ages and had grown up together and shared our life experiences up to that time. Lock was my best running mate from 1950. I met him while we were working in the asbestos mine at Wittenoom Gorge. Harry Murray was the fourth member of the four musketeers. Joe had lodged at Harry's parents' place during his apprenticeship in Perth. Harry had a great zest for life and was an active participant in the escapades we shared. We were footloose and fancy free and looking for some adventure and a change of scene.

There was no shortage of jobs for qualified tradesmen then. Our inventory of expertise in the group comprised the skills of two mechanical fitters, a patternmaker and an electrical fitter. As a team we considered that we would be able to pick and choose where we could earn money and so decided on a working holiday.

Joe and Harry teamed up to buy a suitable vehicle for the long trip. They decided that a ute would be the thing. Their choice was a 1937 Ford with a side valve V8 engine and a canopy over the back. As it had obviously seen better days, they pulled it all to pieces and rebuilt it as a suitable mode of transport to take them around Australia.

Lock and I opted for something big and tough, so we bought an ex-WW2 army Blitz Wagon. Built in 1942, I don't know anything of its military history but we established that it had been an army signals truck. The flat-nosed forward cab design meant there were just two seats in front with the big Chevrolet engine under a hump in the middle between them. It had very large truck wheels and a high chassis and you had to climb a couple of steps to get up into the driver's seat. As our only mode of transport around Perth we drove it everywhere. We enjoyed going to the Embassy ballroom for the Friday night dance as a regular thing and the Blitz could often be seen parked in the street outside. The fun part came if we had been successful in impressing a couple of attractive girls selected from the 'slave market' at the dance, and who agreed to us driving them home. What we found to be a real clincher in winning them over was to tell them that we were the proud owners of a big 1942 Chev. which was parked right outside. Reactions varied when our perfumed companions in the ballerina dresses were introduced to the big, ugly lumbering green army truck. I have to say at this time that Perth girls are really good sports - they would have to be, to tuck up their tresses and their dresses and climb six feet up into the open cab for the wild and sometimes roundabout trip home. With all honesty I can say that not one of them refused - and most had a lot of fun.

We prepared for the big trip by doing some necessary repairs and also building a framed canopy over the back of the vehicle. Adding a touch of opulence we fitted two folding bunks in the rear together with steel boxes for the storage of tools, food and camping gear. We slapped on green enamel paint to cover the army drab and hand-painted a cartoon character on the driver's door. Superimposed on a map of Australia was the image of a black swan in big boots striding across the continent with a swag on his shoulder. We prepared for our adventure with so much enthusiasm, but alas that swan was not destined to stride the full distance.

On our way north at last, the smooth bitumen road surface cut out just north of Geraldton. At about the 320-mile peg the countryside changed dramatically from the green farmland of the South to the red of the Pilbara landscape. We felt we were getting somewhere.

Our truck ate up the miles of sandy, rocky and corrugated road. As it stretched to the next hill and then the next and the next in a long winding ribbon ahead of us I began to feel a surge of freedom. There was something fascinating and seductive I felt about travelling on the open road to God-knows-where.

The urge to travel and see new horizons has stayed with me through the ensuing years, influencing my life and lifestyle in many ways.

Lock and I decided to push on ahead and establish a camp in Carnarvon, some six hundred miles up the coast and look for work. It was winter and we could feel the distinct climatic change from the wet and cold of the South to the fine, sparkling days and cool nights of the northern latitudes.

We were making good time, sharing the driving and soon after sundown, which is a bad time for colliding with kangaroos on the road anyway, we pulled off for a meal. Two piping hot tins of Red Feather Irish stew were retrieved from underneath the engine canopy and were soon satisfying our hunger pangs. The dining table was the top of the engine cover. While travelling, the tins had been simmering on the exhaust manifold of the engine sitting in special holders we had made for just such a purpose. This arrangement worked really well, with a quick hot meal just sitting on the engine and ready to be served at a moment's notice.

Carnarvon, a major banana-growing centre and ex-whaling station is situated on the Gascoyne River mouth and was reached without incident over a pretty atrocious road. The trip to this point however was not entirely without incident. I must confess to a minor crisis, the result of youthful inexperience, which occurred on our arrival there.

Lock peels an orange with a bayonet at a rest stop

One of the left-hand road wheels on the truck needed to be changed as we had suffered a puncture. The mention of the left-hand side is important and not just a matter of insignificant detail – so bear with me. Before jacking the wheel up clear of the ground I attacked the wheel nuts with the wrench with the idea of loosening them and then lifting the truck so we could get the wheel off. Not one of the damn nuts would budge. Lock took over with some disparaging remarks about my physical strength - or lack of it, and slipping a piece of pipe over the wrench handle heaved on the end with extra leverage. Still no go. Then the wrench split and dumped us in the dirt.

The local blacksmith was able to forge a steel band around the end of the wrench to reinforce it. We attacked the wheel nuts again with the stronger tool and renewed vigour. Considerable time and sweat and swearing later when success in this relatively simple operation seemed beyond us, an idea began to dawn. Some small voice was telling us to rotate the nuts in the opposite direction to loosen them. We did – it worked! The nuts were made with left-hand threads but only on the left-hand side of the truck. We live and we learn.

Arrived Carnarvon - Lock, Harry, me

I won't tell anyone about this if you don't.

A tropical cyclone had ravaged the coast at Carnarvon a week or so before our arrival. The dirt airstrip had been flooded by a swirling red torrent, which had uprooted and scattered the embedded landing lights, which were supposed to mark the borders of the runway.

"Your mission, should you accept it" said the airport boss "is to lift out all the heavy concrete landing light foundations, level them and replace them correctly."

So, lifts with a chain block on each one, a bit of shovelling and then drop it back and it's done. The work targets and the payment were set and we began – in the most glorious sunny weather, the likes of which we agreed we had never experienced in our lives before. What a differenced a bit of latitude makes.

Joe and Harry rolled into town soon after us in their Ford ute and we met on the enormously wide main street of Carnarvon town.

The locals told us that the reason for the great width of this road was so that in years past the camel trains and big wagons from the outback could be turned around there.

We established a fine camp on the airport property under an enormous old shady mesquite tree and proceeded to make ourselves comfortable. One of the first modern innovations to grace the camp was a luxury armchair. Its frame was fashioned out of rusty pipe from an old bed, while rubber bands cut from a truck inner tube were stretched across the frame and provided the spring suspension. It was exceedingly comfortable and there was fierce competition for the rights to sit in it. The credit for constructing this piece of bush furniture must go to Harry.

Memories of our stay in Carnarvon conjure up an idyllic lifestyle. Basking in the warm winter sunshine, we caught big fish by the dozen, swam in the warm ocean and lazed around under a cloudless blue sky. Our work commitments on the airstrip were not demanding and with practice we were finished each day's work by noon.

There was entertainment provided too, on this one day and with no prior arrangements. One of the MacRobertson Miller Airlines DC3 aircraft sat on the pad in front of the terminal building and the announcement was made for passengers to board. A very noisy and very inebriated group of about ten revellers weaved out of the terminal, across the tarmac and up the stair into the plane. I could see them bobbing about through the windows, apparently intent on continuing their party all the way to Perth. The engines were started and run up one after the other and the plane began to roll forward to line up for the take off. The next event was apparently due to the failure of a very important and indispensable part of the aircraft, especially when it's on the ground.

One side of the undercarriage collapsed totally and the big aircraft crashed to the concrete like a stricken bird. One thing you could hardly fail to notice was that a big propeller, spinning away under power cannot continue to go around for long when its engine is sitting flat on the tarmac. The combined visible and audible effects were most spectacular. With the propeller doing its best to dig a trench in the concrete, the sound of the heavy fuselage crashing to the ground with bits flying everywhere, I felt privileged to be there to see it all happen.

There was an amusing sequel to all this and it was played out as the passengers were being offloaded from their crippled craft. They had boarded the plane a scarce ten minutes before almost legless and very jolly, but they were not so now. With ashen countenances, steady hands and a kind of robotic walk the party people crossed the tarmac in the opposite direction and filed back into the terminal – without a murmur.

While I must admit to having a fair respect for law and order, I must also say I was pretty astounded as to how far the long arm of the law can reach.

For a short time prior to leaving Perth for our trip around Australia, Lock and I stayed in some cheap digs in West Perth. We had our lumbering Blitz wagon there too, which was a bit of an

Lame duck DC3. This spectacular but inopportune incident spoiled a good party

embarrassment, as we had nowhere to legally park it. There were the kerbside parking bays in the streets of course, but a low form of life better known as parking inspectors patrolled these. Standing out among the little shiny city cars like a dunny in the desert, we considered our vehicle to be a certain target for the boys with the piece of chalk. With only a couple of days left it was decided to test the compassion of the council and see if we would be pinched if the truck appeared to be broken down. We let all the air out of one of the tyres to create this effect. The ruse worked the first time as no little dockets appeared on our windscreen so we tried it on the second and final night too. We bombed – the damned parking infringement was there fluttering under the wiper and then we had to pump up the damn tyre as well. While we grumbled about unfair and unreasonable and discriminatory treatment, it occurred to us that in a day or so we would be in Carnarvon 600 miles away so why worry?

The crumpled parking ticket sailed into the nearest rubbish bin.

We had only been soaking up the glorious tropical sunshine for a couple of weeks and were lazing around our camp on the Carnarvon airstrip when a visitor arrived. I didn't recognize the man, but was sure that I had seen the blue uniform, the cap and shiny buttons before somewhere. We thought at first that we were up on a speeding charge or similar infraction but discounted that, as our Blitz was barely capable of reaching most of the posted speed limits, let alone exceeding them. The policeman was pretty good about it all as he passed over the piece of paper. We were gob smacked – it was the parking ticket from two weeks ago and six hundred miles away in Perth. How they tracked us down we will never know, but the term 'Big Brother' took on a whole new meaning from that day.

A man called Jack changed the fortunes of our party and together with that our plans for the great trip around Australia.

Jack was a small time farmer-cum-entrepreneur with a wheat property in Dongara, three hundred miles south of Geraldton. One of his moneymaking schemes was to obtain land-clearing contracts with farmers who wanted to extend the cultivated acreage of their holdings. He was looking for a suitable team to run this project and reckoned we four would fill the bill.

We had a meeting with him in the Carnarvon pub with him buying of course and his proposal was enticing. It promised good wages, an active and exciting lifestyle and an opportunity to flex our collective technical skills. He said he had purchased, as his heavy machinery, two wartime British infantry battle tanks, stripped of their armour and armaments and had pressed them into service as tractors. He explained that unfortunately, neither one of them was operating at the moment, but they were broken down on farm properties and required repairs to get them mobile again. He explained that the previous operator was a useless character who drove the machines until they stopped and without the skills to repair them, abandoned them in the paddocks.

"I want a good team to repair the tanks, then to continue operating them so that we can all make some money," said Farmer Jack.

It was clear that he saw us as the means whereby he could get himself off the hook. He could have his machines repaired at minimum cost and then reap the benefits from the services rendered to the farmers if we continued on as operators. Ah the clarity of hindsight. We should have twigged that turning south and travelling opposite to the planned direction was a bad omen. Anyway, we fell for it.

Always keenly interested in the technical details and specifications of war machines, and with the chance now to operate one I did some research on the origins of these two.

This British tank was known as the Valentine. A medium infantry tank, it served in North Africa track to track with the General Grants and the Shermans in desert battles against Rommel. It also showed up in the Pacific theatre against the Japanese, which is probably the source from which these came when they were surplused after the war ended.

The Valentine tank in battle trim was powered by a 160 HP AEC diesel engine. Protected against infantry weapons by 2.5 ins. of armour plate, it had for its offensive capability a main gun of either six-pounder or

Harry is doing some repairs to the tank

75mm calibre, a 7.92mm machine gun and a Bren for protection against air attack. Ready to fight it weighed in at 16 tons and had a top speed of 24kph.

We found the first one squatting derelict and quietly rusting, in the shade of a gum tree by a creek. It bore little resemblance to its former appearance as a battlefield warrior. Adapted for use as a farm tractor it had been shorn of its wartime offensive accessories which was a disappointment for me. The greatest threat we were likely to encounter after all out there was the odd cranky rabbit. The turret and weapons were no longer there, having been replaced by a massive rigid frame welded up from railway line steel. This frame supported a heavy pusher bar at the front rising to about ten feet from the ground. Crude decking of heavy timber planks was supposed to protect the driver from falling trees and other miscellaneous debris.

Joe and I set up camp, dragged out the tools and got to work. Shortly after, the roar of the diesel engine and the gout of smoke signalled that our machine was ready for active service. With all repairs done we climbed on board and set about completing the clearing contract on that farm.

Depending on the type of native bush to be cleared, the tanks could be used in a choice of several modes. If the land was covered in dense light scrub and small trees we just drove around and flattened it all ready for burning later. When the two machines worked in tandem we could cut a much wider swathe by dragging a heavy chain linked between them. Towing a large heavy steam boiler between the tanks would provide the greatest destruction of all.

I have to admit that it was great fun.

In a welter of noise and dust and diesel smoke, crashing trees and flying fragments, with a bit of imagination you could see yourself in a role as part of a massive tank

Even a big tree was a pushover as the tank forced its weight higher on the trunk to topple it

offensive charging against the enemy. While the heavy timber and steelwork on the machine provided reasonable protection during operations, there was however danger lurking in an unsuspected quarter.

Using all this brute force and power to push over quite big trees was a real buzz. You rammed the pusher bar against the tree trunk, changed to low gear and opened the throttle. The tank would shudder and bellow and dig its tracks into the soil to push even harder. If the tree did not succumb to the first onslaught, the tank would begin to climb the trunk until it was almost standing on its tail. With all the weight and power now being applied at a higher point the tank usually won and with its roots ripping out of the ground, the tree toppled and fell.

The tricky part of this operation for the driver occurred when the tree and the tank both crashed to the ground locked together. When the tree was felt to be going over, it was necessary to steer the tank off the trunk and down one side to prevent the situation of being 'bellied.' Our earliest efforts in this procedure often ended with a red-faced driver calling for assistance from a team mate. The sight of the machine balanced on a tree trunk with both tracks spinning uselessly in air was sure to bring down a flood of good-natured sarcasm and abuse. It required considerable effort to get the machine back with its tracks on firm ground and mobile again. For difficult rescue operations such as this we devised an effective method of extraction. A stout log was placed across the tank at the back, this was chained to the links of both tracks and then the tank was driven off in reverse. The log would be dragged under, to catch the obstruction and free the tank.

It was hard and dirty and tiring work, but it was different and far from dull and routine – you never knew what was going to happen next. They paid us ten bob an acre and we flattened eight hundred acres of the countryside on farmer Brand's place.

It was time to move on to the next contract and take one of the beasts with us.

Farmer Jack had left a message that we should resume clearing - scrub rolling is a better term, 150 acres of bush on a farm further south at Greenough. This was Clinch's property 70 miles away. Without the benefit of a tank transporter we would have to drive the tank all this distance. It looked like a daunting task.

Unwelcome intrusion. Chopping out a limb that speared into the back of the cab. It gave me a nudge too hard to ignore.

As we quickly discovered, a Valentine is not easy to drive at the best of times. Steering is achieved on the skid-steer principle. Pushing or pulling on two hand levers either powered or braked the two tracks independently. Pulling the left lever back and pushing the right one forward accomplished a turn to the left. You could spin the beast in its own length, which is fun to do, as

164

long as you don't throw a track off its drive sprocket in the process. At speed you try to maintain a generally straight line of travel that more commonly appears as a wild swerving course from side to side, depending on the skill of the operator, while generally aimed at a distant fixed point. With practice you could finish up pretty much at the objective, but not without leaving a pair of serpentine track prints and miscellaneous debris behind.

In its stripped-down condition, but with all that railway steel added we estimated the weight of each machine to be about twelve tons.

Neither of these tanks was licensed to travel on the nation's roadways, nor would they have ever complied in any one of a dozen different ways. They were oversized, unpredictable, lumbering juggernauts with neither lights nor registration, nor could they have ever been insured for public road use – in a fit. It was obvious that an unexpected confrontation one-on-one with the average motorcar would have left same as a flattened, crumpled wreck, while scarcely slowing the tank.

Our strategies for the trip were discussed and agreed on. The seventy miles was to be traversed mainly in the dark hours. Illumination for driving was to be in the headlight beams of the attendant vehicle travelling behind. We would use the back roads and byways wherever possible and take advantage of every bush shortcut on the route. It was all planned with great precision and we were very much aware that we could have enemies – the state police for one and possibly the local population en masse, if we caused too many breaches of the peace.

The route we had decided to take bisected the townsite of Northampton and crossed an allegedly stout timber bridge over the river. While it was no doubt of adequate strength for local traffic, we had reservations about our concentrated twelve tons of tank rolling across it. We decided to clear that hurdle when we came to it – in a figurative sense of course.

Luck proved to be with us up to this point, as we were able to complete the trip, mostly in the dark without causing major damage, loss of life or being arrested.

There was one reportable incident on the journey. Well into the trek I was bowling along in the dark at a steady clip, using the headlights of the attendant car behind me to get a rough idea of where the road was. The straight course stretching on ahead and steady roar of the engine must have lulled my senses. The sudden swerve to starboard was totally unexpected, when the steering gear had a sudden fit of wilful independence. Clearing a ditch in fine style, we demolished a stockpile of gravel and ran down the side of a cocky's fence bordering his sheep paddock. Regrettably a projecting piece of the tank engaged a strainer post in the fence and extracted it from the ground like a large toothpick. When the dust cleared and some light was shed on the scene, the extent of the damage was clear to see. There was quite a collection of those steel dropper posts piled like a bunch of asparagus on the front of the vehicle, representing what had been supporting the last fifty yards of wire fencing.

When we fronted at the farmhouse and owned up to our misdeeds, we were taken aback by the cocky's response.

"That's OK boys, it's easy fixed and I just wanna say thanks for telling me."

It appeared that his fences had been damaged before, allowing his stock to escape but the situations were not reported to him. We were apparently the first ones to demolish one of these and then actually go and tell him. It was almost as if we had done him a favour.

We hove in sight of the town of Northampton, which was going to be the bottleneck with its suspect timber bridge. Hiding the tank out of town a bit - not an easy thing to do, we reconnoitred to try and find an alterative crossing point in the river and so avoid lumbering over the bridge. The contemplation of the possible scenario of a big tank sitting down in the riverbed amongst the wreckage of the highway bridge at sunup was not an appealing one.

The river over time had carved a mini-gorge through the countryside and also had about three feet depth of water in its bed. The riverbanks, although very steep, should not have been an obstacle that would defeat a tank, so it was decided to cross about fifty yards upstream of the bridge, fording the river in the process. All this was to take place after dark to reduce the chances of our being sprung in the attempt.

Our machine appeared to be up to its designed capabilities as it plunged down the face of the river channel and forded the stream with aplomb. Scrabbling up the opposite bank in a cloud of exhaust smoke and flying mud appeared to be going well too, until the dark shapes of human habitation suddenly loomed around us.

The odour of chicken manure and the sight of the black silhouette of a chook house caused us to make a sudden swerve to avoid it. Shedding a track off the driving sprocket of a tank must have been a tanker's nightmare under enemy fire, but being crippled in a stranger's chook house at night seemed even worse to us at the time. This did transpire however and we were forced to wait until daylight to effect repairs.

Dawn light showed the opportunistic chooks roosting on this great new perch that had appeared in the night. We were able to replace and re-tension the wayward track, clear off the dozing chooks and get out of there by sunup. It surely was a sleepy town. We rumbled away in low gear in the gathering light headed for the main road south and relative safety. I reflected on the possible reaction of the chook owners next day upon sighting a set of tank tracks going clear through their poultry run and also their consternation at the possible refusal of the girls to lay any eggs for maybe a week or so afterwards.

In our initial negotiations with Farmer

Driving the tanks was a dirty job -
We had an excuse for looking a bit scruffy

166

Jack we agreed to accept payment for clearing properties at ten shillings an acre. Additional to these payments and particularly as the machines were in such poor condition, it was also agreed that there would be an hourly fee payable for the repair work at tradesmen's rates.

Our relations with him soured somewhat when he announced that we would be paid for the acreage cleared, which was earning him money, but he could not afford to pay for our mechanical services, which were not. The situation became so bad that we were spending more time fixing than driving and we put it to him that without our skills his tanks would not even be operating in the first place. He finally conceded and promised our money would be paid when he had harvested and sold his mythical wheat crop. We deduced that judging by the way things were going for him, the crop would have been mortgaged anyway and that we were way down the totem pole of his mounting debts.

Back in the scrub again we thought that with our collective experience in the business by now, there could be few difficulties or dangers we had not had to cope with. This was not to be we found, as time went on.

It was my turn to drive and I was clearing thick mallee scrub and trees on the ridges overlooking the little settlement of Greenough, south of Geraldton. I had assessed that the big dead tree in front of me was going to be a pushover and charged the tank at it head-on. The trunk was a mere shell however and fractured into several sections to crash down on the protective decking over my head. Feeling pretty safe and snug in the driving seat, I was startled by a sharp sting on the back of my neck. We had encountered bush bees before in ones and twos so it wasn't of great concern.

The weather was cold. My kit was a woollen beanie, a pair of tanker's goggles and an army greatcoat with the collar turned up. Another sting on my neck and then another got me fearing a mass attack, so I prepared to abandon ship in a hurry.

One of the tree sections that had fallen and shattered over my head contained a regular metropolis of angry bees. The horde was pouring out of the remains of their hive in a buzzing torrent, down into the tank and down my neck. Those bees were not amused by the disruption and had found the source of their problem – me, and were bent on revenge and retribution.

Managing to scramble out of the tank I headed back to the truck parked nearby, swearing and swatting as the swarm wreaked havoc on my neck and face and all other exposed body parts.

My caring and helpful comrades back at our camp suggested that the best way to remove such a forest of embedded bee stings is to shave them off with the big cook's knife. This operation took place with me lying on my stomach on a camp bed. I believe that they extracted something like a hundred stings and I felt decidedly unhappy with my situation.

Thankfully, the large dose of bee venom absorbed into my bloodstream, did not affect me too seriously. The swelling gave me the silhouette of a sumo wrestler with no discernible neck. It appeared, as one of them remarked – a little puffy. Apart from this odd shape, which returned to normal in a week, I remained healthy and in good spirits. There have been accounts of people stung by bees or wasps en masse, and they died – the person that is. I didn't and so must be grateful for that.

The sequel to this unpleasant brush with the bees took place a week later in the same area of bush. With a healthy respect now for the rights of all bees, I was driving with due care to avoid smashing any potential bee habitats.

There was another hit on the back of my hand and I saw the bee sink his barb in. It occurred to me that maybe it was a member of the original aggrieved group still trying to hunt me down. Reasoning that just one bee could not possibly be a threat after the previous working-over from all his mates, I brushed off the thought - and the bee, without further concern.

Within a few minutes however, I was feeling dizzy and sick and had to bail out and head groggily for the parked truck again. Driving down the hill to get help, I was feeling worse all the time. My hearing failed – I was in a silent world. My eyes began to play tricks as a yellow haze clouded my vision. Picking myself up after falling out of the truck down at the Greenough post office, it was a supreme effort to stagger up the path and into the building. The reactions of the postal pixies when they saw this bloke vomit and pass out in front of the stamps counter can only be imagined.

Rescue, in the form of a small car to take me to hospital came swiftly and it seems that I appeared to be past caring by that time. There was a dim awareness of a sharp pain in my ankle. While still capable of feeling pain, it appears that to all intents and purposes I was pretty limp and had been left without the power of intelligible speech.

They were trying to stuff my sagging body into the back seat of a small car while one of my feet was still outside the car, and then someone slammed the door on my ankle. The door slammer made the excuse that it didn't really matter, as I couldn't feel anything anyway. How wrong he was. It was akin to being a patient in the operating theatre regaining the feeling for pain, but not being able to tell the bloke with the scalpel about the problem.

While unable to see my surroundings very clearly, from the sensation of travelling on a fast-moving trolley later and with the sounds and the smells of a hospital casualty ward, I felt reassured. Still incommunicado, I was aware of being examined and then began to hear voices apparently discussing my fate. The comment that really shook me was that of a doctor or a triage nurse, who passed the opinion, -

"I don't think we can do much for this one."

Whether this was referring to me or another patient with something really serious wrong with him I shall never know, and I was feeling decidedly helpless and pretty vulnerable until a well-aimed needle put the lights out.

Four days later I was back at work in the bush again.

It has been a resolution of mine to do some research into the effects of bee venom on the human body – in particular, why one grumpy bee could strike me down when a hundred of his fellow bees had failed in the attempt. It is all apparently something to do with sensitisation and the body's violent reaction to even a very small amount of venom after having been geared up to handle such a massive dose delivered on a previous occasion.

Our ongoing association with Farmer Jack and his tanks was not a happy one. The engines suffered major failures and most of our involvement with the tanks was not in driving them, but in doing unpaid repair work. Our resources were very low with hardly enough money for food and other essentials. We called a strike, and threatened to disable his tanks completely if he didn't come good with some of the money owing to us. The hard fact was however, that you couldn't get blood out of a stone. We were down to mere survival rations in our camp and began to live off the land. We still had rifles and ammunition and there was wheat in abundance so the local rabbits and birds began to appear on our menu together with such toothsome delicacies as boiled wheat.

Life can't ever be so bad that it quenches all innovative and original ideas. The countryside was alive with fat-breasted native topknot pigeons. Harry had been eying them off for some time as they were roosting in the trees around our camp and perhaps he imagined them in the camp oven with thick brown gravy and boiled wheat on the side. We knew that Harry's rifle was an army model .303, which all would agree was much too much gun to be used for bagging pigeons. Innovative as always however, Harry began experimenting and one day came back with a couple of fat birds with not a mark on them or even a ruffled feather.

"I just put one past their ear and they fall out of the tree with shock," was his laconic explanation.

Apart from our troubles we were having a fine time, as a group of young blokes is wont to do, but the realities of life had to be faced. Our dreams of enjoying a working holiday travelling around our great country were fading fast. Even a contract to cut firewood for the local pub in Geraldton ended in failure. Having only axes to cut the iron-hard bush wood, the end of a hard day saw just a pitifully small stack of firewood but lots of blisters, sore muscles and sunburn for our efforts.

With still no money and no immediate prospects of being paid and also needing new tyres for our vehicles and other incidentals like food, we decided to quit and run for home. So, defeated and discouraged, but a whole lot wiser, we rolled back down the highway to Perth and home though still with the determination remaining to build up our resources and have another go.

We did eventually get some of the money owed to us by threatening Jack with solicitor's letters, which were sent from Perth. It was too little too late however, as by this time the four musketeers had found themselves engaged in new jobs, new interests and new relationships, while the dream of setting off again on the big adventure gradually faded and has never been revived.

Here I am relating the accounts of those adventurous times, which we experienced fifty plus years ago. We four all together with our wives still stay in touch, with a reunion every year where we talk about the Trip around Australia - and other good yarns.

Episode Eight

<u>Learmonth - Big Challenges</u>

Being a member of the team which discovered the first oil in Australia was a high point in my working life. Experiences in those heady days have left an enormous feeling of excitement and achievement and satisfaction too in having left some tracks in Australian history – albeit oily ones. In the light of these feelings I have dedicated the first story in the first episode of this book to the account of those experiences. The years in the oil search business were times of hard work and drama with the strike as the focal point – but that is not all that happened in the far north fifty years ago.

Two years had passed since my first pioneering stint away from the city lights, which was in the asbestos mine at Wittenoom Gorge. In truth you could say the work at Wittenoom was not really pioneering as the industry had been operating for a number of years, but a new groundbreaking experience it was for me nevertheless.

In the year of 1952, employment for a qualified tradesman was there for the taking. Perhaps it was the country town upbringing that was asserting itself, but city life just did not appeal to me – the noise and the crowds and the artificial surroundings held no attraction at all. I was yearning for some place fresh and unspoiled with new activities – man stuff - in a remote location if needs be. Maybe that was my preference, as I had always loved the bush. Anyway, I was ripe for the picking as the saying goes and was just looking for the right opportunity.

The stories and pictures I had heard and seen of the American oilmen impressed most young men of the time I think. The Hollywood portrayals of their exploits featuring rich oil strikes and the drama of well blowouts and fires intrigued me too. I found myself moved by the drama of the industry and wanted to be a part of it. We had no oil industry in this country at the time, but here it looked as though someone was doing something about that.

I wanted to be one of them - a roughneck on a drilling rig.

The press in Perth had reported that favourable indications were forthcoming from the oil exploration teams, which had been combing the areas in the far north of the state. For some years, searchers using geological, seismographic and other means of assessing the oil potential of the country had been probing the subterranean structures and had come up with some encouraging data.

I asked a friendly geologist just how any one could discover oil below the surface of the ground, perhaps several miles deep.

"We can't," was his answer. "It's not like water divining – there's no way of detecting the indications of oil from the surface. To prove an oil deposit exists you have to drill a hole down into a reservoir – that's if there is one there to find in the first place."

The technology I had learned so far was intriguing. It seemed that searching for oil in hitherto unexplored country was such a massive gamble – a win-or-bust effort. Should no oil be found, the millions of dollars expended in the search were gone. This was, I surmised, offset by the massive profits flowing from a production oil field where the returns were likely to be enormous. Our local press also declared that the establishment of such an industry would put our State on the map with worldwide recognition. This certainly did sound like something worth pursuing.

This oil search challenge was going to be taken up it seemed. The operating company formed for the task was Wapet - the West Australian Petroleum Pty Ltd. It was formed as a consortium of some of the major oil companies in the world at that time. It was to them that I made my job application.

170

The word soon spread – "The Yanks have landed."

Reminiscent of the arrival of American servicemen to our country during the war a few years previously, they were back – albeit in much fewer numbers than before. A specialist team had arrived this time, complete with modern technical oil drilling equipment to begin the search here in our state.

This was the spur for me to apply for a job with the drilling crews, which had already begun the journey north to the exploration area. The most encouraging results from the surveys by the oil search teams were centred on the North West Cape where Exmouth Gulf forms a large peninsula on the coastline. Known as Potshot, it was also the site of a wartime base for submarines and also for military aircraft, which were operating across the sea northwards, raiding Japanese forces in the Islands. Much of the infrastructure was still there in the way of roads and an airstrip with some buildings still standing. This made Learmonth, as it was known, an ideal base from which to mount the drilling operations.

Big red Mack oilfield trucks were offloaded at the port in Fremantle after their voyage from the USA. We had never seen the likes of these monsters, which were specifically designed and built for the transport of heavy drilling equipment. My guess was that the rugged semi-desert country of the Pilbara as I knew it was certainly going to severely test all men and machines sent to operate there in its vast tracts.

Our news services in Perth closely followed the formation of a large road convoy packed to the limit with building materials and oilfield equipment. The National 130 drilling rig itself was to arrive by sea on the ship the Ellen Maersk. The convoy headed north with great fanfare and carried with it the hopes and best wishes of Perth people for a successful venture. These activities so captured the imagination of the public that the newspapers allotted a news team to cover every stage of their movements and the dailies ran photographs and personal interviews from the operations almost every day.

Paul Rigby kept the citizens amused in the pages of the Daily News with his cartoons too. Prominent in the pictorials were images of the American oilmen. To many of us these closely resembled the Hollywood stars of the John Wayne variety. They wore the badge of their profession, the aluminium helmet, almost everywhere they went. To complement the picture, many sported big fancy belt buckles, cowboy boots, big wristwatches and big sunglasses. Cigars often reinforced the oilman character, these clamped between the teeth and jutting out from the side of the mouth. One driller I later became very friendly with named Lee White always carried his small round tin of snuff, which is finely ground tobacco. It wasn't used in the traditional old Victorian way of snorting it up the nostrils but was chewed and parked in the cheek behind the teeth, with regular jets of brown juice splattering a range of targets at random along the way. The characteristic swagger and lazy Californian or Texas drawl always made them easy to locate in a crowd. For better or for worse I was going to get to know these men in the years ahead and was looking forward to the experience.

The first convoy trekked up to Learmonth into the dusty heat of the Pilbara in 1952. I followed soon after.

I flew into the Learmonth base in a Douglas DC3 aircraft, which was the workhorse of the Mac Robertson Miller Airlines fleet on the Northwest service, arriving at the long gravel airstrip shortly after the first road convoy had made the trip from the south. The Company job offer was in my pocket. It stated that on arrival, I was to begin duties as a rig mechanic, which sounded pretty close to being a roughneck anyway. This first attempt in the Learmonth area would be made with a big and powerful drilling rig. The purpose of the first hole was to be purely exploratory – just to see what was down there. The intention was to penetrate right through the sedimentary layers to igneous rocks, below which no oil is found. Should oil be discovered along the way, this would be a bonus. It is almost unheard of in 'wildcat' drilling operations in untried territory such as this however. The anticipated maximum depth required to drill they said would be about 22,000 ft., or more than five miles straight down! I researched as much background as I could on the business and soaked up information at every opportunity.

My first impression on arrival at the Learmonth base camp was one of frenzied but well organized activity with vehicles and men scurrying around in clouds of dust and diesel smoke. Rows of tents stretched away from the airstrip; apparently to house the workforce until some permanence had been established. The mess building, warehouses, workshops and more substantial accommodation were starting to appear as the work progressed.

One of the supervisors answered my query as to the location of the drilling rig.

"It's still on the beach in crates," he said above the roar of engines. "We had just unloaded it all from the ship by landing craft, but before we could move it to the site we were clobbered by this bloody great cyclone. The place is a shambles. Anyway take my ute and go down there if you like and have a look."

There was indeed a picture of chaos on the beach. A large camp complex of prefabricated buildings had been constructed with loving care overlooking the sea. It had been home to a large seismic exploration crew for some time. The cyclone which came in from the sea two days before had ripped it to pieces. All the structures had been lifted from their foundations and strewn as pitiful wreckage for miles inland. Some twisted sheets and panels could be seen scattered on the

Cyclone Vance wrecked the oil search camp just before I arrived at Learmonth

slope of the hills miles distance away. The largest item left intact on the original campsite was a lone porcelain toilet pedestal squatting there amongst the litter. I remember it well and took a photograph, which generated considerable interest as there was someone sitting on it at the time. The 'someone's' name turned out to be Fred Dockray who was to be our radio operator and first-aid attendant in the years ahead.

The drilling rig was there too. Dumped randomly in big sections in crates along the white beach fresh off the landing craft, beach sand was piled against them in great mounds and streamers of seaweed fluttered in the sea breeze. It was hard to imagine on this bright calm day, that the beach was in the grip of a howling tropical cyclone just two days before. Previously painted metal sections of machinery had been sand blasted to a bright, bare finish by the wind and abrasive sand. One of our brand new Mack oilfield trucks parked at the beach had suffered also. It was all bright steel down one side that faced the sea, leaving the original red paint finish on the other leeward side. All glass in the windows of the truck facing the storm was etched to a frosted opaque finish by the blown sand particles.

The newly arrived workforce was extremely busy. Apart from the normal tasks of establishing the camp and the drilling site was the urgent work of salvage and cleanup after the big blow. A drilling rig of this size costs a large amount of money to sustain whether it is working or not so it was a case of head down and get things moving - now.

The site where the rig was to "spud in" was prepared, the machinery unpacked, cleaned up and assembled. Within a very short space of time the tall silver mast stood on the crest of the Rough Range escarpment. Our 'Eiffel Tower' in the wilderness was a spectacular sight on dark nights. Its mast lights could be seen on all the approaches to the site long before a traveller had arrived there.

This first location for exploration drilling in this area was named Rough Range-1. The crew accommodation and other facilities were known as Ranger Camp, or in the words of one of the wags 'The Wapet Inn.' There was a prominent sign over the door of the mess, which bore that title. There was some graffiti added too, which was a bit indelicate, so it won't be repeated here.

Some unusual events occurred during the preparations however leading up to the actual identification and pinpointing of the first drilling site.

The precise spot on the surface of the ground in thousands of square miles of country where the drill is to penetrate has to be chosen very carefully. A mass of information and data culled from exploration and surveys is compiled over several years. This has to be carefully sifted and correlated to paint a picture of the spot where the best chance of an oil strike might be.

In our case the decision for pinpointing this target, on which rested the success or failure of the project was in the hands of one man. His name was Morgan, from the USA - he was one of their top experts in the oil exploration business.

In my job at that time as a general mechanic, I had to drive over wide areas of rough country to service oilfield equipment every day and so became familiar with every inch of the area. My chosen vehicle was a genuine US Army surplus Willys Jeep. Stripped down to the bare essentials, it proved to have the performance of a very fit mountain goat in the hilly terrain around Rough Range.

When the acclaimed Mr. Morgan arrived in Learmonth, he immediately set to work. He conferred with the other experts, scooped up an armful of maps, emerged from the field office, vaulted into my jeep and said, -

"Let's go son!"

The expert and I were in the front seats of my little vehicle and in the back were the maps, a geologists' hammer and a rough pinewood cross. On this marker he had painted the message 'DIG HERE.' It had been knocked up just prior to our setting off.

If I can digress for a moment – both nationalities, Australian and American, were at times highly amused - and bemused, at the others' expressions and idiom. It especially caused a lot of hilarity when one of us Australians used the expression 'knocked up,' meaning feeling tired or having made something in a hurry. Their version on the other hand referred to a girl who was in the family way – this being most unlikely in a single men's camp of course, so their version didn't really count.

Back into the hills on this blistering hot summer day we scrabbled over dry, rocky creek beds and clawed up the scree at the foot of the hills in the jeep. Following faint goat tracks and referring all the while to the geologist's maps, the elusive spot for the first drilling site was finally located. We had halted on a rock-strewn slope near the jagged top of the Rough Range ridges.

The Man consulted his maps for the last time, peered around and swung out of the car to stride a few paces and stand thoughtfully. It was at that spot he drove the DIG HERE sign into the ground with his little hammer.

The gravity of this situation made me think. 'Perhaps I am involved as a party to the historic first discovery of oil in our country – a find that would make us significant oil producers on the world scene'. My companion threw his hammer in the car with a loud clatter and cut my musing short, saying for the last time.

"Let's go home son!"

The Man was soon back in camp, on a plane and gone within the hour.

An interesting sequel to these events began when I was asked to drive the contractor's superintendent of roadwork out into the hills to the chosen location. He was to be responsible for the earthworks construction of the rig site and connecting roads. At that time I realised that I was the sole person who knew where to find it, and I experienced a brief moment of power being in possession of this exclusive knowledge.

The antics of this fellow when we arrived back at the DIG HERE sign were intriguing to say the least. Standing with hands on hips he surveyed the site from different angles and muttered something about 'bloody boffins.' He uprooted the wooden sign from the carefully chosen spot - the culmination of years of exploration and millions of dollars expended, and strode with it several hundred yards distant. This was a comparatively level site apparently, so he skewered it into the ground again. Returning to the car, he declared, -

"Yep, we can put the rig down there alright."

Fully intending to keep this unusual turn of events to myself, I fervently hoped that in moving the sign we would not have got it off target and perhaps changed the course of Australian oil exploration history in the process. The records show that we did that, but not in the catastrophic way that I had feared at that time.

The irony of this chain of events lies in the accounts of the ongoing oil search. The rig, drilling on the altered site chosen by the contractor, did strike good commercial grade oil. I could not help wondering then whether we would have been as successful at the original and official location chosen by the expert geologist. As a matter of fact, fifty years later, the Company examined the exploration data and proved that the first site chosen, if it had been drilled, would have resulted in a dry and unproductive hole.

As events unfolded, further extensive drilling in that immediate area in subsequent months proved there was but one producing well ever to be found. I decided to say nothing of my misgivings about the altered location until the outcome of the whole search was finally established.

The recorded history of the oil exploration at Learmonth shows that some misguided assumptions were made as to future prospects. Given that the first hole drilled in such a vast area had produced oil of good quality and quantity, it was reasonable to assume that not only were the exploratory surveys accurate, but also that there was a good chance that we were operating in an area of vast reserves. Searches in other parts of the world had continued sometimes for years with many holes drilled before they hit the 'pay.' We hit it first time so there must be plenty there – or so we thought. A number of smaller mobile T32 rigs were brought in to step out the searches for the extent of the original successful find. A ring of holes was drilled spaced in a circle, but as the bits went down, so did our hopes sink too. All those holes were dry dusters.

When one hundred men are set down in such a remote area with virtually no support from the services that we come to expect in the cities, some interesting events are bound to take place with interesting outcomes. The work was inherently dangerous and although safe- working procedures were observed, accidents did occur.

We had no doctor at the site. In cases of serious illness or accident we were pretty much left to our own devices. Help in the form of the Flying Doctor Service could be days away, depending on circumstances. We had no police force. Any cases of bad or criminal behaviour had to be dealt with by our own staff. We had no fire brigade on call or emergency rescue teams poised to take over in emergencies. It was up to us to get out of trouble and do what was necessary in our own way. It suited me fine - this was the stuff I was looking for.

There was no place for a slacker in that crew. Anyone who was not seen to be putting in a one hundred percent effort quickly found himself on the next plane out. To give you an idea of the hair trigger judgment of our field superintendent. I recall one incident when we were recruiting new hands during a period of operations expansion. We had just disembarked eighteen new hopefuls up from Perth at the airstrip and brought them into camp. The boss stomped over and looked at the group and then barked, -

"Put those guys all back on the plane and send 'em home!"

A typical day for the roving field mechanic in the trusty Jeep

I was pretty taken aback by this unusual order, but did as he requested. I queried this impulsive and irrational behaviour with him later. It was revealed that he didn't like men with red hair and as one of them was a blood nut, they all had to go. I didn't know that the condition of red hair was infectious and suggested this to him. All I got in return was a glare and a grunt.

There were no women employed in the staff of the field operations of the Learmonth oil search in 1953. This social imbalance in itself was thought to have the potential for creating problems, especially as leave away from the site was very infrequent and short, usually comprising a weekend every four weeks at best. All in all the isolation, the lack of facilities and absence of community services did help to create some interesting situations. Any woman who visited the camp, especially if she was young and attractive would be the object of the longing gaze of a hundred and twenty pairs of eyes – give or take a few.

One of the special skills acquired during my years, as Transport Supervisor at Learmonth was the ability to sleep standing up. It's hardly worthy of a mention you might think, but I have called upon this handy ability a number of times since then.

There was no dock or jetty on which to unload the huge amount of stores and equipment coming in from the ships anchored offshore. We used large ex-military LCM landing craft for the task and the cargoes were dumped on the beach. Most unloading operations were spread over two or three days working around the clock. Part of my job was that of beach master, organizing the trucks and cranes and personnel from the time we started until the ship departed. I discovered that one gets pretty tired after seeing the sun go down and come up a couple of times, so in the short spells of inactivity, while trying to give the impression of still being on the ball, I would grab

175

short catnaps from time to time. It was during these spells that I trained myself to sleep vertically orientated as it were. You can do it anywhere there is a handy surface - like a stack of sacks, a power pole or a truck. It is best to work the boots down into the soft beach sand, lock the knees so they don't fold and thinking nice thoughts, go off to nigh nighs. With the two-way radio close to my ear I could respond to calls and found that the sound of approaching diesels would get the blues to pop open too. One of my drivers approached me about four a.m. as I stood there. He swears he was talking to me for several minutes but was rudely interrupted by my snoring. The long-term benefits of this skill I have called upon over the years and am still able to have a short zizz and wake refreshed. There is no need to be in the vertical position these days, which would give the impression that I was more eccentric than I really am, and besides, if I fell over there would probably be something broken.

When driving down from the camp to the beach landing one evening the two-way radio squawked.

"We've got an emergency medical in the camp and you are needed here pronto."

As the supervisor for transport and a few other things, it seemed that my input was required. Doing a quick U-turn on the gravel road I put the foot down and headed for the camp.

It seemed that our medic had already relayed his patient's signs and symptoms over the pedal radio to a doctor in Carnarvon, three hundred miles south of us. The doctor had explained that we should not try to bring the sick man down to hospital there by vehicle, as the rough ride would most likely finish him. They could not send an aircraft at such short notice, as it could not leave until first light the following day, which may have been too late. The doctor confirmed the medic's diagnosis as severe peritonitis requiring urgent surgery.

A hurried meeting was called to find a way to try and save this man's life.

Our marine supervisor chipped in.

"There's a navy frigate anchored out in the gulf. Maybe they have a surgeon on board. It will take hours to get a boat out there to find out, and we don't have radio contact with them. I could try Morse with a signal lamp from the beach though."

Half an hour later the marine crew was perched on top of a high sand dune with a big spotlight. Focussing on the navy ship about two miles out in the Gulf a message was flashed with the request for help.

In what I thought was typical laconic navy style the answer blinked back.

"Affirmative surgeon on board – hasn't done an appendix since med. school but keen to have a go – advise transport arrangements."

Our patient's chances of survival were looking better.

He was a big man and we lifted him onto a steel wire bed frame and lashed him there securely with rope.

Daybreak found us down on the beach where we had a small dinghy drawn up ready to take the sick man. His bed frame was laid across the gunwales of the little boat and I can still see the look of worry and pain on the helpless man's face as we pushed off into the surf. A big Swedish seaman was on the oars and I scrambled into the stern seat and took a vice-like grasp on the bed frame. The Swede pulled hard to get through the line of surf to clear water while my grip on the bed didn't relax for an instant. The patient was silent through all this. He was a very brave man or he was unconscious – it was hard to tell.

The navy vessel had sent a speedy motor pinnace to meet us, and our patient was transferred for an even faster and more exciting ride - still lashed to his bed frame.

The helmsman gunned the engine, the boat heeled and we described a white circle in the bay. Everyone started to relax in the comparative safety of the bigger boat and while still maintaining my hold on the bed frame it was a hell of a shock when it happened. There was a jarring crash and a steep heeling of the boat that suddenly interrupted the calming effect of the rhythmic planing across the smooth sea. The trussed cook was headed for the ocean depths again as he struggled in his restraints, with good reason to think he was doomed. Our vessel had struck a niggerhead - one of those coral columns, which rise from the sea bottom to just under the surface. They are invisible and a dangerous hazard for boats of all kinds. Men piled on top of the cook, who by now was getting his feet wet over the side. It required all of my strength and that of two sailors, struggling and cursing to drag him back to safety again. We were lucky in retrospect that our patient was healthy – apart from the peritonitis that is, or we might have had a cardiac arrest to deal with as well.

Loading the cook onto my ute after his epic sea voyage and his naval operation - assisted by some of my marine crew.

The cook and bed assembly was quickly hoisted aboard the big navy vessel with the boat davits and he was whisked below to the waiting surgeon with the knife.

Perhaps it was the part we played in creating a diversion for the ship's crew from the monotony of a navy patrol, but the officers treated us most royally and plied us with numerous tots of navy rum. The rest of that day became something of a blur as we awaited the outcome of the emergency operation on our cook. When the good news came that our man was alive and well, it must have been the release of tension, but then we 'tied one on' in a really serious manner. We brought the patient back to shore after a few days on a calm sea and in a considerably more sedate manner this time. He was soon reunited with the relieved kitchen staff. There is a photo here somewhere, which shows us loading the cook back on a vehicle for his return to camp.

177

There was one story around that time about our involvement with the Australian navy again and which I hesitate to tell. After the mercy dash with the cook, we invited the ship's crew ashore for dinner and a picture show. This they enjoyed and they then responded with an invitation to go out to the vessel still anchored offshore for some of the same.

Six of us made the trip from shore just at sundown in our fast workboat. We were soon enjoying navy hospitality again below in the wardroom with a dinner and copious liquid refreshment. It was easy to form the opinion that navies run on rum, as it always seemed to be the standard drink on board.

The subject of the lack of normal social interactions and enjoyment of life was a favourite topic among lonely men like us. After describing our situation on shore where there was an absolute deficiency of the joys of wine, women and song, one of the tars described their life on board as just 'rum, bum and gramophone records.' Now I know why I didn't join the Navy.

The camaraderie was most enjoyable, but it was getting late and time to return to the beach. We weaved across the deck and literally fell into the workboat, fired up the motor, spun the wheel and headed directly away from the ship.

The Transport Supervisor's job was an eventful one at the height of the operations

The night was pitch black and we were soon aware of some strange activity back on the navy vessel before we had gone a mile. A signal lamp was flashing at us accompanied by blasts from the ship's klaxon. Someone suggested that they were bidding us farewell, so we doubled back and circled round under the stern for a last goodbye. An officer was pointing frantically and bawling at us through a loud hailer.

"You are going the wrong way – that's bloody Africa over there!"

Seeing as he was Navy I had to give him credit for the advice, so we waved back and headed on our new 180-degree course again into the night. We knew he was right when we hit the beach – in Australia that is, and so were very thankful for the timely interruption.

The explanation is of course obvious - to seafarers at least, although it took some time to sink in – being in the state we were. Upon arrival, we had tied up alongside the ship on the shoreward side, which was only natural, as we had just come from there. Unbeknown to us, while we were enjoying the navy hospitality a very natural thing occurred – the tide turned. This swung the ship around on its anchorage so it was pointing in the opposite direction. Not that I'm trying to make excuses, but as we could not see Australia anyway due to the fact it was a very dark night, we naturally headed away from the ship, but out towards the Indian ocean. I have always wanted to go to Africa, but not in that little boat - legless, and at night especially!

One warm evening some time later I drove a couple of the lads down to the beach for a refreshing swim after working a hot and sweaty day in the field. I was content to sit on the sand and enjoy the cool evening breeze off the sea. They were silhouetted against the fading sunset having a fine time in the surf.

As darkness began to fall, both men left the water and ran up the beach towelling off. I could not believe my eyes. What appeared to be livid welts crisscrossed their chests and stomachs although they seemed oblivious to anything being wrong with them.

My God I thought – a box jellyfish or Portuguese man-o-war has hit them. The deadly tentacles were still adhering to their bare bodies and would soon be firing the venom into them. We tried to rub off the stinging threads with sand but then the pain hit them both at the same instant.

My ute careered back to camp with the accelerator floored and the men, now doubled up in pain, lying in the back.

After the hectic dash we soon had the two victims on stretchers in the sick bay. Our first-aid man was not optimistic.

"John here is in a bad way and will soon be in serious trouble if he doesn't get oxygen – trouble is we don't have any in the camp yet."

John had stopped yelling and was unconscious, while his mate was still crying out with the pain. This was OK apparently. There was a question that I put to the medic, -

"Will oxygen save him?"

"Probably would," said the medic. "He needs proper resuscitation with oxygen. His chest is being paralysed by the venom."

"We have to do something – will welding oxygen do the job?"

"If it's contaminated we could kill him – we don't know that."

"Might he die if we don't try it?" I countered.

"A possibility, but I have never heard of welding gases ever being used on medical cases, and it's certainly not recommended."

We were damned if we don't and possibly damned if we do try this radical treatment. There was one thing I wasn't going to let happen however and that was see my friend succumb without lifting a finger to help him. Drastic situations call for drastic measures and although I didn't fancy the prospect of standing trial for manslaughter, we decided.

The big black gas cylinder was dragged in from the welding shop and we attached a pressure regulator, cut the hose short, taped it to the resuscitation mask and prayed. As far as I know, John is alive and well today, at least he was the last time I saw him. I prayed my thanks as the gently hissing gas brought back his vital signs and he completely recovered in a few hours.

We were so lucky to have pulled this one off. In retrospect I doubt that full details of our makeshift treatment would have been entered into the medic's records – I hope not.

In later years I attended several First Aid and Trauma Management courses and seminars. A question I liked to put to the lecturer just for interest was - in the absence of the proper medical grade oxygen, whether welding oxygen could be substituted – this in life threatening situations only of course. The look of censure and disbelief shown by the lecturer was the usual response.

There was always a buzz in the audience at my story that we had done just that – in a desperate situation of course and had possibly achieved a good result as the outcome. Then the question was put to the group.

"What would you have done?"

Later on yet again it was a quiet Sunday evening in the Learmonth camp. While strolling across to my office after a good meal I could see that the camp was settling down in the approaching darkness. Through lighted doors and windows there were the typical peaceful scenes of men reading or writing to their families back home with the music from a radio as a backdrop.

Passing by a large Quonset accommodation hut I was startled by the loud report of a rifle shot. Almost simultaneously the figure of a man burst out through the heavy doors and disappeared into the night, yelling his head off with alarm.

With appropriate caution I sneaked up to the side window and looked inside for the source of the violent disruption. There was a collection of startled white faces with men in various frozen attitudes all focussed on the figure in the doorway at the other end of the hut. It was easy to recognise the man with the .303 rifle in his grasp as one of the kitchen staff. He was standing with legs astride, head thrust forward and appeared to be in a very ugly mood.

"Are there any bloody Yanks in here?" he bellowed.

Obviously he had a bit of a down on our American workmates.

There appeared to be no dead bodies inside the building, which was a relief, but this obviously required careful handling and negotiation. The shooter looked drunk and dangerous and out of control, and it could all take a turn for the worst if nothing was done to defuse the situation.

All eyes moved to the far end of the hut. Gus was a North American Indian, a rouseabout with the drilling team from the States. He uncoiled his lanky frame from the bed and slowly moved across the floor towards the man with the rifle. I could faintly hear his soothing tones as he spoke to the belligerent drunk. He spoke to him about his family and his kids. He mildly admonished him for causing such a racket on a quiet Sunday night, meanwhile slowly approaching right up to the muzzle of the rifle.

The barrel gradually lowered. Gus stood close to the man and gently took the weapon from his hands. The standoff was over. It was a vast relief and gave me a new impression of the quiet, unassuming Gus. It was a privilege to have seen such a display of cool courage and fearlessness in a situation most men would shirk – particularly if he happened to have an American accent - which indeed he did. You can read another account of the lifesaving actions of the dependable Gus at a later time.

The full story emerged. Learmonth had strict rules and restrictions on the use of alcohol and as a result, inebriation was almost unknown amongst the men. It was a mystery how this one had got into the state that had contributed to his dangerous behaviour. Investigation homed in on the kitchen staff where there was often good reason to suspect that the lads there were into the turps. The giveaway was the discovery of the generous commissary orders they had put in for flavoured essences, which of course are of high alcoholic content. It must have taken a considerable quantity of this to send this fellow off on his rampage and I assume he would be heartily sick of vanilla and cochineal flavours for some time afterwards.

Apparently an argument during the evening meal with one of the American drillers had flared into the situation that we later witnessed. Yanks, in his opinion were fair game. Firearms were common in the North at the time – throughout all the country areas in fact. We thought it a good idea to collect all those in the camp that night and remove the bolts in case others were tempted to try this stunt as well.

The shooting scenario in the hut was described to me later. The major casualties were Gus's alarm clock and then the wall, which took the bullet - in lieu of Gus presumably. At the call "Any bloody yanks in here?" the unsuspecting Indian had sat up and responded, much to his surprise and regret.

Two of our biggest bystanders grabbed the culprit. He soon sobered up to the point where he could see the error of his ways and was blubbering his apologies.

My car was refuelled and brought around and we bundled the sorrowful and penitent offender into the front seat with a very big man on the outside to deter any escape plans. He was followed by all his possessions thrown into the back.

Jack was our American field superintendent, also cast in the John Wayne mould with both image and attitude to go with the character. I sent for Jack to come to the car where we waited for him to pass judgement on the wrongdoer and tell me what to do with him officially.

My messenger returned.

"We can't find the boss anywhere now, but someone saw him hightailing it into the dark just after the shot went off."

A search for the next senior staff man also came up with no result and then it dawned on us – there was not one of Jack's countrymen still in the camp. They had all 'gone bush' and disappeared into the night. After reading the prisoner his rights, or rather that he had just lost them all, we drove out of the camp and headed for the nearest major town which was Carnarvon, three hours away. We let him loose there in the main street. No charges, no police paperwork or other action was taken and we didn't ever hear from him again - not even a postcard.

Rigging up. Raising the mast - National 130

The full story emerged when I drove back into Learmonth camp with my 'deputy sheriff,' just as the sun rose next morning.

They had located our Field Superintendent behind a warehouse with others of his staff. He had been told that the man who fired the shot was disarmed and harmless, but he still refused to emerge until he saw the lights of my vehicle heading out of the camp for Carnarvon.

Poor old John Wayne lost a lot of his glamour for me after all this, especially when I saw him in later years starring as the hero in Hollywood oilfield dramas

A Mix of Cultures

The interactions occurring between the Americans and the Australians in the oilfield at Learmonth were interesting at times. Beginning on my arrival there at the start of Wapet's drilling program in 1953, I worked alongside them for two good years. There were many remarkable situations occurring over that period, some of which I would like to relate.

Hollywood has a lot to answer for in my book, as the image of the tough, resourceful, cocky and a heroic screen hero was firmly implanted on our minds. Throughout their formative years I think all boys were impressed and even overawed by the feats of daring and bravery we witnessed in the picture theatre. We were grown men now though, thrown together with representatives of the American culture that gave us these heroic figures. Most impressive in my boyhood memory were the oilmen. These were they, in the flesh, direct from Texas and California and here to show us how it is done, with perhaps a few superhuman acts thrown in for good measure. Yep, we reckoned they were genuine all right – the good oil in fact. This team represented the Brown Drilling Company from Long Beach California. They numbered about fifty, comprising the engineers and superintendents down through the ranks to roughnecks and rouseabouts.

A roughneck racks drill pipe on a T32 rig drilling in Shot Hole Canyon. His coating of drilling mud is

At the outset, it has to be conceded that they really knew their business – drilling for oil. We Australians, about the same in number, knew little or nothing about this new craft then, other than what we had seen at the pictures Saturday nights. Many of the visitors, I discovered, had apparently known little else than this in their working lives. One driller claimed he was born under the floor of a drilling derrick. Knowing the man, I could well believe him.

We were young and fit and highly competitive. I was twenty-five years old and relished the situation, the excitement and the hard work. There were new experiences and new challenges

182

to be enjoyed and a lot to learn. It suited my needs to a tee. It was not long before we picked up the physical and the technical skills necessary in this new industry. The Australians were willing students and were soon matching the performance of the American drillers in many ways. I will say one thing about the newcomers without fear of contradiction – they knew the meaning of hard work and despised any one who wasn't up to it.

As you can imagine, the mix of cultures that occurs when two different nationalities are brought together in a remote location like ours is bound to have some interesting outcomes. We were all suddenly thrown together to work and to live in that remote spot and pretty soon the individual and cultural differences began to show. We found for instance that despite the drawl and the tough oilman image they projected, they were really just ordinary blokes like us. There was much amusement among the visitors on hearing the wide variety of the idiom and slang of the Australian language. One American who heard us describe something, as 'bloody beautiful' was so intrigued that he went around saying it for days. Sometimes you would get a blank look until you explained the meanings of expressions like 'you little beauty,' 'she'll be right,' 'no flamin' worries,' 'no wuckers' and others. 'Crikey' was adopted by the Americans too.

When a tired Aussie remarked that he was 'knocked up' it raised a lot of American eyebrows. One Aussie expressed his need for a drink by saying he was 'as dry as a dead dingo's donger'. That one took a while to explain. In a short space of time everyone was everyone else's cobber or mate which was encouraging.

On the downside in this mix I detected a hint of superiority in one or two of the Americans in their dealings with us. I put it down to the fact that they had worked in developing and backward countries where the quality of the local help left a lot to be desired both culturally and technically. It appeared to me, especially in the early stages of the project, that they considered Australians as inferior. One American driller called a big Aussie a Peon, which is a Mexican peasant. Everyone thought it was a great joke at the time, but the Aussie went and looked up the word in his dictionary, became very unhappy at the slight and had to be restrained from punching the joker's lights out.

The drillers brought with them not only a lot of their culture, but their favourite foods as well. The first American breakfast I ever had was composed of three very large plate-sized hotcakes, with three fried eggs - sunny side up, thick slabs of ham, some beans and grits. This pile of protein was swamped with maple syrup and to a hungry man it tasted so good. One of the drillers would take over the kitchen occasionally and in a short time would come up with a whole mess of chilli beans. Was it hot! One welder there used to chew the small Mexican chillies all day. I tried one – they weren't hot, they were incandescent! A good antidote for the chillies and a wonderful thirst quencher on a hot day was iced tea. A beer jug was filled to the top with ice cubes and fresh brewed tea poured in. With a dash of lemon and limejuice it really hit the spot. A cold beer was something we all enjoyed at the end of the day too and it was found to be the best thing for overcoming most cultural barriers. Our Emu and Swan brews were judged to be a lot stronger than theirs. After sampling their Budweiser and Schlitz from the US of A, I would have to agree. In the food line there was one tasty Australian favourite which was abhorrent to the Americans. 'That black muck you Aussies eat' was of course Vegemite. The look on the face of a driller when he tasted it for the first time, said it all.

My trade skills were utilised in the early stages of the operation of the National 130 rig at Rough Range. One of the largest and most powerful of its type in the world, this rig was specially designed so it could be dismantled and moved from site to site around the country by the purpose-built Mack trucks. It featured a folding 'jack knife' mast 140 feet in height and was capable of drilling deep test holes to 22,000 feet. Always interested in fine machinery I was mightily impressed. With my trade qualifications, my sights at that time were set on the rig mechanic's job.

Al, the drilling superintendent and a thorough gentleman, also a golden gloves fighter in his time, gave me some advice.

"To be a good repairman on a drilling rig like this you first need to know how it operates. The best way to learn that is to sign up on a rig crew and do some roughnecking for a while." This fell in with my long-term plans, so I did.

There is a warning given to 'boll weevils' - new hands on a rig, in typical jargon by the driller in charge of the crew on the drill floor.

"If you-all ain't careful, that goddam iron will eat you up!"

Literally translated this means that if you don't exercise due care you may be caught up in the rapidly moving equipment and suffer an injury. It really was a potentially dangerous job though and I found at a very early stage that you had to be alert and keep the wits about you. On one graveyard shift I didn't get my hands away quickly enough and one finger was mashed between the heavy tongs and the drill pipe. We were 'tripping' at the time, a process of pulling drill pipe out of the hole. It is a procedure that cannot be stopped until completed. For me to go off the floor would

Capable of drilling down 21000ft the National 130 rig was one of the largest jackknife rigs in the world at the time. My schoolroom as a roughneck

have disrupted the whole operation. Gingerly easing the glove off, I showed the driller my mangled finger. His reaction was that if the finger stayed inside the glove when I removed it, it was probably worth treating – so I mopped up the blood, put the glove back on and went back to work for another three hours until the trip was finished. The finger still bears the marks of the night when the 'iron tried to eat me up.'

The drilling operation continues around the clock seven days a week while they are 'makin' hole.' There is a continuous need for maintenance on a drilling rig, which means that nobody is ever idle. There's a saying in the oil patch, -

"If it moves, grease it – if it don't, paint it!"

I revelled in this full-on strenuous activity. It was an interesting thought, I mused, to grab a city bound fitness junkie out of some gymnasium and throw him into a roughneck role on a big rig in the northern summer and watch how he fared. A solid eight hours continuously wrestling heavy drill pipe and tools in the mud and the sweat with seldom a break or a breather would see him as lean and as fit as most of us were in no time at all. At my young age it was good to feel the benefit of these compulsory all-day workouts. Keen to show

Chuck Kemp of Brown Drilling Co. prepares to run the hole opener, which is a significant stage of 'spudding in' a new well. We had a lot of VIPs witness this ceremony

a competitive spirit and to try my newfound fitness I put myself to the test one day. There were twenty 44-gallons - 200litre, drums of oil lying in the warehouse compound and I had driven a small truck there to take them away. There's a bit of a knack to begin with in standing a large number of full drums up on their ends. While not seeing myself as particularly powerful, it was done without too much effort and then alone and unaided, I got them all up and loaded on the truck as well. Those were good days.

There is always humour to be found, even in that world of grime and hard yakka. I can picture now the lights from the tall mast casting an illuminated circle around the rig on a graveyard tour - midnight to 8 am. If you were busy on the floor but felt an urgent need to answer nature's call, it was a case of having to 'go bush' for the purpose. Heading away from the rig you

185

could relieve yourself or squat at a considerate distance out among the tufts of bull spinifex. This by the way required great care, as spinifex spines can penetrate a boot and the damage that could be inflicted on an exposed and naked posterior and tackle, doesn't bear thinking about.

This particular roughneck, at about two in the morning, as regular as clockwork, would apparently feel the urge and make the trip. He would get the OK from the driller and go out for a squat in the spinifex. So regular and so dedicated was he in this habit that we decided to pull a stunt on him. True to his routine one night, he grabbed his toilet roll and shovel and disappeared just beyond the lighted perimeter, followed silently by another crewman with another shovel - a long-handled one. Obviously enjoying his ritual, the victim squatted to unburden himself. Unbeknown to him and masked by the noise of the rig engines, the stalker very cautiously slid his shovel under the victim's rear end. When the job was done and the squatter reached for the paper, the stalker just as carefully retrieved his shovel and disappeared back into the darkness. The victim was unaware that he had been robbed.

Standing to hitch up his overalls, our man did the most natural thing on earth and looked back to where he had been – nothing there! We could vaguely see the figure in the dim light. He had consternation written all over him searching in vain for that which was missing. He was certain he had done it – but where was it now? We saw him drop his pants again and search around inside. He even looked in his socks and at the soles of his boots – still nothing!

The bemused and worried look never left the victim's face for the rest of that shift back on the rig. When he thought nobody was watching he frisked himself again and even patted his pockets – all to no avail. No one ever told him that he had been conned and I suppose to his dying day, he will have this nagging thought that what had gone missing would turn up at the most inconvenient and embarrassing moment.

Thrown together as we were in this isolated spot, working and living so close and depending upon each other in all situations, it was inevitable that the Americans and the Australians would form bonds of friendship and mutual respect for each other. For all their bluff and bluster however, they were in some ways naive and easily fooled.

Those of us who were there at Ranger Camp at the time will surely remember the legend of the travelling brothel.

The word flashed around the camp - among the Americans, initiated by the Aussies that a travelling knock shop had recently arrived nearby. The report also implied that the 'girls' would be ready for business that very night. To convince the sceptical among them, we pointed out the two vehicles with caravans in tow stopped by the roadside at the foot of the ranges. They appeared to be camped there for the night. Lines of washing were strung out between the trees and figures could be seen going about the business of settling in.

The yanks after several months of hard work and having been deprived of even the sight of a female of any kind - as were we all, decided without hesitation to grasp this heaven-sent opportunity.

The reek of after-shave and deodorant wafted through the camp as a dozen sex-starved American oilmen prepared themselves for the encounter. They buffed the cowboy boots, pressed the shirts and slicked down the hair and in a body, headed for the car park.

We watched the drama unfold with great interest from the top of the hill. The cars flew down the road trailing plumes of dust, heading for the little encampment. We could see in the failing light a scene reminiscent of an Indian attack on a wagon train as it had been in their own Wild West on the Santa Fe Trail.

The travellers, whom we discovered later were two retired couples, must have got the shock of their lives when the fast-moving convoy of randy oilmen descended upon them and began to circle the camp. I recall the images of frantic figures snatching laundry from the clotheslines, bundling up their camping gear and hurling it haphazardly into the vehicles and hightailing it for somewhere more safe and quiet.

The mood of the would-be lover boys on their return to the camp was like a gathering thunderstorm. Not a peep, a chuckle or even a smirk could you see on the faces of their reception committee. We felt that to even look slightly amused would have been akin to signing your own death warrant.

Standing with the Drilling Superintendent, Al one day, on the hill by a water tank overlooking Ranger camp, we were discussing a problem. The tank was new and put there as storage to supply fresh water to the drilling operations. The still-empty tank had a capacity of two hundred barrels – about seven thousand gallons and there was a steady stream of water just starting to fill it from what were known as the Billy Wells - a remote pumping supply.

Al said, "I reckon we'll have to station a man up here to turn off the supply when the tank is filled up."

Probably hoping that I would not be asked to do that job, I suggested to him what I thought was a better idea.

"We can measure the rate of flow of the water coming in and we know the volume of the tank, so we can calculate the time when it will be full."

Al seemed impressed. "Can you do that?" he queried.

With a piece of chalky rock and a few numbers scrawled on the side of the tank it was easy to estimate the time which would see it full - that was to be at 5 am next day.

."You better be right," he warned, "We don't want a goddam flood down the hill during the night."

To make the calculation I had to know the rate at which the water was flowing into the tank. Using an empty oil drum as a receptacle, the flow was diverted into it and the time taken on my watch for it to fill the drum. The rest was just arithmetic. The upshot of this exchange was that the man sent to check the tank water level at five the next morning reported it just one inch from the top rim. I was pleased and reckoned we could chalk that one up for the Aussies - in a manner of speaking.

In Al's office next day he expressed considerable surprise that anyone short of a qualified engineer could work such miracles. I stopped short of telling him it was a problem that could have been solved by the average primary school student here. Further to all this he said that they were looking for a good foreman for the field operations and was I interested. Always on the lookout for a new challenge and direction, I accepted.

Next day they presented me with a new Dodge Fargo pickup truck with a two-way radio, a white hard hat to wear and the title of Transport Supervisor. This job was retained until I left the field operations there in late 1954.

This field commission was a surprise to me, but then I had always liked the idea of organizing and running things. It was the first opportunity to work in a supervisor's job and I hoped I wasn't going to blow it.

Our operations had expanded considerably after the oil strike with the importation of a number of smaller T32 mobile drilling rigs to prove the extent of the Rough Range oil field. This meant we had a bigger work force too and my department grew in size and workload. Simply put, it was my responsibility to direct almost everything that moved. All road transport and haulage, ship unloading and marine operations as well as mechanical maintenance and workshops were to be looked after. This was proving to be a full-on and very interesting life.

The job had its less desirable side as well with some weighty responsibilities. A tragic accident occurred early one morning on the access road to the top of the Cape Range where one of the drilling rigs was operating. An urgent radio call came in reporting a head-on smash between a crew car and a loaded Mack oilfield truck. At the scene later, I could see that the truck had literally cut the crew car in half down the middle. One man had died and others of the crew were seriously injured. The accident was to have an overwhelming effect on morale in our close-knit community.

The day following the fatal smash will haunt me forever. The field superintendent issued instructions that when we brought the mangled wreck of the crew car down from the accident site, it was to be placed in full view of the camp in front of the mess. It was to stay there, still with the stains of the broken bodies of our workmates, as a warning of the dire consequences of careless driving. I disagreed with his judgment and said so, but he insisted. My opinion was - how could we come up with such a draconian idea as a remedy, when we did not even know the causes of the accident.

There was a sullen undercurrent of simmering anger in the mutterings of the group of men which had gathered at the forlorn wreck. This mood turned to one of disbelief, as the battered vehicle appeared to take on a life of its own in a most peculiar and horrifying manner.

Suddenly, with no one near it, with a screech of metal and a sudden lurch, the front of the car fell in a cloud of red dust as the last leaf of a shattered spring finally gave way. As if this was not enough, it began moaning like a demented ghost with a wavering strangled sound that made my blood run cold. Everyone looked a bit pale and backed away. It only stopped moaning when I went forward and disconnected the battery cables. The horn wires had shorted, making it sound like demented ghosts. The weaker current from the smashed battery caused it to emit the weird moaning sounds we heard. It is certain that some had trouble sleeping that night – I know I did.

The big rig on Rough Range-1 ran into difficulties while drilling down through the limestone strata beneath the hills. Alarm bells sounded when the mud stopped flowing back up to the surface one day. This is a serious situation that could shut down the operation.

None of the attempts to seal the hole worked. Now at their wits-end the drillers headed for the biggest water source they could think of - the Indian Ocean. A big diesel powered pump was set up on the shores of the gulf and when it was started, gave a steady supply of seawater up the hill to the rig. Halleluiah, drilling was resumed. Not out of the woods yet however, the big pump ran a bearing and seized up. More head scratching took place with the solution being a call to the USA for a new part. This, we were told would take two to three months to get on site. The drillers were desperate. One of my fitters and I had a look at the damaged pump with the failed bearing. I was able to tell the Field Superintendent that we could arrange to have the part manufactured in Perth and have the project back on line in a week. A drawing of the part was sent to Tomlinson's Engineering. They manufactured the bearing and air freighted it up. We hand-fitted it and the pump was soon running again providing the vital water for the rig. Actually it took ten days.

That was another brownie point for our tradesmen, and Al commented that the Aussies were 'real smart guys.'

The Rough Range exploration area had been marked as the most likely place to start drilling and the first strike was made at Rough Range-1 in 1953. With this encouragement for further effort it was decided to begin the search further a field to see if there was oil below the Cape Range near the Gulf. This range forms the hard spine of the North West Cape and is extremely rugged and difficult country to operate in. It is over one thousand feet in height and heavily gouged and fissured by a multitude of rocky canyons and ridges.

I was with Rab Bell, one of the Bell brothers, well known for their expertise in large earthmoving projects standing at the foot of the range. He had the unenviable task of pioneering a road from the sea, up through the foothills of the range and across to the prospective well site on the crest. The Eastern flanks looked forbidding to say the least from where we stood. With my lack of experience in these things, it seemed there was no way for a mountain goat to reach the summit, much less a road works crew.

188

The team boss called forward a little wiry bloke who introduced himself as their bulldozer operator. When the steep razorback ridge was pointed out to him he was asked if he thought he could get his bulldozer up and follow the narrow crest, which would eventually be flattened and widened sufficient to carry a road for heavy trucks and big loads. Shading his eyes against the glare with a greasy hat, squinted quizzically at the route indicated, shrugged his shoulders and said, -

"No probs Rab, I'll be up and across before you know it."

Having seen a lot of dozer work in my time on construction and mining projects, this fellow still astounded me with his skill and daring.

Clanking his way on his machine up to the spine of the ridge, he reached a point so narrow you could not walk on it, with steep sides falling into deep canyons on both flanks. No longer able to maintain a grip with both tracks, he slewed the dozer at right angles where it tipped forward at a crazy angle sending showers of rock down into the canyon. His next move was to slew his machine and reverse until he was teetering at a crazy angle backwards. And so he went, tipping forwards and then backwards perched over the sheer drop, all the time crabbing sideways, flattening the crest of the ridge as he went. Apart from the comment that this man was one of his better operators, the boss seemed to think that what we had just witnessed was pretty routine stuff in his line of work.

The last we saw of the man and his dozer was a couple of hundred yards further on after he had traversed that section of the ridge and headed for the next one. He disappeared from sight with a wave of his hat, partly obscured by a pall of dust and diesel smoke.

When completed, that road, carved into the rock up to the summit where the rig was to operate, was known as the Charles Knife Road. It still bears that name today and can be found in the tourist brochures. Following the track of the pioneering bulldozer operator, the route climbs the range by way of the crests of the razorback ridges. Most access roads to drilling sites are temporary by nature, so if the site is abandoned, so is the road. This one was narrow, steep and precipitous and required the utmost care when negotiating its loose gravel surface. It is much safer now having stood the test of fifty years and with much needed attention is a favourite route for sightseeing tourists.

We had just taken delivery of a number of brand new Ford F800 oilfield trucks. I had made it a practice to road test all new vehicles myself before they went into service, to detect any faults. Heading up the Charles Knife Road I proposed to bring down twenty-five tons of drill pipe and test the truck at the same time.

All went well until half way back down the hill. That was when my troubles started. These bigger trucks have their brakes operating on air pressure. No air – no brakes. An audible alarm sounded and glancing at the pressure gauge I got a shock – no air left. With still a mile or more to the bottom of the hill it seemed likely that I was in for a wild ride if I stuck with the truck. Bailing out was not an attractive option at all with the sheer drops on both sides of the road. With the twenty-five tons of steel pipe nudging the back of the cab it was logical to assume that if the truck was to hit anything solid I'd be speared by the load. I couldn't change down a gear either – if I missed the shift and wound up in 'Angel Gear' - neutral, then the angels would be the next people I would be seeing. It seemed a simple case of hang on and try to keep the speeding, bucking truck from plunging off the road.

As you will have already guessed, the truck and I both survived the wild ride. After negotiating a few bends the truck picked up speed on the last slope and rocketed to the bottom of the range. There was enough air built up in the brakes after that for me to stop before going on and into the ocean. The most exciting part was when it was doing seventy-four miles and hour and the rig was actually airborne for a spell as it crested a ridge in the road. We found the fault in the brake system later and all those new trucks of ours had it. It boiled down to a case of shoddy pre-delivery checks. If I had had a lucky rabbit's foot that day it would have been a case of saying - 'it works for me.'

'Pappy' Newcombe was an old shellback oilman from Texas who had been pulled off his retirement ranch to take charge of the Wapet drilling operations at Learmonth. Pappy was short in stature and powerful in physique with a bark like a bad-tempered Rottweiler. He did not walk like ordinary people, he stomped on his short legs around the site, firing orders at all and sundry with dire threats of what would befall you if his orders were not followed to the letter. Pappy and I got along reasonably well although I didn't hold much with the yelling and arm waving bit that most of the Americans seemed to go on with to get a normal job done. While he was at times an object of fun, especially among the Australians – it seems we are by nature an irreverent lot, it did not pay to let him suspect that he was the butt of the humour. There was no doubting his vast oilfield experience, but like most of his countrymen he claimed that there was literally nothing he could not do. I can still hear him bawling.

"Hey you guys get the lead out – let's go, let's go, let's go!"

He could never shut up! The effect of all this bluster on the Aussies was minimal – we had heard it all before - in the John Wayne films.

Cargo vessels from overseas brought thousands of tons of equipment and supplies to Wapet's drilling operations at Learmonth. They anchored about three miles out in the Gulf while we ferried the cargoes ashore in LCM landing craft. The marine section of the project was my responsibility too after the promotion. It was staffed and crewed by mainly ex-professional fishermen from the coastal fishing fleets and they knew their business. Laurie was the foreman and a good friend.

Pappy announced out of the blue one day that he was going to check on our marine operations. I suggested at the morning meeting that he should ride on board one of the landing craft as they plied back and forth between the ship and the beach, so he could gain a first hand idea of the way things were being done.

Of the three landing craft, I chose to put Pappy on the one with the most experienced of our skippers, the marine foreman, as I knew the ageing American could be a troublesome and interfering old coot.

The lights of the departing ship were disappearing over the horizon behind us as we ran the three craft in pitch darkness, loaded to the gunwales with the last of the cargo. We were heading for the beach and in particular the big tidal creek, which was our harbour. It incorporated the wharf for unloading the barges, which we would complete on the following day.

In the beam of the spotlight on my boat, we could see the skipper at the wheel of the lead boat with Pappy at his side. In the still night over the mutter of the engines we could hear Pappy's barking and see his arms waving around. He seemed to be advising the skipper, with all the gesticulations, on the rights and wrongs of operating the craft. Typical, I thought.

It was very dark and the sea was an oily calm as we idled in towards the beach. With no shore lights or line of surf to guide us, it was a matter of dead reckoning and probing the shoreline to find the opening to the tidal creek. I had every faith in the skippers to go unerringly to the opening as they had done many times before.

There was some activity on board the vessel in front. It appeared to have had a change in command. Pappy had taken over from Laurie, shouldering him out of the way and was proceeding to operate the ship 'Texas style.' The twin engine exhausts bellowed as he opened the throttles and a creamy wash shot out from under the stern. He was spinning the wheel from port to starboard and the boat responded to the rudders by pivoting one way and then the other. We nearly ran him down however as we were still coming up on his stern. We had to reverse the props, back off and lay in the water. There was no doubt at this point that despite all the revving of the engines and the wheel spinning, Pappy's boat was not going anywhere. It was obvious to his amused audience that he had run the craft gently but firmly up onto a sand bar. The stern was still free and swinging in response to the helm and the wash was still streaming back from the props, but due to the lack of any reference points in the pitch-blackness, he thought he was still making knots.

He must have been twitching his throttles and spinning the wheel for about five minutes before we came up under his stern and above the racket, shouted that he wasn't making headway and was in fact on the beach – like a stranded whale!

It was a precious moment for my lot and me and unforgettable. He panicked a bit and ordered his crew to jump over the side in a vain effort to lighten the vessel. Not too effective with a fifty-ton load on board, we thought. He allowed us to finally tow his boat back off the sandbank with him in a fit of pique and bad temper, while we plucked his crew out of the cold sea and wrapped the cursing men in blankets. Laurie took over command again and they headed unerringly towards the creek entrance and home.

Many of the humorous situations arising from time to time on the job were lampooned in the cartoons I liked to draw and pin up on the notice board in the mess. This subject was too good to pass up; even though it depicted the hoary old Texan in what must have been one of the most embarrassing moments of his long career.

His familiar figure came stomping across the yard next morning to where I was refuelling my car. I'll swear he had sparks coming out of his eyes and smoke from his ears as he jabbed a stubby finger in my chest.

"Goddam it son" he spluttered "did you draw that X$#%* picture of me that's on the notice board?" I was pleased to see that he had recognized himself in the drawing at least.

This was a sticky situation all right and I countered by saying that he had provided us with a lovely piece of comic drama, which we all enjoyed and felt that it should be recognised in some way.

He was still breathing fire as I tried to smother a laugh. He eyed me off with a withering glare from under the rim of his helmet and then collected himself to deliver the verdict on this threat to his standing and authority.

"Goddam it son, if you-all hadn't come clean and owned up to that like you-all did, I'd have fired your ass!"

There remained a bit of tension between us for a time, but it seems he must have reconciled himself to having to work with these crazy Aussies, who seemed to have absolutely no respect for authority and so he just had to wear it. It would be hard to say whether this situation could have been repeated back in Texas, but it was a fine bit of humour and remains one of my fond recollections of the oilfield operations at Learmonth and of the characters that worked there.

By the way, the cartoon depicted a landing craft balanced on top of a sand hill by the beach. Its propellers are spinning in air and smoke and sparks are jetting from the engine exhausts. Pappy Newcombe's likeness at the wheel wears a Lord Nelson cocked hat and he waves a big sword. The figures of his hapless crew can be seen abandoning ship just as he had ordered them to do on that black night. I was a bit miffed when Pappy did not ask for an autographed copy of the drawing.

There is still one here if he ever changes his mind.

Thar She Blows

The classic tale of Cap'n Ahab's fatal encounter with Moby Dick, set in the early days of whaling in sailing ships, was a source of great fascination for me as a growing boy. In fact all stories of men sallying forth to hunt big and dangerous game on their terms and on their home turf would set me daydreaming about African safaris and hunting the giant whales from a plunging longboat.

Whaling was still a thriving and profitable industry when I worked on the Learmonth oilfields in 1952. Western Australia had three whaling stations on its coastline - at Albany - Cheynes Beach, at Carnarvon and one near the Northwest Cape in the Pilbara region of the state. This last one, known as Point Cloates was only a couple of hours' drive from our base camp at Learmonth.

These whaling stations consisted of a large shore-based processing plant or factory, which produced the whale products and had a small fleet of whale chasers. These small nimble vessels combed the migratory tracks of the humpback whales some miles offshore during the season, to capture selected individuals and tow them back to the shore station to be processed.

Our oil drilling base camp at Learmonth often played host to other residents and small communities from the surrounding district. Inviting local graziers' families generated considerable goodwill. Also fishermen and other local residents shared a good dinner, a few drinks and a picture show with us in our mess. Included in our guest list were the staff and their families from the Point Cloates whaling station. This social interaction was an important and enjoyable part of life in an unkind and remote part of the country. They in turn would return the compliment, giving us a welcome break from camp food and looking at the same faces around the table every day.

An invitation to visit our friends at Point Cloates would have been difficult to refuse, especially as we were also offered the rare privilege of a trip out to sea on a whale chaser on the second day.

Alan was a friend and colleague, and was the chief engineer at the station. We had helped each other on occasions with the sharing of engineering maintenance facilities between our two sites, so it appeared that a good time would be assured on our visit with them.

Early one morning a party of us drove our Landrovers up and over the craggy spine of the Northwest Cape and descended towards the Indian Ocean in the West. We rattled over rocky outcrops and then skirting the sea, hummed along the sand tracks of the coastal scrubland. The whaling station appeared etched against the blue of a peaceful bay ahead of us. The plumes of steam and the fleet of small vessels riding at anchor off the dazzling white beach pointed us to its location.

Alan woke me roughly at three am.

"The chaser has to up-anchor soon to clear the reef and be out to sea by dawn" he said. "Here's a coffee. I'll take you down to the anchorage - the skipper won't wait for you."

The scalding black brew helped clear my head from the effects of the middling-to-severe hangover, which was a legacy of the previous night's hospitality. Dressing in the dark I followed my friend down to the beach in the pre-dawn darkness. The fresh salt tang in the air rapidly cleared my head and restored the faculties.

The winking lights from the whale chaser moored out in the bay gave the workboat skipper his direction and we soon rounded its stern and clambered up the rope ladder, over the rail and onto the deck.

This was an ex-navy wartime Fairmile boat built originally for minesweeping and other coastal duties. Alan commented, -

"They are a good boat for the job - but long and narrow-gutted so they roll like a bastard in any kind of a cross sea."

From the bosun on the chaser who had taken me under his wing, I learned that whaling was conducted under Norwegian seafaring principles and traditions passed down over hundreds of years. Some of the skippers were Norwegian and were practising the skills and knowledge and seamanship handed down by their whaling master forebears. Among the beliefs that were still observed was the superstition that taking a woman on board a whaling vessel would bring bad luck. This possibly accounted for the remarks from the wives the previous night, that none of them had ever accompanied their men on a whale hunt. At worst, it seems that serious misfortune could befall the vessel and crew, or at best the vessel would return to base with no catch - or so it was believed.

Still in darkness I felt the throb of the twin diesels, sensed the surge of the propellers and watched the shore light tracks swing in a rippling arc across the stern as we headed for the gap in the reef.

The crew was on deck preparing for the hunt. Two thick manila ropes were run forward from the winches squatting back against the bridge. These were harpoon lines converging to the powerful-looking gun on a pivot mounting on the forepeak, now just becoming silhouetted against the lightening dawn sky.

Out in the deeper ocean on the whales' migratory course, the deck underfoot lifted and tilted to the run of the long swells marching in from the west. Detail by detail the ship revealed itself - high sharp bows swept back to a narrow waist, trailing to a low stern. The bosun was at the wheel on the bridge and further aft the exhaust stacks left a greasy smoke trail above the white curving wake. The lookout swarmed up the ratlines and dropped into the barrel at the masthead. There was apparently no 'crows nest' on this vessel as in the Moby Dick tale - perhaps this familiar term had been lost in the Norwegian translation.

A voice alongside me explained, -

"The lookout will be up there until we sight some whale - after that you can go up yourself and get some photos of the action."

I had almost forgotten the little Leica camera that was buttoned into my shirt pocket as I looked up with some misgivings at that lofty perch which was describing wild arcs across the sky in time with the roll of the boat.

The lookout in the barrel soon yelled at a sighting and indicated the bearing to the helmsman on the bridge. It wasn't too clear exactly what it was he shouted, but I was disappointed - no cry of 'Thar she blows' carried down from the masthead. No 'Crows nest' and no 'Thar she blows' were certainly detractions for me from the accepted whaling tradition I had expected to find, but then this was modern times whaling and the year was 1953!

The low stern squatted into the vessel's wake, the bows lifted and I had to grab a railing as we heeled to the bite of the helm and headed for the distant sighting of the quarry. There, halfway to the horizon we could see the telltale twin jets from a pair of whales blowing. The white plumes hung in the morning air for a time and then disappeared in a fine mist to rise again as they ploughed southwards on their steady course.

A crewman raced to the gun on the bows. He swiftly loaded a heavy harpoon into the muzzle. A fat brass cartridge containing the propellant was rammed into the breech and locked, while a blond seaman pulled the end of one of the thick harpoon lines under the gun. I was intrigued to see this fellow make an eye splice with the rope to the shank of the harpoon - it took him less than ten seconds. They stood away - the gun was ready.

I asked the nearest crewman, -

"Who fires the gun?"

"The skipper does," he grunted.

"Where's the skipper?" - I hadn't seen him as yet.

"Sleeping below still - I'll get him up now." Grabbing a steel bar he pounded on the steel deck twice.

"That's the signal that we have a fish in sight," he explained. Some fish I mused.

Sure enough a figure soon emerged yawning and stretching, appearing from the companionway under the bridge.

The skipper seemed to take in the whole scene in an instant, but moved leisurely forward to his position on the gun platform. He appeared to check that everything was in order, pivoted the gun a few times and shaded his eyes in the direction of the targeted whales.

My adviser was back beside me. He explained, -

"There's a big bull humpback and a cow with him on the port bow. We have to run up close behind them, match our speed with theirs and without spooking them, try to close up for a clear shot."

The broad backs of the two mature whales broke the surface ahead of us while two blowholes blasted and two jetting plumes of water drifted back over the boat. It was an exercise in stealth on a grand scale - the hunter and the hunted in a stalk across the sea - but there was no time for idle daydreams.

"You had better get up that mast and into the barrel if you want to see the action," said a seaman as he indicated the ratlines converging up to the swaying masthead.

We were now running diagonally across the long swells as I climbed with a life or death grip up to the lofty perch. It was necessary to swing out over the boiling sea to secure a footing on the rungs of the rope ladder and then at the top, haul myself over the rim of the barrel. I gratefully dropped into the relative safety inside - all without looking down lest I lose my nerve and join the sharks.

The view was magnificent. It was possible to see deep into the green ocean with the still unconcerned whales running close under the bows with leisurely sweeps of their enormous tails.

It was big disappointment for me when the skipper said I wasn't allowed to fire the whale gun

Through the viewfinder of my camera the crouched figure of the skipper was training the gun - he had screwed his cap around backwards on his head for an uninterrupted sighting. The atmosphere was tense and electric.

It seemed we were about to run the boat right up on the backs of the whales when at last the gun fired. The harpoon flew with astonishing velocity for such a heavy weapon, with the line snaking out behind it. The head of the harpoon with its deadly folded flukes was suddenly buried out of sight into the broad back. The whale checked and shuddered at the impact. The ship lost way and reversed the screws to heave to as the winch took up the slack to haul in the secured catch. I was saddened, witnessing the death throes of this magnificent animal and hoped that a merciful end would come soon. My pleas were answered as a muffled explosion sounding deep in the whale's vitals signalled the detonation of the grenade on the tip of the harpoon. It is designed to deliver the coup-de-grace and end the suffering quickly. The great beast then relaxed as though defeated and rolled over dead in the sea.

The bosun later remarked, -

"That's a copybook shot and one we always try to get. That bull was finished in less than a minute and though we always try to make it a one-shot kill it sometimes takes another, but they never get away to suffer unnecessarily."

This was some consolation I thought, as it seemed to be about as efficient and quick and humane as the skill and equipment employed could make it.

One of the interesting things about whale behaviour is known and exploited by the whalers when they are hunting. When a bull and a cow are being pursued running together as these were, it is customary to take the bull first. After the bull is killed the cow usually stays close by and can often be taken as well. If the cow is killed first, her mate seems to think better of it and he usually disappears

My precarious perch was in the barrel on top of the mast. It's a great place to be to watch the action

195

leaving her to her fate.

This is not very noble of male whales in general, I thought at the time.

Both of those whales were taken within a half-hour that morning. The carcasses were brought alongside, inflated with compressed air so they would stay afloat, marked with a yellow flag and set adrift to be retrieved on the return trip to the whaling station. A chemical shark repellent is discharged into the sea around the floating whale, leaving a green stain in the water, with a taste supposedly effective in deterring those predators. The bosun was not convinced that it did the job.

. "The bloody sharks get to recognise the smell of the stuff after a while," he grumbled. "Any time they pick up on it they reckon there's a feed somewhere around and start looking for it - as a shark detergent it ain't worth a damn."

My adrenalins were pumping and my nerves were jangling after the explosions. The witnessing of the death throes of these great beasts and the sheer enormity and the scale upon which a whale hunt is conducted left me with such mixed emotions that it was as though I had seen it all now and was unprepared for further drama.

The scene had returned to relative peace and calm down on the deck as the crew went about their tasks preparing the dead whales and releasing them to float free astern. The blond seaman had already ducked under the muzzle of the reloaded whale gun to splice the line to the harpoon and the vessel was moving slowly under the thrust of engines running slow ahead. The skipper seemed relaxed beside the gun showing his obvious satisfaction at having secured two whales so early in the day. A seaman's yell broke the silence.

"Fish on the starboard bow!"

There he was, with a broad gleaming sea-washed back appearing suddenly from the depths to surface about twenty yards from our boat. He seemed to display what appeared to be just idling curiosity to see what all the fuss was about. So close was he that I could make out the crop of barnacles on his back and the unblinking eye staring up at us. The swells sluiced over his black hide as he swam alongside us. He had appeared like a huge surfacing submarine.

The gunner was galvanized into action. Yelling at the blond seaman to jump clear, he swiftly pivoted the gun in a sweeping arc and depressed the muzzle to its maximum angle. It was an instinctive bid to line up on this new prime target, which had appeared so unexpectedly right under our noses. This was a standoff on a grand scale I thought, peering down from my vantage point and I steeled myself for the shot. The concussion soon shattered the relative tranquillity again. It looked as though the gunner was having difficulty with the extreme depression at which he had to train the sights of the weapon, as the whale was so close, the deck was rolling and now the whale was beginning to submerge.

It was a snap shot. A whale gun fired from the hip as it were in a scene from some bizarre Wild West film. It was an instinctive action on the gunner's part but the harpoon missed! The whale won the contest and sounded down into the safety of the green depths without suffering a scratch. This without a doubt would make an extraordinary story of the one that got away. After witnessing two of his fellows killed, it was hard for me to see him go without taking with him my good wishes for his survival and a long life.

As the big fellow sank from sight, instead of meeting a target of hide and blubber, the harpoon encountered a protective layer of seawater. The velocity of the shot caused the projectile to glance off the sea surface like a skipping stone on a lake!

Completely forgetting my photographer's role I followed the parabola of the airborne missile as it rose level with my position at the top of the mast, its flight being traced by the curve of the trailing rope. It had not occurred to me to take into account the burning fuse in the harpoon warhead, which was ticking off the seconds in its flight. I watched, mesmerized by this unexpected turn of events.

At about fifty yards range the cast iron grenade on the tip of the harpoon exploded in air like the shell from an anti-aircraft gun. The bright flash, the white smoke and the concussion caused me to duck instinctively below the rim of the barrel. It was at that instant as though I was a RAF tail gunner with his bomber surrounded by flak on an air raid over Germany. This image was made even more real as the shrapnel from the harpoon grenade buzzed past my head and ricocheted off the sides of the steel shelter.

There was no shame in admitting to the experience of a dry mouth and a fit of the shakes at that time. After reassuring those anxious faces peering up from the deck, that I was unscathed by giving them a nervous thumbs-up signal, I decided that there must be safer locations for a landlubber like me.

Never was there a mast so high or a sea so rough, or a pair of knees as weak as on that day as I descended to the relative security of the deck below. They reassured me that what I had just witnessed was not by any means a common occurrence - quite rare in fact, and I was just so lucky to have been in a position to see it at such close quarters. I did not feel all that lucky at the time and decided to reserve my opinion on that one.

Before that day was over, I was to witness one of nature's seemingly brutal acts of survival, which I felt that few would have seen in their lifetime on such a great scale.

The floating whale carcasses from the chasers had been marshalled together just outside the channel in the reef and alongside our boat. They would soon be towed into the station and hauled up onto the flensing deck to be processed. Suddenly one of the huge bodies shook with a massive impact to its flank. All forty tons of a nearby whale carcass began to roll and jerk spasmodically as if it was once again in its death throes. Down there in a swirling mass of shredded blubber and blood and debris I could make out slashing jaws and thrashing bodies engaged in a mad feeding frenzy.

The bosun cursed, -

"They are the great white pointers – the big blokes from the deep ocean attracted by the blood in the water."

It was both fascinating and bizarre to watch the enormous grey streamlined shapes of the sharks streaking in from the depths beneath our boat. Each of these great predators attacked with gaping jaws to strike and bury their fearsome teeth into the flank of the dead whale. Then the ripping and the tearing began as each one whipped its body back and forth and thrashed until it had torn out flesh in a giant bite. Appearing in the whale's side, I could see cavities the size of a garden barrow exposing snow-white blubber and torn red flesh.

There was some comfort in the thought that the whale was past caring, but I still recollect the sight with a feeling of awe and a healthy respect for nature's savage ways.

"A pack of those big bastards," growled the bosun, "can tear off so much meat in a short time that they leave the whale in tatters so it sinks and we lose it."

I discovered later when my films were developed, that one of the pictures from the barrel caught the gun firing and the harpoon in flight. Such split second timing can only be attributed to the fact that I had my finger on the shutter button and with the concussion of its firing, the gun had virtually taken its own photograph. It was a shame to have missed the spectacular anti-aircraft harpoon incident with my camera though. That would have been one to brag about and would have served to settle the doubts of any unbelievers.

It is unlikely that another account of these experiences will ever be related in these enlightened times. Laws have been passed which virtually eradicated any possibility of the whaling industry being revived in this country. For this reason I feel privileged to have witnessed the sights and the sounds and the drama at first hand, even though they were packed into just one memorable day.

Land Voyage to Dirk Hartog

After the 1953 strike Wapet decided to drill in some of their oil exploration leases further a field than those in the Learmonth area.

Dirk Hartog Island lies off the West Australian coast at Shark Bay where the little townsite of Denham is situated. This was to be next on the list of sites to be drilled in the extended search for the elusive oil deposits.

My transport department was ordered to move a complete National T32 drilling rig with all equipment and personnel from the Learmonth base down and onto Dirk Hartog Island. The geological surveys had suggested that it had likely indications as a favoured drilling site. A landing barge was required to be available there to affect the transfer from the mainland.

In a manner that seemed typical of many American team leaders, who saw themselves as all knowing and all-powerful, Jack the Field Superintendent made an executive decision. This was that one of our landing craft would make the sea voyage from Learmonth down the coast to its destination at Dirk Hartog Island. After consulting with Laurie, my Marine Supervisor, I had to report back to the boss that this method was impractical and dangerous. Laurie had said, based on hard experience, -

"Landing craft are designed for the short trip from a ship to shore and are lousy sea boats – no keel. With the strong sou-westerlies this time of the year you would be on the rocks in no time."

Jack didn't take too kindly to having his orders taken in vain and issued the warning that if the boat was not on site at Dirk Hartog on time there would be hell to pay. He even intimated that my job would be on the line if we failed. This I took with a grain of salt, as it wasn't the first time he had been heard sounding off with this kind of bluster.

My teams put their collective heads together and came up with a plan to pull the craft out of the water with a bulldozer and transport it by truck down to a proposed launch site some two hundred miles south at Carnarvon. This had not been attempted before and with the local roads little better than narrow tracks it could be seen as quite a challenge.

Loading the barge on the Mack lowboy trailer for the road voyage to Carnarvon

Our boss said it couldn't, and therefore shouldn't be done, which was of course a great incentive for us to do it anyhow. The American built landing craft, many of which were a familiar sight in the Pacific war involved in beach landings, was ideal for our purpose in the Exmouth Gulf. With a cargo of heavy equipment from ships standing offshore, they would run up on the beach, drop the big gate in the bows and unload without the need for a jetty or a wharf. About sixty feet long and weighing in at about fifty tons, they were however a high and awkward load to move around the countryside on a truck. Nevertheless, with the aid of a bulldozer and a crane, early one Sunday morning, we soon had the craft sitting on a heavy-duty lowboy trailer and secured. The load was hauled by one of our oilfield Mack bobtail trucks.

Jack was having a bit of a sleep-in that morning as we sneaked the secured load back into the camp. When he finally got out of bed to head off for breakfast he found that the door of his hut was difficult to open. It seemed there was something solid wedged up against it. The obstacle actually weighed about a hundred tons all up, consisting of a big red Mack truck and the lowboy trailer with the barge on top. We had been able to quietly idle the rig up against his hut door as he slept peacefully on. So far - so good!

Jack, as we had discovered, had a fragile sense of humour, especially when he realised that the joke was at his expense. Eventually he stood looking up at the towering load and muttered something like, -

"Goddam, this ain't gonna work."

It did work however and I managed to keep my job as well.

We sent a team ahead on the narrow gravel road to prepare for the passage of our big load. They cut fences, widened gates, reinforced cattle grids and lifted telephone wires all the way to Carnarvon. Another team brought up the rear to put everything back the way it was. Our big boat was soon making its stately way southwards over the spinifex plains of the station country towards the ocean at the port of Carnarvon.

We took along the boat's crew of experienced hands and skippers on the convoy to operate it on arrival. One of these old salts decided that rather than travel, jammed with the rest of us in a hot truck cab, he would climb up on top of the barge and relax in the wheelhouse and enjoy the cool breeze and the smooth ride. He appeared to be rapt with the view from his lofty perch when we set off and was obviously enjoying the gentle swaying of the heavy load as the truck rolled on its journey southwards.

We had not travelled a great distance when a hail of nuts and bolts rattled on the roof of the truck cab. Our seasoned sea dog on the top of the boat was signalling us to stop. He was in a real state heaving and retching over the side of the vessel. We thought he had food poisoning or something. It was revealed that due to the unaccustomed swaying of his elevated perch, he became violently land sick. This was the cause of much hilarity among the crew, especially as he was reputed to be a veteran sailor who had never been seasick in his life before. He had succumbed out there a hundred miles from the sea in his boat on the top of a truck. I had never seen him as embarrassed as he was on that day. It took him a considerable time to live it down.

With the aid of the powerful truck winches we worked the craft down into the water over the fascine at Carnarvon and were soon running southwards on a calm sea for Dirk Hartog Island. The drilling equipment and supplies were ferried off the beaches at Denham with only a short run across Shark Bay to prepare for the drilling program beginning on the island.

The unusual, the interesting and the amusing incidents seem to stay in the memory no matter how insignificant they may have appeared to be in the scheme of things at the time.

On board the landing craft anchored in a bay close to the island we had to wait for the tide to turn and take us over the sand bar and onto the beach. Much of the time spent on this kind of job is whiled away in idleness waiting for something or other to happen. Thinking to make myself useful I spooled out a new fishing line to try my luck. The line, with its new hook was hardly in the water when the tug of a big fish almost pulled it out of my hands. It was a big fat Northwest Snapper, which had apparently been attracted to the bare shining hook. This seemed to be a rare fluke, as there was no bait on it. Looking down through the crystal clear water to the bottom, we saw an amazing sight. The floor of the ocean at that spot was a seething mass of fishy fins and scaly backs. Never had I seen so many big fish in my life before - and they were hungry. With only pieces of rag used as lures, we landed about a hundred pounds of fat and succulent Pink and Blue Spangled Emperor Snapper in a flapping heap on the deck in the space of half an hour. We were confident in making a hit with the team cook and the crew at dinner that night - and we did.

Once used as the site for a sheep station, Dirk Hartog Island still had a long stone building that had served as the shearers' quarters. This was to provide the ideal base for our drilling operations. The big wood-fired cooking range was soon alight and giving off delicious aromas. Rodger, the camp cook was busy with his pots and pans and was soon sweating profusely over his hot fire while we lounged at the long wooden tables enjoying a well-earned cold beer.

After a while, as the crew relaxed and talked over the day's events, we noticed that Rodger, who was singing away tunelessly to himself, had begun to sway perceptibly back and forth. When asked if he was all right he waved an egg slice nonchalantly in our direction and mumbled something unintelligible. A glance over his shoulder confirmed the cause of his unsteadiness and the reason for the definite lisp in his speech. A near-empty bottle of Bundaberg Rum there confirmed that we had here a cook who was well under the influence and should never have been driving a stove.

In a voice with an unmistakable slur, he turned, flapping his arms around and waving an egg slice, announcing, -

"Dinner ish ready."

He was a comical sight with his baggy shorts, skinny legs, greasy apron and cook's hat set at a crazy angle on his wispy hair. We watched amused and expectant as he closed one bleary eye and lined up to launch himself at the doorway in the concrete wall. He apparently wanted to go outside to "take a pish."

The heavy wooden door was wide open and had swung back inside flat against the wall. Although the door aperture to the outside was clear where his unsteady rush should have taken him, he mistook the door against the wall to be his exit. Grabbing the door handle before anyone could blink he pulled it, and with a wave to the crowd, cannoned straight into the wall headfirst. Down he went in an untidy heap – like a stunned mullet. After the laughter and the cheering had died down we attended to his injuries and then stretched him out on his bunk to sleep it off.

The fresh fish, chips and eggs he had cooked were superb.

A big patch of sticking plaster adorned the lump on our comedian's forehead for the rest of the week. He was always on his feet before dawn however to cook a hearty breakfast for the crew and didn't let us down once.

I have seen many camp cooks in a similar state over the years – it seems to go with the job. There was never a nasty one however and in many cases they provided us with the source of some good entertainment along with some great tucker. One memorable event of this nature involved the heavy imbibing of the flavouring essences in the kitchen store, but that is another story.

Episode Nine

The Derby Years

People have asked me why I would choose the town of Derby in the West Kimberley as a place to live and raise a family. They remind me of the heat and high humidity there, of the lack of decent amenities and services in those days and the general outback appearance of the place as it was then. The fact is that my employer had transferred me there in the search for oil , but when their operations wound down I stayed on. Together with my wife and our growing family, I lived there for eight years and this probably calls for an explanation.

Aged twenty-seven and single I was working at Learmonth when the oil search fizzled out in that region. West Australian Petroleum, my employer, then moved its teams and equipment to the Kimberley to extend the scope of the search. A job is a job and while I was initially unimpressed with the town I enjoyed the life that required that I travel extensively by vehicle and aircraft into the outback maintaining a service to the drilling sites. The country with its ranges and rivers and lush grasslands captivated me. I was intrigued by the reversed weather patterns too as it rains mostly in the summer time with some large destructive cyclones and flooding. It was so different from the impressions you get in the south of the state and the Pilbara where I had lived and worked before. This was before the bitumen roads, before the tourist invasion and before the pill. The country was secure in its relative isolation and pristine. I enjoyed many wonderful trips into the backcountry exploring and hunting, some in the company of aboriginal friends who related stories of bygone days. That is in another tale. I have tried to give in some detail the more memorable and interesting events experienced during our time there. Life in the Kimberley was sometimes hard and uncomfortable but never dull.

Bev and me on that special day -
21st.September 1956.

While sitting, musing in an aircraft on a flight to Perth I felt I would like to give in to an ever-increasing need in my life. I decided that I wanted to get married. Beverley and I had maintained a steady relationship over several years, this however being restricted by my brief and infrequent trips to Perth. I had a really deep feeling that I wanted to settle down with her and have our own family. On that stay in Perth I talked to her dad Bill first and then popped the question, which is in the proper order of things.

We were engaged then and married three months later.

On our return to Perth from the honeymoon two important events took place. The happy event was Bev's announcement that we had started our family already. The second, a sad event, was the death of my beloved dad - aged fifty-five. He did however live to see me, his eldest son get married.

Still currently employed by Wapet, I organised some family housing and began our

201

married life together where I was based back in Derby.

As a city girl Bev had no idea what she was letting herself in for. As loyalty is one of her greatest attributes, she settled down and endured the harsh living conditions, the unforgiving weather and some hard times with steadfast determination, helping me through the rough patches and having our four children as well.

To those who asked me why I chose such a place to live with a wife and family, I would like to say that I felt that the place chose me. Have you ever found yourself in a situation which in retrospect you would never have initially selected from a list of options? There are people living and working in places much more difficult than this all over the world, purely because over time they have become a part of it. They see less and less of the downside and more of the satisfaction and contentment that is in it for them. I found that I was becoming involved in the community and forming many friendships and associations. Also, for all life's rough discomforts I felt I was becoming a part of the wild countryside as well. There was freedom from restrictions, which I liked, but some excitement and challenges that city people know little of and experience even less. As Bev wryly commented on more than one occasion, -

"It's a man's country."

It is true that I loved the bush for its vast spaces, spectacular scenery and for its solitude. Crowds of people were never my scenes.

Once again in the Kimberley, the oil search discovered only minor traces of oil of no commercial value and our operations began to wind down. By this time I felt committed to the town and the people in it and left Wapet's employ to start a small business. I was offered a going concern – the town's taxi business running three cars and later combined with it a small engineering shop as well. Operating the two conjointly seemed a good idea and viable, as one could support and complement the other.

Those people who share the isolation and difficulties of an isolated country life are generally drawn together in the spirit of mutual support. In contrast, I lived at one time with a family in the city who had been in their house for twenty years and never taken the time or felt the need to meet their next-door neighbours. We had to make our own fun and entertainment there while our growing family kept us busy with their needs. There was no TV of course but there was the local open-air cinema where you could recline in the canvas deck chairs and contemplate the stars, and lots of social interaction and activities of different kinds. I formed a photography club, which I felt was a contribution to the community and stood for and won a seat on the local shire council. You could depend on it being hot almost every day of the year, so the family Mecca was the town swimming pool - you couldn't swim in the sea, and then of course we could sometimes be found bending the elbow at the one local pub. It was named the Club Hotel if I'm not mistaken.

Climatic cycles throughout the year were strange to the newcomer. Winter was actually a misnomer in the accepted sense as it was really warm to hot all year. Somewhere in those winter months however you might experience a coolish day and this became a talking point in the town for some time

There was a lot of travelling to do

202

afterwards. One resident complained bitterly, saying he had missed the winter that year. It appeared on one Sunday he said and the reason for him missing it was that he had slept in late.

As a young bride Bev's first house there was a little wood-framed galvanised iron cottage situated on the edge of the salt marsh. We had settled in to married life as happy as clams when an old horse adopted us. During the day while I was at work, the horse would lean on the house and follow the shade around looking for relief from the heat. Bev got a shock the first day the old fellow pushed his head and neck into the kitchen through an open window. They became firm friends and she seemed glad of his company, which took her mind off the few hundred giant cockroaches which were in residence too.

Upon starting the business and more or less joining the town community we were allocated one of a street of Housing Commission homes. Built up on stilts they were designed for the tropics with large windows devoid of glass with storm shutters to close when the weather turned rough as in a dust storm or a cyclone. The household amenities were very basic with a wood stove, a kerosene fridge, a chip bath heater but no hot water system. The water pipes baking out in the sun for most of the year generally made the last item redundant.

The wet season living conditions were very trying as we attempted to cool off with fans, which only moved the hot humid air around, usually failing to dry the perspiration. I wandered from the bedroom one night in the early hours, leaving a sweat-soaked bed and found myself in the living room. The thermometer on the wall showed it was still 30 degrees C, with the humidity around saturation point. It was also the time of the year when all the termites took wing and our insect screens were a seething mass of rustling bodies.

The shower recess was a small cubicle with a door opening onto the verandah but with no interior light. I was standing naked in the dark under the shower one night and unhooked the face flannel from the wall to work up a good lather. Feeling a strange prickling sensation on my bare skin, I opened the door to shed some light on the problem. Smeared across my chest and belly were the mashed body and hairy legs of an enormous spider, which presumably had been camping in the folds of the flannel during the day. As he didn't bite me I assumed that he had been caught napping. Lizards and snakes were a common sight around the house too, with fat green frogs regarding us solemnly from windowsills, and stick insects over a foot long climbing up the screens. Bev announced that she had encountered a snake on the front lawn and I enquired whether it was alive or dead. She said she presumed it was dead, as she had whacked it with the spade. It was dead all right; she had chopped it up into little pieces like macaroni. A local snake expert who was brought in to make the identification shook his head sadly and pointed out that there wasn't enough snake left for him to be able to do that.

Our livestock inventory around the house exploded as the kids grew older and adopted pets. There was a goat, too young for positive ID when we got it, which they initially called Suzie Wong. Before long she grew a fine pair of knackers, a respectable pair of horns and a beard. Without wanting to create an identity crisis for the beast, we modified the name to Suzie Gone Wong. There were dogs and cats of course, also a pig, and a kangaroo joey and a big scaly blue-tongued skink called, imaginatively – Bluey. This last creature was a really low-maintenance pet. It partook of a meal of milk and mincemeat regularly – once every month.

When Bev disclosed her love of ballet and her experience as a dancer, the local ladies roped her in to help stage a concert in the town hall. It was considered a nice touch and a way of bringing a little fine culture into the social life there. To add some novelty and also due to the lack of real ballerinas, the dancers were all hairy men volunteers. Bev taught them the rudiments of the choreography and routines so it would look as though they knew what they were doing. With apologies to the professionals who have performed Swan Lake over the years, their rendition was a huge success. It brought the house down - almost literally.

No attention to detail was spared in the rehearsals or the reproduction of authentic costumes either. With each dancer appearing as an average ninety kilos of suntanned rippling muscle, the Dance of the Little Swans in tutus, makeup and feathered headgear was a sight and must have challenged the structural integrity of the stage too.

Speaking of feathered headgear, one of the leading lights in the town matrons' forces was stumped for a while in sourcing a quantity of suitable feathers. Resourceful as usual she grabbed a rifle, went out into her back yard and shot the requisite number of white corellas out of the river gums. The two wings of a parrot, she asserted, when secured with elastic and encircling the brow of a prancing bulldozer driver were the perfect solution. Left to dry in the sun so they were a bit less odoriferous, it is alleged that a prowling tomcat was attracted to them and ate half of this portion of the costumery just before the opening night. It was reported too that the Little Swans were, despite the temporary costume malfunctions, completely costumed on the night, but were not necessarily pleasant to be close to.

We had a very clever butcher in town who as it happened, was the only butcher in town. He must have been clever as he could find at least fifty pounds of fillet steak on a beast – that is he charged fillet steak prices for most of the meat regardless of where it had been cut from. His antiquated cold storage system in what passed for a shop was a bit dicky and was sorely tested in trying to cope with the heat. Bev opened a parcel of chops in the kitchen while preparing for a tasty meal. I had just purchased the meat that day from said butcher's shop. They were a bit limp and somewhat slimy and the smell nearly drove us out of the house. His advice when I dumped them indignantly back on his counter was this, -

"Aw hell, these are OK. Just pour some vinegar over 'em – that's what I always do." The old sod would not take them back either.

Other traders in the town were more agreeable and accommodating. Mary was a chubby woman who operated the soft drink kiosk in the main street. On a stinking hot day her icy lemon squashes were much sought after. For her favourites among the male population she had special rewards in store. When asked for a big lemon squash she would breast the counter, lean over and in a conspiratorial whisper, ask, -

"With, or without?"

'With' meant the addition of a generous shot of gin or vodka from the bottles secreted under the counter and she didn't charge extra for it either. Mary was understandably very popular with all the thirsty blokes in town and I am happy to report that I became one of her favourite customers. She was never busted for selling sly grog either, as she apparently had others of her favourite men in high places.

The Call of the Kimberley

Our company had constructed a base with offices and a warehouse and large workshops on the fringe of the salt marsh close to the Derby town centre. On arrival, I presented myself to the superintendent and reported for duty. With the main focus of oil exploration now centred in the Kimberley, the Company had moved in a large amount of heavy machinery with which to carry out the task. It was my responsibility to oversee the repair and maintenance of this fleet with the operation based in these workshops. I had a staff of twelve skilled tradesmen and helpers for the work.

There was a shortage of unskilled labour there at one time and I found it necessary to recruit from the local casual workforce. The first job I gave Ernie who was a jack of all things, was to clean and degrease the engine of my ute. He started well and set to with a will so I left him to it. A scant half hour later there was a deafening explosion out in the yard. My personal transport - my pride and joy was there without a bonnet and with the engine a mass of flames. Ernie fell out of the cab and bolted for the gate, refusing to come back and join the team of fire fighters battling the blaze. This was probably a good thing as the heavy engine bonnet crashed to earth where he had been standing. We managed to salvage the vehicle, albeit with some blistered paint and a crumpled bonnet. Ernie fronted up at the office next day to collect his wages. Quite methodical, he had itemised the work done that day to total one hour. Item one was cleaning vehicle Rego. Number KW 714, which I thought was a bit of a cheek considering the mayhem he had caused. To arrive at the full hour he had noted thirty minutes 'getting the sack.'

To give him credit it took him a while to explain the causes for the explosion which were - wash down the motor with petrol, close the bonnet, and then get into the car and start the engine!

The exploration geologists who spearheaded the oil search looking for likely drilling sites had roamed all over thousands of square miles of mostly rough Kimberley country to pinpoint the spots where they thought drilling should be initiated.

There was a saying among the field personnel who had to go out into the bush dragging their heavy equipment with them into really rough terrain - they declared that these 'rock doctors' as they were disparagingly called, only recommended certain areas of country in which to drill. These locations, it was maintained, were typically in the

Jailbirds rescue the stranded geologist's Landrover bogged on Cable Beach Broome - below high tide mark

roughest, the boggiest, the hottest and the most inaccessible parts of the country to be found anywhere.

Our collection of equipment comprised oilfield trucks, earthmoving equipment, and four-wheel drive vehicles and of course the drilling rigs. All were operating under the harshest conditions so breakdowns were frequent. Geologists, I found, were some of the worst in their treatment of vehicles and we had to retrieve their cars from some sticky situations. The excuse was that their work took them to very remote parts of the country, which was true, but then I found they would rather ride than walk and so they pushed their Landrovers and Toyotas to the limit. Being bellied on big rocks seemed to be a favourite pastime, although getting bogged on the ocean beach was high on their list of stunts as well. One geologist sent an urgent call to me to extricate him near Broome. He requested a rescue party to arrive before six o'clock. This was puzzling, as we did not usually run these salvage operations to a timetable. The reason for urgency became clear when we found him and his vehicle on Cable Beach bogged to the bumpers five hundred yards out and well below high tide mark. There was a spring tide on its way and due to submerge his car, with its ETA at about six o'clock. We scouted around town for some helpers, but the local labour forces were all at the jetty unloading a ship. With no ready help available I thought I would try the Broome jail for some recruits. There was an understandable air of enthusiasm as about twenty inmates streamed out of the gates and climbed onto my Commer truck. The prison warder stood by the gate and handed each man a shovel to do the job with as they boarded.

"I suppose you want to come too," I said to him "to keep an eye on them?"

"Nar," said the warder, "they will have a good outing and anyway I've counted 'em."

Never have seen a bunch of blokes enjoy themselves so much. They worked with a will and screamed with laughter when someone fell on his face in the mud. We beat the tide by half an hour as they dug the Landrover out and practically carried it bodily up the beach to dry sand. This happy team of crims was rewarded with a cold carton of beer, which we polished off in the sand hills. The warder counted them back into the jail when we finally returned. He thanked me for easing the monotony of a screw's life by giving him back his guests who he said would be cheerful for at least a week. Even the geologist was happy and waved to the crew as he drove away in his muddy vehicle with seaweed streamers fluttering behind.

There were few decent roads in the Kimberley. Even the main highways were dirt and gravel and subject to flooding at many of the river crossings during the wet season. A real danger known to all the locals was the problem of having a river rise just behind you and then encountering another one impossible to ford further on. One diehard had this experience while travelling with his family. It was imperative that he crossed the flooded river for some reason or other, so he floated his kids over - each trip had one of them crouched in the bottom of an empty 44-gallon drum.

Out in the cattle country conditions were bad in the wet season. The odd cyclone or two did not seem to faze the station families though, as over the years they had developed the experience and the resources to cope with the isolation. They prepared by stocking up on groceries and essential items and then battened down to sit it out for the duration. Many of their children had been getting their schooling over the Flying Doctor radio network, so that was not interrupted. I experienced first-hand a bit of station life at the height of the wet at Gibb River when involved in the search for a lost aircraft. Medical emergencies could cause problems. The Royal Flying Doctor Service was always available to ferry a patient from a remote station into the district hospital to receive urgent treatment in normal conditions. In the wet however, many of the small station airstrips became waterlogged and could not take the weight of an aircraft. The highest rainfall figure I can recall in the Kimberley during the time I was there was that recorded as dumped by a rain depression following one of the cyclones – it totalled twenty-two inches - or five hundred and sixty mms in twenty-four hours!

As I was never one who liked to sit on his hands when there was some action going begging, I was always on the lookout for a reason to get out in the field and do my bit.

We picked up a request on a radio sched from one of the remote drilling locations, reporting a breakdown on a prime piece of equipment. I decided that this would be a good opportunity to relieve the boredom a little and elected to do the repair job myself.

One of our most capable off-road vehicles in handling the worst conditions was the American ex-military Dodge weapons carrier. It was also known as the Dodge Power Wagon and we had several of them. It was decided that to reach the drilling site, we should follow the telegraph line for about 150 miles. It also ran close to our destination. Not really a road as such but a well-beaten track used by PMG linesmen over the years, it served as access to the backcountry and was used by all and sundry. Telegraph lines, as most people are aware generally go by the shortest route from point A to point B in a straight line across country. They do not necessarily try to skirt the worst conditions of the terrain on the way.

I asked one of my team, a North American Indian by the name of Gus Myer, if he would accompany me on the trip. Gus was a member of the original group of drillers from the USA and had decided to stay on in Australia for a spell. I felt that there was no more loyal or dependable an offsider for what promised to be an eventful trip. Gus had a reputation for guts. He was the man who coolly confronted a half-crazed drunk with a loaded rifle and disarmed him in the camp at Learmonth a few years back.

Gus gave me a big toothy grin and in his typical slow drawl said that a break from routine work in town was just what he needed.

The rain had been heavy and persistent all week and we knew there would be some obstacles in our path into the backcountry. We loaded the vehicle with some food and water and tools and spares for emergencies, fuelled up and headed eastwards into the bush.

Gus Meyer tended the winch line on our Power Wagon while we were crossing the black soil country.

Our difficulties began about fifty miles out when the track entered the notorious black soil country. From a distance the black soil plain looks pretty innocent. Fairly flat and studded with spinifex clumps and box trees, in the dry season the ground resembles rough concrete – in the wet it is a different story. While I was easing the power wagon down a bank to where the track had dissolved into grey mud, we instantly sank to the differentials. Gus and I looked at each other and then peered up along the line of the telegraph poles for a distance of about a mile. We could see, etched on the face of a distant red sand hill, the wheel tracks emerging from our swamp. That was where we would have to aim to get onto solid ground and dry footing to continue our trip out to the rig site. We discussed the pros and cons of continuing or backing out and returning to town. The need was urgent however and like the famous mountain climber when asked what was the driving force which made him tackle his mountain, he replied "Because it was there," – or words to that effect. Anyway we would never have lived it down if we had given up and returned to base so we decided to forge ahead.

Our truck had a powerful winch up front. It had fifty yards of steel cable wound on the drum and was driven from the engine crankshaft by a chain. This is a very reliable system and one of the reasons I chose this vehicle for the trip. To attempt to drive through this bog even in four-wheel drive with the differentials locked would only dig us deeper into the mire, as black soil mud is bottomless. The tough spinifex tussocks growing out of the slop became insurmountable obstacles, impossible to climb over and all four wheels were soon wedged in the slime between them. It looked like being a winching job every inch of the way.

As we all know, a winch is useless in dragging a heavy vehicle through these conditions without a solid anchor point on which to hook the cable. There were the box trees, scattered randomly across the plain, which we would have to press into service. I asked Gus if he would put his native Indian cunning and backwoods skills to good use in plotting a course so we could winch from tree to tree until we reached hard ground. He could take a joke fortunately and my scalp was safe for the time being at least. We knew there was a pretty tough job ahead of us and much of it would be in darkness as the sun was now low on the horizon. There was one other detail. It wasn't just the power wagon we had to get through the swamp – there was a Landrover on a towrope being dragged along behind as well.

We were sprawled on that far bank twenty-four hours later, with only our eyeballs and teeth showing through the coating of grey muck and laughing with the relief of our win against the swamp. Gus was talking about the bit of drama we had experienced about halfway across.

During one particularly hard pull, with the engine roaring and the winch line creaking with the strain, there was a loud bang and the cable went slack. I diagnosed the problem to be the failure of a steel pin in the drive designed to shear off if it's overloaded. This is good for the gear, but not good for us, stuck in a swamp in the middle of nowhere.

The toolbox in the truck yielded a good strong Allen key I could use to replace the broken shear pin. Grabbing the front bumper I slithered on my back, half submerged in the mud, under the vehicle to fix the problem. There was little room to move as I replaced the pin and twitched it with wire so it would not fall out.

Soon there was even less space under the truck as I could feel the axle housing touching my chest with no room left to manoeuvre now. With a final twitch on the wire I yelled to Gus to grab my feet and drag me out. This he did with a will, scraping my face on the oily bits under the truck and nearly taking my ear off. Had we not acted quickly to repair the winch we would have remained a permanent feature there and if Gus hadn't acted quickly I would have become part of the swamp under the vehicle. Standing up rather shakily I watched with interest as the vehicle sank a bit further into the mud and the track left by my body began to fill with water. I shook the hand of my Cochise and thought how well you get to know the finer qualities of your companions when you are in the bush.

We had rules about travelling in remote country, one of which was to never go it alone, even in the case of an accident or an emergency. Without trying to make excuses for foolhardiness, here is an account of the situation I found myself in somewhere between a remote rig site and the sanctuary of town.

The wet season was over for that year and although the storms had subsided and the sun was beating down again, much of the country was still treacherous and boggy. An oilfield Mack truck empty weighs in at about ten tons. With a powerful Cummins diesel engine, a good range of gears and power steering, it is relatively easy to drive, even in rough conditions. It is specifically designed to haul heavy drilling equipment even on basic unmade roads.

Feeling a little bored with life at the base again I jumped at the chance of driving one of the Macks back into town for repairs. It would involve traversing the telegraph line track again but most of the country had dried out so I had no qualms about doing the trip. The mechanical fault suffered by the truck was the failure of the power steering system. We knew from experience that without the aid of power steering, the driver, using all his strength on the wheel could guide the truck roughly in a straight line on even ground. I decided to attempt the trip back to Derby via the telegraph line as reports suggested the track was open to normal wheeled traffic again.

Perhaps I was becoming a little too complacent after travelling reasonably well as far as the point where the black soil flanked the road, when the truck dropped a front wheel into a soft patch. This tore the steering wheel out of my hands and the rig nose-dived into the muck.

The Mack has a winch behind the cab, which is rated at 64 tons pull. No problem I told myself, I'd just latch onto something substantial and drag the truck backwards onto the road again. Searching for suitable anchor points on which to secure the winch line, I lined up several possibilities. These, when chained together in series, would hopefully provide the necessary resistance to pull my ten ton vehicle out of the mud hole.

Beggars can't be choosers they say. The only suitable fast points within range were, in order of availability, a small tree, a big termite hill and further out, the steel poles of the telegraph line. Singly, I doubted that any of these would serve my purpose, most likely risking being ripped out of the ground by the pull of the winch. Chained together however, one to the other as it were, their combined strength might just serve my purpose as an anchor.

Sitting in the cab, I took up the slack on the winch cable and engaged low reverse drive gear too. The tree shook, shed most of its leaves and its roots were ripped out of the ground. Pulling some more I saw the anthill shear its foundations and begin to glide across country, while the telegraph pole I was connected to sent distress calls down the wires to all the other poles stretching away into the distance. When the dust and the smoke cleared the truck was thankfully back on the road again.

The tree was shattered, I could see the termites were panicking and tearing around in the sun as their apartment block was disconnected from its foundations and then I peered up along the telegraph line. I have to say it looked different. The steel pole selected for my purpose as an anchor was uprooted. When it toppled, it brought down the next one and the next, until like dominoes they were felled or sloped for quite a distance. Not stopping to count the phone line casualties, I just collected all the gear, threw it up on the truck and departed. Knowing that it would have been nigh impossible to cover my deep tracks and the line of destruction, I decided to go back to town and face the music.

Entering the outskirts of Derby I spotted a familiar red truck festooned with ladders and cables and gear, travelling at speed in the opposite direction out of town. My friend the line foreman Graham pulled up and asked had I been on the linesmen's track and if so had I noticed a break in the wires anywhere. He was a good bloke and deserved my help so I said yes to both questions and told him approximately where to look for the 'fault'.

One of the major units of currency in the culture of the Kimberley is the cold carton of beer. I felt that I should own up to my part in the incident and try to make amends for destroying government property and creating so much work for the line staff.

Arriving at Graham's house about dusk a few days later, I hitched my peace offering onto my shoulder and banged on his front door. The place was really rocking and by the din I could hear there was a good party in full swing inside and the reason for celebration? – The great windfall the team had collected in their pay packets for a major line repair they had been called out to attend to a few days previously.

It's an ill wind they say...

Crossing the Fitzroy at Langi Bridge in the dry season

Langi Bridge after cyclonic flooding – 7 miles wide in places.

Barramundi Territory

In the cool green depths of the Fitzroy River pools in the dappled shade of the river gums, lurked one of the most courageous and incidentally the most succulent of all the Kimberley fish. Marine life of all kinds swam and sported in the lower reaches of the great river. Here the fresh water floods surged brown and turbid in the wet season. Then the salt water came upriver with the sea tides when the flow had exhausted itself in the dry winter months

Two distinct species of crocodile lived there in large numbers. The large and dangerous estuarine or salt-water croc moved freely from the sea and up to the headwaters of the river, sometimes visiting billabongs along the way.

The Johnson River or fish croc, a smaller cousin which grew to a maximum of about two metres, was timid in his habits and relatively harmless to humans. There were catfish and swordfish too, but the prize catch of them all was the 'barra.'

Thick, juicy barramundi steaks, fresh from the river, were the highlight of many a fine party or celebration in the Kimberley towns, stations and camps. Partaken with congenial company on one of the many warm, steamy tropical evenings and washed down with an ice-cold lager, it was fare 'par excellence.'

Common to all of the barramundi I ever captured from the Fitzroy, I remember the flesh had one outstanding quality. No matter how large the fish or how correspondingly thick were the fillets, the flesh was always the same. It appeared snowy white in colour with faint marbling and was always tender and succulent.

The further downriver towards the sea where you threw in your line, the bigger and sweeter were the fish you could catch. I struggled one day with a fighting barra on a heavy hand line, sinking up to my knees in grey mud on the bank of Telegraph Pool. With a supreme effort he was finally dragged up from the river. They are an exciting fish to catch due to their strength and fighting spirit, and I was amazed at the size and bulk of his silver-scaled body.

Fitzroy River pools are the home of the barramundi and the two species of crocodile

We lifted the big fellow across the bonnet of my Landrover and I noted that while his head rested on the top of one of the front mudguards, his tail covered the opposite one. To prevent the loss of

211

the fish by spoiling in the extreme heat and humidity on the long drive home, we gutted him and stuffed him with green grass, then wrapped him in wet bags out of the sun. That particular fish weighed in on the butcher's scales at sixty-four pounds, or nearly twenty-nine kilograms.

Due probably to its warm and benevolent climate, the Kimberley attracted a particular kind of character. There were men who, tired of the city life, with its noise and frenetic pace, -and possibly with some debt to society unpaid, opted for the peace and the quiet beauty of the bush. They came, they stayed and some lived there for years. Many will see out their last days in the Kimberley I am sure, preferring a solitary bush camp on the bank of a river or by a billabong in the shade of a big rivergum.

Such a man was Nugget Moore. We knew little of his background except that he once served in the army on active service. Never far from his side were a battered old army .303 rifle and a rusty bayonet. Nugget's livelihood came from the Fitzroy. His rudimentary lean-to camp was hard by the river at Langi Crossing where he fished for barramundi and sold his catch on the bridge to tourists and passers-by.

We travelled up from Broome on the dusty highway one day. I nosed the car off the gravel and down the slope to the timber decking of the Langi Bridge, which is on the main road to our destination at Derby. Nugget Moore's figure in the middle of the bridge was unmistakable. With leathery skin burnt to a deep mahogany by the sun, he was short in stature and stocky, with legs like trees. He was wearing his trademarks, the old diggers' hat and a tattered singlet, but the most distinguishing feature was his baggy shorts. Called 'Bombay Bloomers' by the troops who served in tropical climates, they reached halfway down his stumpy legs to where they disappeared into old army boots minus socks and laces.

Nugget straddled the middle of the track halfway across the bridge and held up a hand commanding us to stop. This we did for fear of running him down and also because he was an entertaining character and a friend to boot. He strolled to the side of the vehicle and slouched against the driver's door.

"Do you want a bloody fish?" he rasped in my ear. With a lot of fillets in the fridge at home already, I tactfully declined his offer.

"Come on ya miserable bugger, why don't ya buy one of me fish?"

This was his livelihood after all and most had a soft spot for the old battler. There was no doubting that his fish would be fresh too.

"What have you got?" I asked him.

It was not obvious at first as we stopped there yarning, just where he kept his fish, but closer observation revealed the source of supply. Secured to the balustrade, were fishing lines spaced at intervals for some distance along the bridge. They all disappeared into the waters of the big pool alongside where we had pulled up. All the lines were moving, as captive fish secured to the lower ends swam around in the river below like dogs on a leash.

Nugget grabbed the nearest line and hauled it up onto the decking. He held up the flapping silver specimen for us to inspect.

"Twelve pound this one and pretty fresh too eh? That's five quid - do you want it?"

The price was fair, but twelve pounds was more fish than my fridge could accommodate at this time.

"Nugget, have you got a smaller one?" I asked him.

"Miserable bugger," he snorted as he hurled the flapping fish back into the river and selected another line and pulled it in. His stock inventory system could not be faulted as he showed this next one to us.

"This one's eight pound weight for four quid," he offered and held up another, smaller barramundi.

The deal was struck. I paid him his four quid and with a grunt he swung the fish into the back of the Landrover.

A cloud of dust further down the road signalled the approach of another potential victim. The stocky figure in the baggy shorts stumped back in the middle of the deck with his hand held high, his image now receding in my rear-view mirror. His voice could be heard more faintly now,

"Do you want a bloody fish?"

Of all God's creatures, there is one species, which was Nugget's sworn enemy – the Fitzroy crocodile of either kind. They sneaked in to feast on his tethered catches by the bridge and at times were literally tearing his livelihood to pieces. It was an unforgettable sight to observe his reaction when one of them attacked his captive livestock.

On another day, we had pulled up and were yarning with him again in the middle of the bridge. Nugget had six lovely big barramundi circling in the pool, ripe for retrieval, to be offered to travellers in the cars he would bail up. One of his line tethers suddenly zipped through the water, pulled almost to breaking point and humming with the strain of the load on it. Nugget reacted instantly and bellowing with rage, turned and pounded back along the bridge to his tent up over the bank. Shaking his fists like a madman he swore foul curses on all crocodiles and their ancestors.

The comical stumpy figure with his flapping baggy shorts bounded back to where the gyrating fishing line was tied. He had murder in his eye and his rusty bayonet clamped between his teeth.

"I'll fix you - you bastards," he gritted, and plunged headfirst into the deep pool.

Pulling hand-over-hand down the taut line, he disappeared from our view to meet the challenge of the crocodile, which had had the temerity to seize one of his fish. We watched in fear and fascination as the mud and the blood and the bubbles boiled up from the bottom of the pool. There was no way of telling what kind it was or just how big or how fierce his adversary might be, but this did not seem to faze him in the slightest. I can guarantee that in the gouts of air that rose from the depths and burst on the surface, there were strings of Nugget's swear words rising from the murky water.

One of Nugget's sworn enemies

When all went quiet in the river by the bridge after what seemed like an eternity, the head of our gladiator broke the surface. His hair was plastered down, he was bloody, but he had a gleam of triumph in his eyes.

"Got the bastard!" he spluttered.

The question occurred to me later whether he was able to sell the front half of the fish, as that was all that remained after the epic battle when he reeled in the line again.

Nugget's 'crocodile charmer' was a fearsome looking weapon. His old .303 service rifle had had the butt sawn off to form a one-handed pistol grip and the barrel was docked to about eight inches in length to make it 'handy.' One of his mates commented after firing a shot, -

213

"I don't mind the thunder too much – it's the lightning that scares me!"

Summoning all my courage just one time I fired the thing one-handed with full-powered military ammo. I still vividly recall the mind and body-numbing experience.

Our fishy friend could often be found sitting at the top of the steep riverbank with his lines in the water, a bottle of beer and his 'croc charmer' close beside him.

Late one afternoon along the river, just as light was fading as it does so suddenly in the tropics, we heard a great commotion down at the big pool. Nugget's unmistakable string of oaths could be heard through the trees. He was obviously unhappy with the crocs again. He was calling down all the fires of hell on his foes, disturbing the peace of the river and merging with this racket were the bellows of an angry croc competing with his tirade.

The scene of the standoff is etched on my mind for all time. It appeared that the clay of the steep bank had become wet and slick from the fish he had dragged from the pool. He was having serious difficulty staying put on the slippery slope with a big barramundi pulling hard on his line. I am sure he would normally have been able to dig in his heels and handle this situation, had it not been for the extra weight and ferocity of the ten-foot saurian with its jaws clamped firmly on the fish he was trying to land.

The tug-of-war became more desperate, with the croc refusing to let go and Nugget trying to land them both. To further complicate matters, he was now starting to slide down the steep bank towards the water and into the jaws of the beast that had his catch in its jaws.

Nugget must have then decided it was time to get serious and fight his foe with firepower. It wasn't easy to load his bolt-action cannon and fire it at his adversary while being dragged down into the river. He was able to get three shots away which caused the beast to lose interest and he eventually won the day.

Can I move now to a more recent time sitting in a picture theatre watching Paul Hogan's 'Crocodile Dundee?' As you will recall, with the heroine in dire peril, he dived on top of the tinsel-town croc, armed with his 'this is a knife,' knife. Much to the consternation of the closer audience at this exciting juncture I started to laugh. Nugget Moore's image flashed before me, recalling the time I saw him fight a real croc in the mud at the bottom of a deep pool and then wage war on them on their own terms whenever he could. There were a few hard stares and muttered comments at my involuntary outburst of merriment. It seems I was the only one there who saw the humour in the dramatic situation being portrayed on the silver screen.

You see, I had witnessed first hand and in real life, a tough man of the bush that had challenged crocs on their own terms, in their own element, with only a rusty bayonet in his teeth - and won.

As a comment on not only Nugget's hatred of crocs, but also his scornful disdain for them, I witnessed his sleeping arrangements at first hand one night also.

As the river shrinks to its natural bed in the dry season, small, shallow billabongs are created along the banks, where some of the water has been trapped behind sand bars.

Another species of pest that irritated Nugget in his bush camp was the mosquito. His method of escape from these stinging insects was simple but effective. Come bedtime, just after sundown, he would kick out his fire, take his last drink for the day and amble down to the river. Wading out into a shallow billabong, he would lie down contentedly, submerging himself in the water. The spot he picked for the night's sleep had to have a handy tree root on which to rest his head so he could breathe. There he would spend the night and as he claimed, -

"Mozzies don't swim too good."

We were aghast at what we considered to be the very risky practice of sleeping the whole night in a crocodile-infested river, but Nugget reassured us.

"Nar, I'm OK – those buggers are scared of me!"

His claim must have been valid. He was to my knowledge never taken.

The Secret of the Fitzroy

There is an old-timers' prediction that if you drink the waters of the Fitzroy River in the wild Kimberley, you will be drawn back to her for the rest of your days. True or not, this mighty river with its watershed in the upland ranges, swells and bursts its banks and floods every year in the wet season to disgorge into King Sound where it meets the sea. The river, over many years has influenced the lives and fortunes of many those who lived and worked in the country that is part of its domain. I will not easily forget the experiences and the dramas of my years spent in her territory.

The Great Northern Highway, which is the link with the southern parts of the state, intersected and crossed the river at Langi Crossing. From here the road swung westwards to reach the town of Derby on the coast. In the Kimberley dry season, which is the cold, wet winter in the south, the Fitzroy is a sleeping watercourse, a chain of green pools lying torpid between high banks shaded by overhanging trees. At this time of year in the 1950s the traveller would cross the river on the decking of the road bridge, which squatted on thick timber piling set in the sandy river bottom. The bridge spanned the riverbed, bank-to-bank for about 150 yards. Standing in the middle of the bridge, you could follow with your eyes the line of pools and billabongs stretching east and west, separated by river stone bars and silty sandbanks. If you were on foot and quiet, you could make out the languid forms of dozing crocodiles sunning themselves - never far from the secure depths of the pools.

Our drilling operations south of the river required that we had to maintain the road transport of equipment and supplies throughout the year and through all seasons. In the drier winter months the roads and tracks were hard and firm and a trip to the river crossing at Langi Bridge was just a pleasant scenic drive. When the monsoon thunderstorms emptied their black bellies onto the upland ranges and the plains, a foaming red torrent filled the watercourses and charged headlong toward the sea. With little or no rainfall evident at the crossing on several occasions, the word was out that the inland had received ten or twelve or more inches of rain in a single day. The townspeople knew there would be a period of grace of several days before the charging flood came down taking all in its path and hit the Langi Bridge. You could see feverish activity in and around the river as the flood was approaching. All vehicles and machinery were evacuated to higher ground and the final hasty crossings were made with last-minute supplies.

The arrival of the flood was imminent. In less than an hour the landscape at the crossing would be transformed. Where once there were tranquil pools and trees and sandbanks, it was now a swirling red torrent. Where the bridge stood on its stumpy legs, it was now a rushing surge of swirling eddies peppered with smashed trees and the detritus of the land. The bridge disappeared under the surface of the torrent, the only evidence of its existence being the south road dipping into the river on the far bank some miles distant.

One Saturday morning it was decided to send a convoy of our trucks along the highway northwards to Derby. Reports suggested that the level of the flooded river at Langi Crossing had subsided considerably after some heavy rains and that the trucks should be able to cross the bridge there. To remove the tangle of flood debris still festooning the bridge, a small team of men was transported out to the crossing to clear the way for the trucks to pass safely. The water over the bridge, while still flowing strongly, had fallen to the height of a man's waist above the decking.

The task of clearing the bridge was proceeding well as the team worked their way across in single file, throwing the accumulated build-up over the side. A shout went up. The leader of the operation turned to see one of his men behind him being swept off the decking. Turbulence in the deep hole scoured by the current beside the bridge was still fierce and the unfortunate man was quickly being carried away downstream. The supervisor dived in securing a hold on him, but then everything started to work against them. The man could not swim. Additional to this were his large size and strength and the weight of the heavy work boots he wore. The watchers were sure they were in serious difficulties when the big fellow had his rescuer in a fearful grip and was pulling him under. Fresh water of course does not have the buoyancy of the sea and coupled with the fact that he was now in a state of panic, their situation became desperate. The foreman appealed to the man he was trying to save to calm down and allow himself to be towed across the current to the shore, but his efforts were in vain. Finally breaking out of his rescuer's grip, the drowning man disappeared under the surface of the murky water. His head briefly reappeared among the brown whirlpools and eddies and then disappeared from sight. His hand was seen to reach up in a final gesture of desperate appeal and then he was gone.

A message was relayed to the Derby police from a local cattle station and they, with a team of helpers, arrived later in the day.

The search continued for a week downriver, along the banks and in all the likely spots where a body might be found, with no result. It was thought that it could have been carried along in the current all the way to where the river emptied into the sea, in which case it was almost certain that no trace would be found. A theory put forward by one of the local identities was that he could have been taken by one of the resident estuarine crocodiles. There were some very large ones in their territorial habitats along the river. One in particular had been seen in Telegraph Pool, not far downstream from the Langi Bridge where the accident occurred. We had spotted this big specimen lurking out on the surface of the pool on several occasions and his great belly track was evident in the shingle of the riverbank.

In order to complete the death certificate for a missing person and allow the whole situation to be finalised and put to rest, especially in the interests of his grieving family, some tangible evidence of his passing was required. Our departed workmate appeared to be gone without a trace and we decided to hunt the big crocodile, after all other efforts to find a body had failed. As long as there was even the remotest chance of finding some remains as evidence, we had to pursue it, even if it was to be found in the belly of a crocodile.

In the context of this book, some may think that the following account is somewhat macabre or sensational in nature. It is my intention in writing to relate events exactly as they occurred, so for what it is worth this is the way things happened. Life away from the city in these remote parts of our country, especially in those early years, was not always pleasant or comfortable, but more likely to be harsh and even dangerous at times, especially in our line of business there.

I was often involved in hunting in the Kimberley in my spare time. There was a need to cull certain species of feral animals, which interfered with the local beef industry and this, together with some like-minded companions, was what I did. Crocodiles were not normally on our list of quarry and in subsequent years became protected anyway, but in this instance we felt justified in taking some action that might give us the forensic evidence that was so badly needed for evidence of death.

Bruce, one of my close friends and an outdoorsman and hunter, was an officer of the local council. We decided to team up to try and take the big croc at Telegraph Pool. One of his duties was that of dogcatcher, which we considered to be of benefit for our purpose. Big crocodiles love to have a dog for lunch – or a person for that matter. Big ones have been known to pull down a bullock.

Soon after the accident we sat quietly together one hot night behind the cover of a fallen tree on the riverbank at Telegraph Pool with a high-powered rifle, waiting patiently for a glimpse of our elusive target out on the water in the bright moonlight. We knew that we were up against a cunning old veteran of the river – they don't grow to be the size of this one if they are careless.

Down on the beach I could see the shapes of four dogs tied to stakes. After pegging out the canine baits on the water's edge, we felt confident that their movements would entice the big predator to approach within rifle shot.

My mate growled, -

"The damn dogs aren't moving - they have all gone to sleep."

As some movement of the quarry is desired for the croc to be interested, they would have to be stirred into action somehow. He slid down the slope of the riverbank and woke the dreamers with his boot. A few minutes later the hounds were all slumbering again.

Bruce had a secret weapon he said, to rouse the dozy dogs into action. Within seconds of his next visit to the bait line, all four animals were whimpering and tugging at the stakes. I could see them describing circles at the ends of the ropes, round and round, dragging their rear ends through the shingle. Back at the hide, he chuckled and showed me a small bottle of dab-on mosquito repellent called Kokoda. The label had a warning that this powerful stuff should not come in contact with the eyes, mucous membranes or any other sensitive area of the body, as some pain and discomfort would result. I had to assume this applied to dogs too. The dab of this stuff under the tail of each dog was working wonders.

"It will wear off soon," he reassured me, "so don't worry about it."

There was no doubt that our big reptile was interested now. The noise and the vibrations of the distressed dogs had reached his sensors and we observed him closing in from the middle of the pool. His knobbly snout and eye cases were visible making a vee track on the surface. I closed the bolt on the .30-06 Mauser and slipped off the safety catch silently. Our target moved closer on a beeline for the dogs. They seemed to sense the danger stalking them, as they became even more agitated. Bruce was ready with the big spotlight and we waited for the opportunity of a clean shot. There would only be one chance – they don't hang about after you have had a crack at them. Shooting from a hundred and fifty yards across water from our vantage point was being very optimistic, but we had gone to a lot of trouble for this one opportunity and we felt that there was a lot at stake with only one opportunity.

Our target was stationary now and wary. The white beam of the light lit up the pool and the eyes of the croc glowed like two bright coals against the black. With the crosshairs of the scope centred on the red orbs, I fired once. There was a tall spout of white spray where the bullet had struck, obliterating those eyes. The turbulence and the mist cleared, but the croc was gone. A good hit would have been evident at once with a big one like this in his death throes, but it was a miss, probably an inch high or low and with a ricochet off the water. He was a lucky crocodile. All there was left to do was concede defeat, let the dogs loose and go home.

A sure sign that the Telegraph Pool croc had lived to hunt another day was the evidence seen on the beach at that spot a week later. A great belly track was visible in the soft beach sand. He had emerged from the water and investigated the place where the dogs had been tethered, circled our dead campfire, had lain for a while in the sun and then submerged back in his pool again. An idea of the size of this specimen can be gauged by the comment by one of our party, who likened his track to one made by dragging a full-sized bathtub along the beach.

Despite all of the efforts of those who scoured the lower reaches of the river for months after the accident, no trace of the missing man was to my knowledge, ever found. All hopes faded for the discovery of evidence that might lead to conclusive proof of our friend's demise. That will always remain one of the many secrets of the mighty Fitzroy River.

The Spice of Life

The title says it all for me, as opportunities for finding variety in our lifestyle were abundant during the years we spent in the Kimberley.

Settled into operating the two businesses - the engineering shop and the taxis, I enjoyed a close interaction with most of the townspeople and had the chance to play a part in the determination of some of the community issues. Having a seat on the Shire Council opened a lot of doors for me and I felt useful.

My defacto position was the anointed town photographer with a lot of interesting requests forthcoming for my camera. A lot of the more remarkable assignments came from the police and the medical departments. The WA Newspapers and the Daily News also had me on assignment to file pictures and stories of newsworthy events in the district.

There were two memorable photo projects I was asked to cover at the town hospital. The doctor rang and asked if I could do colour coverage of a post-mortem examination on a tribal killing victim to gather forensic evidence. As the procedure was getting under way I stood apprehensively just inside the door of the morgue. As this was a new experience I feared that I would not have the stomach to do the close-up shots of the various bits and pieces that the doctor requested. It seems that it's not so bad when you get used to it however and I kept reminding myself that some people do it for a living, so who was I to be squeamish? All in all it was very interesting to witness the procedure and while not breaking my neck to stand in a queue to see another one, it was all part of the variety of life in those days. You won't find those pictures in our family album or in this book either.

The next request from the same doctor specified some detailed pictures of a victim's head, which had apparently suffered the ravages of a bit of mayhem. Asking would I have to report to the morgue again, he said that would not be necessary, but to come to his office - he only wanted pictures of the head. I felt a bit uneasy about this one and wondered how he could justify parking a cadaver in his office.

"As I told you" reiterated the doctor, "I only want pictures of his head. I've got it here under my desk - in a bucket."

This was becoming pretty bizarre and there were visions of the film I saw which told the story of Salome's betrayal of John the Baptist and the shock of seeing his severed cranium on a platter. Then the doctor unwrapped a parcel and dumped this poor bloke's noggin on to the table. He was considerate enough to close the surgery door first I noticed, probably due to the risk of giving the tea lady a heart attack.

"This was where he was clobbered with the four by two," he said as he indicated the depressed fracture at the back of the skull.

"And this is the major damage causing death on the other side. It's the effect of the compressive forces of the blow which splits the skull – a bit like what happens to a watermelon when you drop it."

While I found all this information and the images quite fascinating, I took my shaky shots in record time and departed in haste.

There was some interesting stuff I did for the police too, including taking pictures at sites of serious accidents, at crime scenes, and also of weapons and other forensic material.

It was reported that a member of one of the aboriginal communities had gone missing from a camp near Fitzroy Crossing. On arrival with the police party, everyone sitting around the campfire assured us that they had no idea of where the missing indigenous person might have got to and maybe 'he just gone walkabout.' I was thinking that my presence was not required and that the long trip was all for nothing, when one of the constables called from behind a copse of trees close by.

My trip was not in vain it seemed as he indicated two black toes protruding from the earth. These belonged to two black feet, which were the property of the body in a shallow hole. When they dug him out there was some discussion as to the possibility of foul play. When his head was examined there remained little doubt. The bloodstained ends of a nulla nulla or throwing stick were protruding from both sides. It looked pretty suspicious to me.

One rather unusual and even amusing murder trial was held in the courthouse - if you can stretch the point to call murder amusing. It was alleged that an irate spouse had hurled a jarrah stool at her partner while chasing him out of the house after a domestic argument. The heavy missile impacted with the back of the victim's head and felled him on the spot. He was deemed deceased soon after.

A flock of legal eagles had flown up from Perth and were seated resplendent in gowns and wigs in the Derby courthouse. The judge presided pontifically and the courtroom was packed. The first expert witness to be called was our local doctor who had conducted the post-mortem examination on the victim. When asked to describe the injuries suffered by the unfortunate party he did so in the prescribed medical terms. He was then asked his opinion as to the cause of death.

"Fatal heart attack your honour."

This was a death as a result of natural causes, so the court was cleared and everyone filed out and went home.

In a town with a small population but with a busy courtroom there was one thing you could depend on – you will be summoned for jury duties sooner or later and more than once. I got the guernsey for this civic responsibility three times. All the cases were interesting, though one was a little more so than the others. I was elected jury foreman for this one.

The defendant stood accused of first-degree murder by gunshot. The prosecution alleged that the shooter had hidden behind a tree and ambushed the victim from some distance. The defending counsel on the other hand maintained that there had been a struggle over possession of the firearm, which had discharged accidentally, causing the aforesaid fatality.

We twelve good men and true - actually eleven and one lady, listened carefully to all the testimony and then retired to consider our verdict. The exhibits were laid out on the table in the jury room. These included the rifle and the bloodstained jacket the victim was wearing when he was shot. There was a half-eaten Granita biscuit on the table too. This wasn't listed as an exhibit in our proceedings, so it was assumed to be left over from a previous trial and was therefore deemed inadmissible.

The pivotal point of the debate on which hung the fate of the accused seemed to be the angle of penetration of the fatal bullet. Trying to be helpful, I slipped on the deceased's jacket and we noted the relative locations of the entry and exit holes in it to determine the bullet's path. It had entered low in front and come out through the collar at the back next to the Dryzabone label. It was conceded then and there that the shot had not been fired from a great distance. If it had, it would have followed a horizontal trajectory, suggesting that the shooting was a premeditated act. We concluded that it was not.

The man in the dock was eventually found guilty on this evidence, of the lesser charge of manslaughter.

Our eldest son Peter, who was aged about one, was pictured on the front page of the Daily News newspaper in Perth. I had learned a bit about photojournalism over the years and knew that the editors are always on the lookout for human-interest stories. They are popular with the readers and help to boost circulation of the newspaper they said.

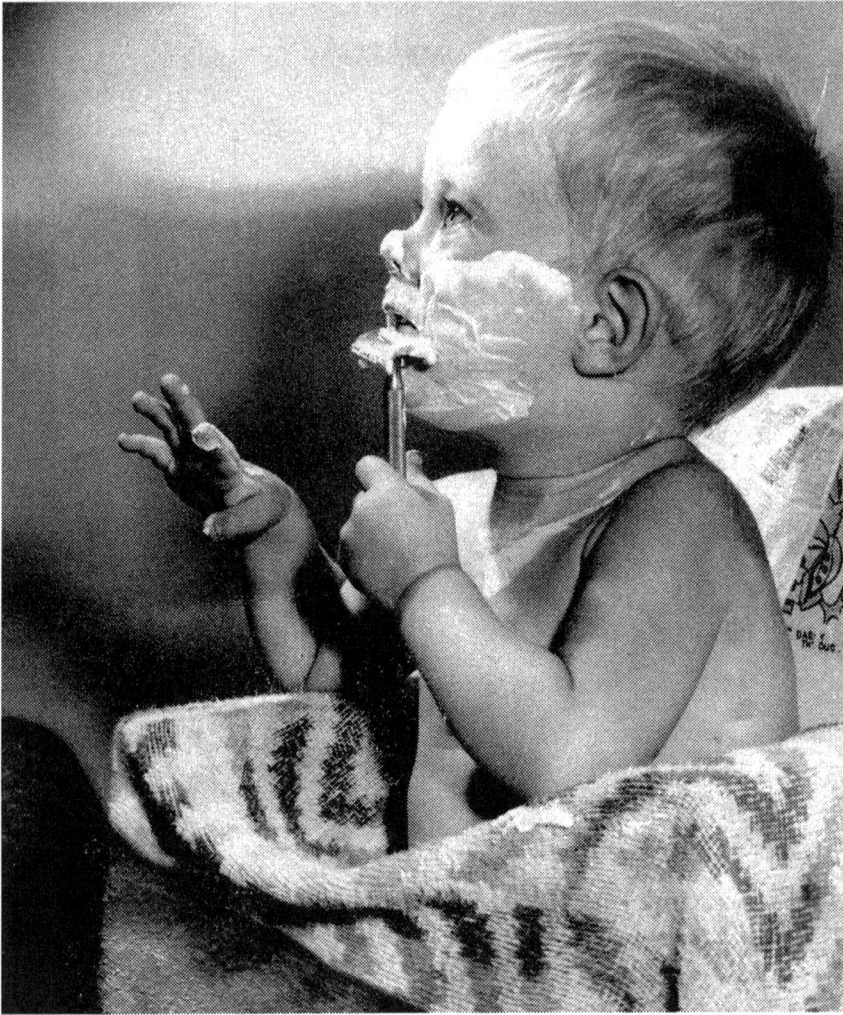

'Young Shaver' was the caption on this photo of son Peter that I sent to Perth newspapers. It featured next day.

Son Peter showed a keen interest in family activities, one of which was his father shaving. I set him up in his high chair with a towel around his neck, lathered his face and told him he could shave – just like Dad. I removed the blade from my Gillette super shaver of course and in the way of all kids he imitated my shaving technique faultlessly. The editors liked that picture and ran it the next day. I had entitled it, 'Young Shaver.'

A really good cyclone and severe flooding and other reportable phenomena provided newsworthy material that I sometimes submitted to the papers. It was interesting being 'their man on the spot' and at a distance of over three thousand kilometres it was unlikely that I was doing any Perth photographer out of a job.

On the subject of tropical cyclones, we had one pass close to Derby one year but learned that it had smashed up Broome a bit. Communications were poor, as the landlines were all down, so we surmised that they could do with some help in restoring order in that town. We rounded up a team of technical and trades people and in three vehicles set off on the two hundred-kilometre rescue mission. Our aim was to help the townsfolk restore power and telephones services and assist in any other way we were able.

The drive down the Broome road, which was normally just gravel and dirt and bulldust, was a bit of a nightmare due to flooding after the storm. After grinding through the mud and winching out of numerous deep bog holes we finally arrived at Broome, weary and muddy, and straggled down the main street of the town.

The whole place seemed eerily quiet and deserted. We took note of the structural damage that was the visible effect of the destructive winds, but there were no people to be seen. We finally found them all gathered down at the soggy sports ground. They had hauled a few kegs of beer down from the pub and were enjoying an impromptu cricket match.

Singling out the Shire President we explained the reasons for our presence there and expressed our noble intentions to assist them in getting back on their feet again.

"We are OK," he said, "although a couple of dopey buggers got themselves killed last night. They crawled under a house for shelter and then the wind picked up and the house fell on 'em. We are waiting for the insurance assessors to come up from Perth, so don't unpack your gear and do come and have a drink with us."

We all went to a 'cyclone party' that night, which is a tradition in the North in these stressful circumstances. Most of these dos rage on inside while the storm rages outside. The president stood up by the barbecue and publicly thanked our Derby team for the well-intentioned heroic rescue bid, which brought a round of applause from the townsfolk. All drinks were on the house and the festivities continued unabated for most of the night. The road back to Derby was drying out rapidly as the sun shone next day, so we bade our bleary farewells and headed north for home.

There was much variety in the work we took on in the engineering shop too. It was difficult to predict what we would be asked to do next, or in fact what surprises the day might bring. I had just finished repairing a clock one day when a tourist's caravan pulled up. They had a broken suspension spring on the van and had it wired up with a block of wood jammed in. We had them back on the road in an hour. They were Rolf Harris's parents and we got a card from them from Sydney a few weeks later. They said their trip was completed and the spring was still holding OK, and they had had a lovely time as well.

The driver of this Landrover failed to see the deep washaway in the long grass

A big Norwegian ship came into the port and tied up at the town jetty. The chief engineer came ashore and approached us about doing an urgent repair job on board. He explained that there was a bad leak in a large pipe, which was open to the sea below the waterline and therefore was a risk to the seaworthiness of the vessel. The work he requested us to do was to go down there and weld repair this pipe. The catch was, it had to be completed between the time when the tide went out leaving the ship high and dry, and the point when it flooded in again and submerged the pipe.

Two of us worked down in the hot, stinking bilges of the ship going flat out to complete the welding in the time available. The engineer said that if we did not finish before the tide came in again the sea would pour in and the ship would sink – or should that have been fail to float? I felt sure he was kidding, but it was a big incentive for us not to tarry.

221

I had not enjoyed the privilege of drinking fine cognac before, but the three or four glasses offered by the skipper before we went ashore again more than compensated for the rigours of the job. They gave us some money too, which was a bonus.

The district hospital called upon our engineering expertise frequently for their equipment breakdowns and to manufacture all sorts of medical gismos they needed.

We could actually be credited with some direct input into patient care as well. A young lad had poked his finger down through one of the holes in the metal strainer of a sink drain. They had to wreck the sink to free him and brought him to the hospital still with the heavy brass plate wedged immovably on his digit. On the phone the doctor said that their available options were to remove the finger, which was his line of business, or remove the plate, which was more in my line. They wanted to try my method first - which was to remove the plate.

The finger was a bit of a mess and swollen and they had given him a needle in the bum to reduce his level of pain and panic. Perhaps the sight of the hacksaw I pulled out of my toolbox might have been a contributing cause for that. I clamped a big engineer's vice to the surgery table and locked the brass plate in it. Slipping some thin metal strips between the foreign object and the finger, and giving the patient a reassuring pat on the head I went to work with the hacksaw - on the plate.

The grateful lad wanted to keep the chopped up sections of the sink strainer as a souvenir. Perhaps he has them still.

There was this unfortunate character that dived into the river in a depth of water insufficient to accommodate his head. This resulted in a broken neck as a consequence of this near-fatal and foolish act. I stood at the head of his hospital bed while the doctor explained their predicament. They wanted the patient out of there and on a plane to Perth as quickly as possible.

"You can see we have tied his feet to the bottom of the bed and that sling under his chin is attached to a rope which goes over a pulley at the top. On the end of the rope is a bag of sand. The weight of this is stretching his neck and keeping the damaged sections of the spine separated so there won't be any more damage there. If we stopped stretching him he could die or be a quadriplegic at best. Your job," said the doctor "Is to make a strong metal stirrup to go down over the top of his head. This we will fasten to his skull and then we'll attach the rope to it for the sandbag. This will allow us to transport him safely on the aircraft in traction."

You would have felt sorry for the poor bloke as I got my rules and callipers out of the toolbox and set about measuring his head. By the initial apprehension on his face it occurred to me that he thought I might have been measuring him for a casket.

The interesting part for me came when I asked the medico how he intended to attach my creation to his patient.

"That's easy," he said. "Those two big screws I asked you to put in the ends will go into the holes I bore in each side of his head so it won't all slip off."

When I returned to the hospital with the engineering masterpiece ready for its final fitting, it looked good, as I had painted it a nice green colour so it wouldn't appear too agricultural. As I was leaving I looked back to see the doctor with a black felt pen marking two crosses - one on each side of the patient's head where he was going to drill the holes for the screws. I didn't stay to see that part, suddenly remembering an urgent appointment elsewhere.

The hospital rang a few days later to say that 'Doctor Mac's Patented Stretching Machine' as they called it, had worked a treat and that the wearer was doing fine. I wondered if it was pressed into service again afterwards, but dismissed the idea, as it was after all tailor-made for a custom fit. Perhaps it rests in a display case somewhere in a museum of medical science. Come to think of it, it would not be out of place in a collection of relics from the Spanish Inquisition either.

A strange-looking shiny instrument, obviously from the hospital, was dropped off at the workshop one day with a note from the doctor. It read, -

"Can you make this work properly Ivan?"

It was a long narrow stainless steel tube with thumbholes in one end and a fine wire loop protruding from the other. The doctor cleared up the mystery by explaining that it was an instrument for performing tonsillectomies.

"You just slip the loop over the tonsil and pull on the wire. It's supposed to snip off the offending tonsil nice and cleanly but it doesn't."

I looked at the thing with a bit of a shudder and went to work. I set to and sharpened the cutting edge and returned it to the surgery.

I had already road tested the instrument on a stick of asparagus. Not being sure of the size of your regular tonsil, I took a punt and it did a pretty clean amputation.

The doctor and I sat one on each side of his surgery desk and tested the tonsil snipper on a range of vegetables from the hospital kitchen. We chopped up enough asparagus and celery and baby carrots between us to make a nice healthy salad. He seemed satisfied.

A report later confirmed that the next set of tonsils that was excised in the hospital with the snipper came off without fuss, just like asparagus in fact.

Norm, one of our drivers, greets a stockman while on a trip.

Because of the lack of any kind of regular public transport in Derby, the option for those who did not own a car was to ring for a taxi. I purchased a going business on leaving the Wapet employ as the oil search wound down and began in the new role of Taxi Proprietor. This enterprise was run operating three cars on a twenty four-hour call basis. We also had free telephones around the town and a PBX switchboard at the house. There were two-way radios in the cars and a base station in the house too. The cars purchased with the business were past their retirement date, so I bought two new Australian FC Holdens.

If you are a cabbie, particularly in a small town, you get to know a lot of people and also get to know a lot about this lot of people. Many of the passengers I found, like to unload on the cabbie, telling him many of their closest secrets in the certain knowledge that the information given remains confidential. Almost like a doctor and patient, or a priest and parishioner relationship, what was revealed in the taxi was sacrosanct and my lips were sealed. There were some surprises too, especially during the night hours with regard to who was going where and why and with whom. I could write another book about clandestine meetings and the like, but think that I should finish this one first.

Aborigines like to ride around in taxis and also they like to be seen riding around in taxis. Their pension day, which was Thursday, was always the busiest and there was a roaring trade for our services until their money ran out. Our Austin mini bus was the preferred ride as they could pack half a tribe into that and they split the fare. They would ask to be taken to Broome, to outlying stations and around the town mostly to see relatives and have a fine time. They were a happy people with a great sense of fun, who loved and cared for their kids and revered the elderly. They regarded taxi drivers as good blokes too.

The taxi service also catered for tourists visiting the area. With even the best of our roads being rough and rudimentary and affected by flooding in the wet season, there were few tourists from the south who would venture to bring their own cars so far north.

Our main source of tourist business was from the State Shipping Service. Their vessels carried cargo and passengers, stopping at the major ports where most of the people on board welcomed a day or two ashore and a look around. We had established standard tours that proved to be very popular. These took in some of the outlying cattle stations, Myall's Bore and the Prison Tree and some of the beauty spots in the ranges and along the rivers. One of the most interesting trips for the visitors was the run to Windjana Gorge. This part of the country is steeped in historical interest, as it was the haunt of the aboriginal outlaw known as Pigeon, who caused havoc among the early settlements there in the late 1800s.

As their bus driver and knowledgeable tour guide, I would tell the visitors the story of the outlaw Pigeon and his exploits as we bowled along and showed them the very country where he roamed. Mostly they were consumed with interest.

We were following the track along the Napier Ranges one day with a full complement of gawking passengers on the bus and I was expounding on my local know-how, when two figures appeared walking along the road ahead of us.

Frank was a tall glossy aborigine with a fine physique whom I had seen before doing odd jobs in town. As many of his fellows did, he would strip off the trappings of the white man's world down to an old pair of shorts, grab a handful of spears and go bush for a spell. The figure out in front of us was indeed Frank, striding along like the warrior he was, while trailing him further back was his gin Nancy. Hers was the traditional role of general carrier of everything and she was loaded down with all they could possibly need for a stretch out in the wild. I could see that she had their lunch with her too by the kangaroo leg sticking out of the sack on her back. There were no baby aborigines with her that I could see.

Overtaking Nancy in a cloud of dust I pulled the bus up alongside the warrior who loped over on his big bare horny feet and stood talking. I was congratulating myself on this great opportunity for the passengers to see the real thing – a Myall black in the setting of his traditional hunting grounds. I don't know if he had rehearsed it or not, but the tourists were delighted and gave a lot of oohs and ahs as he assumed the classic native bushman's posture. He stood on one leg, leaning on his long spear, with the other leg crooked and the foot on his knee. The tourists' cameras were going off in rattling volleys.

Deciding to cash in on this unique opportunity to impress my clients in the bus, I took it a bit further when I returned Frank's greeting.

"G'day Frank. You're a long way out – you goin' walkabout?"

Frank's beetling brows beetled a bit more and he seemed puzzled by my question. By this time Nancy had plodded up alongside to stand with her mate. She was giving us a shy smile with lashes downcast over her big brown eyes but saying nothing.

I repeated my question, "You goin' walkabout Frank?"

He was obviously still confused, shrugged his broad shoulders and turning to Nancy fired something off in tribal dialect. She confided something to him that made him throw back his head and roar with laughter.

"What's dis Walkabout Ivan?" He chuckled, "we are jus' goin' on Holiday!"

I wished the earth would swallow me then as the whole busload erupted with mirth. So much for the great tour guide – my balloon was burst.

In a similar vein I cannot help but repeat the one about the Bedouin and his concubine striding across the desert somewhere near Cairo. A white tourist stopped and remarked that the woman was walking ahead of him and wasn't this against the custom, as the proud warrior is usually in front. His reply was short and to the point. He simply said, -

"Landmines!"

Myall's Bore is about ten miles out of town and is reputed to have one of the longest cattle troughs in Australia. It earned this distinction in its role as the final staging point for drovers moving mobs of cattle from the inland down to the port for shipment. It also has the famous Prison Tree, which is a mighty boab with a hollow centre and with the front door an opening in the bark. It is told that police patrols returning with prisoners rounded up in the backcountry would camp there. As the name suggests, the felons were often secured inside the tree for the night.

Myalls Bore was also the place where you could find a local celebrity. Living in a bush camp close by was a real Kimberley character that could outride, outshoot, outdrink and beat any man in a fight as well. Her name was – in the local idiom, The Apache Kid, although I never did get to hear her real name. Her stamping ground there was known as Indian Territory and yes, this was a young woman. She was tall and tough and good-looking with the body of a thoroughbred. Her job at The Bore was to look after and water the mobs of cattle as they were passing through. She had the appearance and the skills to make you think she was born to ride a horse. Raised on a remote cattle station as the only daughter in a family of men, she had learned to look out for herself and to all accounts, did so admirably.

The Apache loved Bundaberg Rum and I knew that if I responded to a call for a taxi out to Indian Territory I was in for a session with her at her camp with a bottle of the preferred brew. The most common reason for her to call a cab was to have us bring some bales of hay for her mounts and any other needs from town. Though she was friendly to a fault and interesting company, I decided that it would not be prudent to upset her, particularly as her .44 Winchester was never far from reach.

During a local plane flight out of Derby one day, I noticed the Apache sitting in a seat in front of me. She always wore moleskins and a bush shirt and carried a big bag she had stitched up from a black and white spotted cowhide. I had accepted her offer of a drink to ease the boredom, from the bottle of Bundy she produced from her bag – to refuse might have been risky, when she spotted a target in the seat in front of hers. The light reflected off the shiniest bald head I had ever seen and it seemed that the Apache was winding up for a bit of devilment. He looked pretty uncomfortable as she leaned over and began chiacking her victim and then turned beet red as she pulled the bandanna from around her neck and started to polish the top of his head with it. He stood up and turned around to face his tormentor, took stock of the situation and promptly sat down again. The vigorous polishing was resumed and the cabin was in an uproar of laughter. She was afraid of no man.

It was a hot and sultry night when I was on night call and had just got home and into bed in the early a.m. I had one of the taxis, the newest Holden parked in the driveway in case I should need it. The unusual sound that woke me was that of a car engine being cranked with a weak battery out in the roadway. Looking into the street I could see my car with its headlights on, apparently all lined up for a quick getaway. The would-be thief was having trouble getting the engine started, which gave me a little time to get out to him. You could not blame me for feeling a bit upset as I was hot and tired and very grumpy and someone was trying to steal my livelihood. I was clad only in underpants and the gravel hurt my bare feet atrociously.

It flashed through my mind that if the engine did start he would be gone in a cloud of dust, so I resisted the impulse to run around the front of the car to the driver's door. Pulling the passenger door open, I dived across the seat and caught him a beauty just above the ear. That punch came all the way from our front veranda. Since the other side of his noggin had also struck the door pillar as a consequence, I reckoned he would be a bit disorientated, so I took the opportunity to pull the car keys from the ignition.

What to do now was the question. As things were becoming clearer – for me at least, I locked the passenger door and went back to the house for my pants. Pushing him across the seat, I barged in behind the wheel, started the car and headed for the police station. So there was no misunderstanding as to what was expected of my reluctant passenger, I pointed out that if he tried to abscond I would kill him. How grumpy is that?

Telephoning the police posed a bit of a problem, as I didn't want to give my would-be carjacker an opportunity to escape. Pulling up alongside a telephone box so close he couldn't open the car door gave me the opportunity to duck out and call the sergeant. I saw my bed again an hour later, having delivered my passenger into the waiting arms of the law and then slept the sleep of the just and righteous.

We were all together again in front of the magistrate on Monday morning. The villain was asked if he had anything to say before sentence was passed. Pointing an accusing finger at me with great indignation he shouted, -

"That bloke didn't have to hit me so hard!"

I could not help feeling a slight pang of remorse as our would-be thief got six months in the cooler and I assume a bad headache from the lump I could see just above his ear.

1 Delawar Street Derby - our family home there.

226

Aircraft Down in the Wet

The wet season in the summer of 1956 in the Kimberley region of Western Australia had set in with a malevolence of black storms with towering thunderheads and drenching tropical rain. Most of the dirt roads and tracks linking the outlying cattle stations and towns were flooded and impassable to wheeled traffic. The inland was turned into a quagmire with barriers of black mud and swollen watercourses, while the inland communities resigned themselves to months of enforced isolation. Travel across some of the land was virtually impossible.

The major river, the Fitzroy, usually bursts its banks in the summer wet season to spread in a swirling brown flood over hundreds of square miles of country. Rising water levels inundate the Langi Crossing, which was the connecting bridge on the main arterial road southwards. Would-be travellers can be seen waiting, blocked by the torrent of debris-filled water flowing down to empty into the sea. Woe betides those caught on the road between any two flooded rivers at this time.

Townspeople resigned themselves to the discomfort of the steamy heat while many of the station people just broached their stores of emergency supplies and prepared for the inevitable stretch of confinement and isolation.

Our business of oil drilling and exploration in the Kimberley was not made any easier by the onset of the wet seasons. As the foreman responsible for ensuring the availability of transport for supplying our drilling camps, I discovered that the Wet was to impose great difficulties on the task of servicing the remote sites with essential food, materials and machinery spares. Distances of up to three hundred miles had to be traversed by trucks and light vehicles from our base at the port of Derby over roads and tracks, which could best be described as rudimentary. Some of our access roads were simply trails bulldozed through the bush.

As a large company operating these resources, we tried to ensure that we did not interfere too much in the lives or business of the local people and the pastoralists, but rather strove to assist them in any way possible. This community spirit was tested on several occasions, but none more so than in the event of a call for help, especially if it was a dire emergency.

News travels fast in a small country town like ours - especially bad news. It was the 4th of February 1956 when an aircraft was reported missing while on a mercy flight to the interior. Grave fears were held for this Flying Doctor plane with five occupants, which was well overdue on its return from a remote outback station.

The police sergeant was on the phone with a note of urgency in his voice, describing the situation and asking for my Company's cooperation and help.

The patient, one of the children of the manager of the station, was seriously ill, apparently needing urgent hospital care. While extreme weather conditions were prevalent at that time of the year, a special mercy flight contingency was raised. Pieter van Emmerik, who was allotted the Flying Doctor pilot's duty in the region, had exceeded his allowable flying hours for the month. This meant that he had to contact his office and DCA in Perth for dispensation to fly this mercy mission. He had been trained as a fighter pilot and had a distinguished record as a flight instructor with the reputation of a skilled and careful airman. He headed out inland at the controls of an ancient Avro Anson aircraft VH-MMG, to pick up the sick child and bring it in to the Derby hospital, a distance of two hundred miles. Hardly a modern state-of-the-art machine, the construction of this plane incorporated a framework covered with doped fabric. Its two engines had to be started with the use of a hand crank handle, a job I had done more than once.

At the station, the weather was fine, and an immediate return flight seemed to be a safe proposition. While he was on the ground however the pilot picked up a message from Derby that there was a large and rapid build-up of severe electrical storm activity to the West and was advised that he should delay his departure until the next day. The record suggests that he was reluctant to take off after receiving this warning of the bad weather ahead and also, to compound the risky conditions, darkness was falling. He reluctantly gave in to the pleas of the parents of the sick child however and loaded his five passengers. They took off from the station strip to begin the return flight.

Radio contact with the plane was lost soon after however and the authorities feared the worst as the flight had literally disappeared without trace.

At the point when the fuel duration time of the plane would have expired, some time after the radio became silent it was decided to bring emergency responses into action. A search was to be mounted as quickly as possible, but with little immediate hope of finding the plane and its passengers, as darkness had descended and the intense lightning and heavy rainstorms continued to lash the backcountry. This was later recorded as one of the most severe wet seasons to strike the Kimberley in years.

I was asked to attend an emergency meeting of the police, representatives of the airline and members of other authorities to plan and carry out a search and rescue mission.

Few of the townspeople were untouched by the grim news, as almost all had some connections with the passengers on the missing aircraft. One of the young nurses from the district hospital had been sent as escort for the sick child. To provide company, another nurse, her close friend, volunteered to go too. The baby's father, who joined those on the plane for the return flight, together with the pilot, made five in all. We had almost nothing to indicate the location of the plane or the fate of its passengers.

The emergency committee decided initially on a strategy to fly the flight path of the missing plane, leaving at first light next morning with weather permitting. A Douglas DC3 freighter aircraft was selected, chosen because of its long range and the capacity to carry a fully equipped ground search party. This was all in the hope that there would be a need and even an opportunity to deploy one.

The police asked me to provide a small four-wheel drive vehicle together with a trailer that would fit in the belly of a DC3 and also to join the search team as an observer, mechanic and to drive the car. After some quick measurements I confirmed that a short wheelbase Landrover would indeed fit into the plane, but it would have to be stripped of mudguards, its top and other projections. I met the rest of the search party at the airport. Two policemen, a surveyor, two black trackers and I joined the aircrew there where the plane was fuelled and waiting. I saw that it was a DC3 Air Freighter, one of the trucks of the airline. We loaded a number of packsaddles, the Landrover and its trailer, together with rescue and survival and medical gear and rations. With some difficulty the big door was closed and we took off into the gloom of the rainstorm clouds heading eastwards.

Hope springs eternal they say, but the mood was gloomy in the plane as we flew out into the atrocious weather with extreme air turbulence and poor ground visibility. Compounding the difficult situation was the lack of information as to the location of a likely landing site for the missing plane. Another possibility put forward was that the un-pressurised aircraft might have strayed off course attempting to circle the worst of the weather, which they most certainly would have been experiencing. The crippled plane was down there below us somewhere in the thousands of square miles of green waterlogged landscape – but where? Its fuel range from the station as plotted on the map at the base gave us an enormous area to cover. Not even a valid radio signal or a Mayday call had been picked up. The search attempt seemed hopeless, but we had to do something.

All windows and the floor hatch had been removed from our aircraft before take-off. Each man took station as a lookout at these vantage points. As we gathered speed a wet gale whistled through the apertures and all eyes peered through binoculars sweeping the ground passing below. To avoid the low cloud we flew at almost treetop height, across the plains and through the gaps in the ranges. I doubt that this would have conformed to the normal civil aviation flying regulations, but apart from the grim task at hand, I found the whole experience to be very challenging and exciting though tiring and stressful.

Hour after hour we flew and we searched – to no avail.

The opinion of our aircrew was that given the near-impenetrable thunderstorms blocking his path to safety, the Anson pilot might have headed for the coastline. Smooth beaches and shallow water are apparently considered preferable alternative sites for a forced landing. To encounter a mass of rocks and tangled vegetation inland would surely have spelled disaster.

As time passed, we heard that several other aircraft were dedicated to the task of combing the vast area of the territory too.

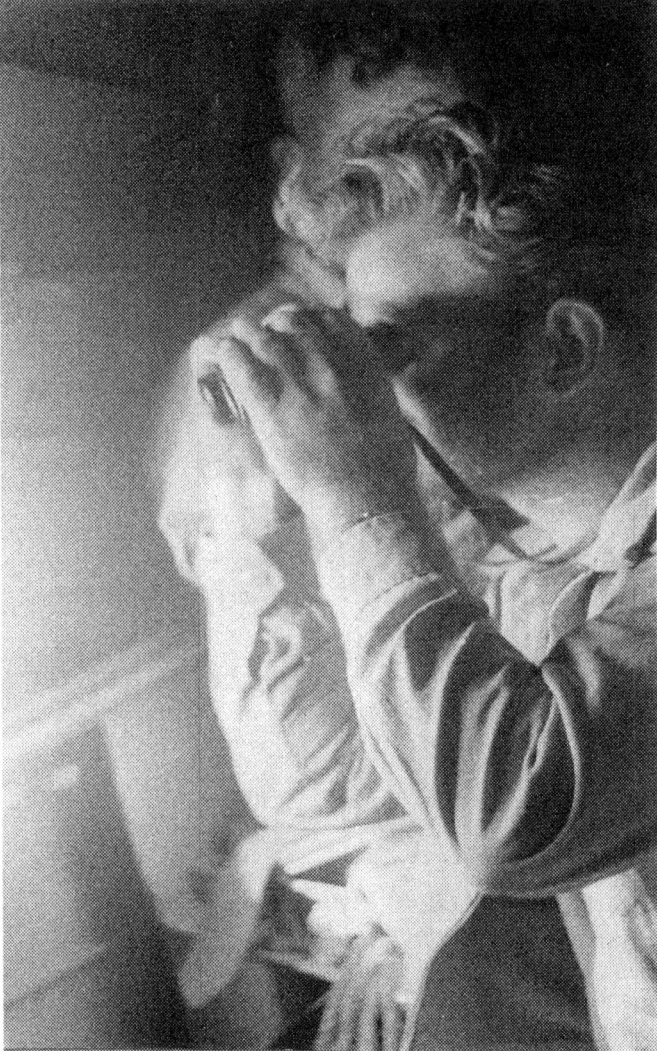

I spent hours spotting from the open windows of the search plane but failed to get results.

We returned to Derby after a long and tiring search, to refuel and then took off again immediately, this time to follow the coastline. Our destination was Darwin and we would fly that great distance with the intention of searching every inch of the way.

It is difficult to describe my feelings especially as the hours wore on. All of us were crouched at the open windows and hatches searching, searching with ever-dwindling hopes of sighting our target. The sheer vastness is so intimidating and discouraging with the dark green countryside flashing past intermittently obscured by the low cloud. In the attempt to recognise what it was we were looking for, I tried to form a mental picture of what I could expect to see. Would it be just a few pieces of wreckage perhaps? - Or the aircraft on the ground intact? My God I thought, it probably wouldn't even look like a plane anyway, so what chance do we have of stumbling across it? Another thought crossed my mind and that was, if on the slim possibility there were survivors down there, the chances of them lighting a signal fire in that drowned landscape would be almost nil.

Our pilot showed consummate skill as he nursed the big plane at just a few hundred feet altitude, banking and twisting to maintain a good view of the beaches on the tortuous coastline of bays and inlets and rocky headlands. Seabirds rose in clouds from the cliffs while the mangrove creeks and the mudflats were crisscrossed with the tracks of bolting crocodiles, stirred from their solitude by the thunder of the plane's engines.

As our quest dragged on there was still no sighting of the missing plane or its passengers or even the smallest sign down there that would have given us some hope.

While we were refuelling at Darwin it was reported that two remote radio stations in the outback had picked up faint broken carrier wave signals. The timing of these transmissions was about right for the period when our lost aircraft would have been in trouble. Crossed bearings from these points plotted on the wall map suggested a possible target area in the vicinity of the Gibb River Station. The weak nature of the signals could have been an indication that the transmissions were from a plane on the ground, possibly with failing battery power. Our hopes were buoyed once again and a course was set for the Gibb River homestead. This was well off the planned course of the Anson, but it had to be checked out. We were clutching at straws now. There was a landing strip there too where we could put down and unload all the gear, as it had been decided to fan out from the station on a ground search.

Our search party unloaded all the equipment at Gibb River station. That's me backing the Landrover out of the plane

Later that day we stood in the mud and watched as our DC3 freighter roared down the sodden station strip. With the props blasting back sheets of spray it lifted off to climb and disappear into the low cloud ceiling. The Landrover and all the gear had been unloaded and covered before the threatening clouds could dump more rain on us.

Fred Russ, the station owner greeted our party and soon organised a virtual tribe of aborigines to help us set up our base camp in the machinery shed behind the homestead. Our arrival had created tremendous excitement in the lonely station community, which had already battened down to sit out another dreary wet season. One of our black trackers suggested they should organise a corroboree with the station natives to mark our visit, but it had started to rain again and so what's the point in planning a corroboree if you can't have a decent fire anyway?

We were up and ready for action at first light. It was still raining - but heavier now. The packsaddles were cinched up onto the station donkey team. Fred remarked that if the donkeys couldn't get out there nothing could. I loaded the Landrover and its trailer to the hilt with gear and we straggled in single file out through the yard gate and into the bush. Without any clear plan to identify a likely search area or form any other strategy, we decided to pick the most negotiable track and go as far as possible looking for a hill from which we could scan a wider sweep of the country.

Driving the Landrover, I took the higher route along the ridges where there was some wheel traction to be found on the stony ground. Looking down I saw that the gullies between the ridges however were a morass of sucking black soil mud through which the donkey team was floundering under their heavy loads.

It was all over in a distance I estimated as just short of a couple of miles.

The donkeys bogged down to their bellies and refused to go further. My Landrover sank between the rocks with all wheels spinning at about half that distance.

Then the rain really set in with a vengeance.

Back in the shed that night we slept like the dead in sodden clothing with the steady roar of the Kimberley wet on the tin roof.

The grey dripping morning revealed a depressing sight. A swirling torrent of red water had invaded our building a foot deep and had carried away all that had been on the dirt floor. Clothing and boxes and personal gear bobbed around in the flood and were floating purposefully toward the far doorway.

I could make out that the station owner's figure was knee deep in water as I saw him standing in the doorway of the shed surveying the depressing scene.

"Well I'll be buggered," said Fred. "I've never seen this happen before as long as I can remember - and I was born here."

An adjacent creek had burst its banks. Why should we be so lucky?

It took ten long days for the rain to ease and the floods to subside. When the airstrip had dried out enough to take the weight of our DC3 again we saw it circle for a look and then land to pick up our bedraggled and dispirited search team and its gear. The whole exercise had been an abject failure for us. The lost aircraft had not yet been found and hopes of rescuing any survivors were all but gone. I am glad we didn't know it then, but it was shown later that we had been looking in the wrong part of that vast area.

Some weeks later, intensive air searches were called off and the wet season had moderated to the point where some land travel was again possible. Pilots were still on alert to look for any significant indicators on the ground as they made their routine flights. One of the airline pilots I knew very well told me the fascinating story of the events leading up to the final identification of the crash site.

He explained that it all took place as one of their aircraft was heading out on a scheduled flight inland. It was a DC3 freighter on its way to one of the larger cattle stations. An aboriginal helper was sitting in the plane, perched on the cargo with his keen eyes scanning the ground passing below, hoping to see some sign of wreckage. He yelled excitedly to the pilot that he had spotted something down there that did not belong in the pattern of the bush. He could not say what it was, although he was certain it didn't fit into the natural scheme of things.

Banking for a low pass back over the spot and searching the ground with binoculars, the crew finally zeroed in on what the spotter had seen that caused him such agitation.

They could make out what appeared to be a long ribbon of green fabric. It was trapped in the branches of an ironwood tree and was fluttering in the slight breeze. Another pass and a closer look also revealed what looked like scraps of aluminium embedded in the soggy ground.

The area was a creek bed or wash, which during the recent wet season had seen the passage of great volumes of floodwater. There were no clear signs of anything resembling the shape of an aircraft, but the aborigine insisted that the tiny movement he saw should not have been there. This was an oddity in his eyes that did not conform to nature.

As soon as it could be organised, a land party, guided to the area coordinates given on their map, reached the position indicated by the pilot who had called it in.

The ground party organised by the Derby police, who were the first to arrive at the crash site, had a difficult time reaching it. Driving first to Kimberley Downs station, they then went on to the abandoned Napier Downs homestead. From there they continued on foot towards the objective. The country was still awash from the heavy rains and they found it necessary to use an inflatable dinghy to cross the swollen creeks. After two days of hard slogging, they were finally standing on the spot, which appeared to be the impact site of the aircraft and also the grave of its occupants.

It was indeed the place where the ill-fated mercy flight had so abruptly and tragically ended.

Both engines were buried in craters in the ground and smashed. They saw that most of the main wing had been reduced to splinters of airframe and tattered fabric. A bauhinia tree close by had a vertical score in its bark, confirming that the crash had been from a direct dive into the ground.

Their report stated that the plane appeared to have struck at high speed, driven apparently by a vicious downdraft spawned in one of the violent thunderstorms. The entire aircraft and all its passengers had disintegrated on impact. The virtual explosion which destroyed the plane had scattered small debris far and wide. One of these remnants was the essential clue - the ribbon of green fabric of the kind used to line the inside of the cockpits of Avro Anson aircraft. Most other small movable remains had been swept away in the floods of that wet season. A major part of the aircraft and its contents were entombed in the earth.

The clue that led to the final discovery was that tiny signal marker impaled on the ironwood tree. It had been seen from a considerable height and from a fast-moving aircraft by the sharp eyes of one of our indigenous hunters of the bush.

The search party could find only small remnants, which were seen to be the only evidence of the existence of the human occupants. A few tattered scraps of clothing recovered there too were solemnly interred under the bauhinia tree.

This was the tree that bore the vertical scar in its bark - thus inscribed to serve as a mute epitaph and a symbol, which pointed the way to that lonely resting place in the bush.

My young wife Bev waited for me to return from the search for the missing aircraft.

My Friend Albert

There was a special character whose company I valued highly in Derby. He was very different to most of my other friends in the town but was a staunch and respected friend nevertheless. Albert was my senior by a few years and was a full-blood aborigine from one of the local Kimberley tribes.

The mission station was not far out of town and Albert lived there with his clan He carried the role in the community as one of the tribal elders. He was a wise old fellow with a calm, serene way about him and with a natural dignity that befitted his status among his people. With quite a striking appearance, he was well built in the way of many of the Kimberley aborigines and sported a shock of hair and an enormous grin showing perfect white teeth, which contrasted sharply with his coal black skin. He was usually seen wearing a stockman's trousers and shirt and an old battered Akubra hat.

We made frequent runs in the taxis out to the mission and I began to look forward to seeing him and enjoying his company. Albert's spoken English was good and he gradually opened up and began to relate stories of his life and also gave me glimpses into the world of his forbears. He would go back in time through his tales passed down to him, especially those of the years when the white man arrived and spread and occupied their tribal hunting grounds.

One of my greatest interests lay in travelling out into the backcountry on exploration trips and to hunt. In those years the places of historical interest and scenic beauty throughout the

Albert and I unload the ute to make camp near Tunnel Creek.

Kimberley were virtually untouched and unspoiled by the flood of tourism and all it brought with it, which came in more recent times.

Albert and I arrived at a mutual agreement. He would accompany me on trips out into the backcountry where he would show me some of his special secret places and relate the ancient stories of the free and nomadic life his people had led for thousands of years. In return he would have the pleasure of revisiting the land of his youth and we would hunt kangaroo for him to take back to his family.

Albert's grandfather had apparently been caught up in the revolt and the battles between the warlike native tribes and the white settlers. The pioneering pastoralists had established homesteads and run their flocks of sheep over much of the land between Derby and the Napier and Oscar ranges. A massed aboriginal force stopped the incursion of the whites at these natural barriers. They were organised and led by a native known to the whites as Pigeon. Although this part of Kimberley history is well known, Albert seemed keen to tell the story as it was passed down to him and to show me the sites where the pitched battles and killings had taken place. Albert was a little vague in his information as to which side his grandfather was aligned in the conflict, so I did not press him.

Windjana Gorge is the gateway through the jagged limestone ranges to the broad expanse of country beyond. In the wet season the flooded river churns between the high walls. In the drier winter months you can see the cliffs mirrored in limpid pools lined by shingle banks. Anyone remaining still and quiet there for a spell will be rewarded by the sighting of the shy Johnson River crocodiles coming to the surface for a look at you.

In my A model Ford which I favoured for these trips, Albert sat with me and began to talk about the man they called Pigeon. To most of the blacks he was a hero with magical powers who fought and eventually died for the cause in protecting their tribal lands against the white invasion. To the whites he was a dangerous criminal, a will-of-the-wisp who raided and killed indiscriminately and evaded capture for many years. We sat in the shade of a big rivergum sharing a billy of tea, as my friend talked of some of the dramas that were played out along these very ranges more than a hundred years ago.

Pigeon was the name given to the black man Jandamarra, as he was small and

My preferred transport was the Ford A safari wagon.
Absolutely reliable and tough you could seat four in the front.

quick. His people were the Banuba tribe. Born in the late nineteenth century he grew up on one of the sheep stations established by a white pastoralist. Adapting well to the European way of life he soon acquired great skills as a stockman, a shearer and a marksman with a rifle. He found himself rejected by his own kind and later also became alienated from the whites too when he went bush. Reports later filtered in of a band of aborigines who were spearing station stock and causing

235

trouble on some of the more remote properties. One of their members was identified as Pigeon and as a consequence he was declared an outlaw and became a wanted man.

In the year 1889, in league with another renegade named Ellamarra and a group of followers, he began to ravage the Kimberley countryside. Four thousand sheep were speared on Lilmaloora Station alone and after each raid the attackers simply melted away into the hills. All efforts to bring Pigeon to book were unsuccessful until he was finally cornered and captured with his lieutenant at Windjana Gorge. Ellamarra was returned to gaol to serve out a previous sentence from which he had absconded. Pigeon on the other hand, due to his skills in riding, tracking and in the use of firearms, was given the option of being drafted into the police force.

He proved to be invaluable in his new role as a member of police patrols scouring the country for renegade blacks and earned the trust of his employers. In one pitched battle Pigeon was paradoxically instrumental in saving the life of a white policeman in the party named Richardson.

While it would have normally been considered inadvisable to recruit an aborigine into the police force to assist in hunting down his own tribesmen Pigeon, due to his unique skills found himself in this situation. While on one patrol, a police party led by Constable Richardson and including Pigeon camped overnight at Lilmaloora Station. With them were sixteen black captives in chains and several of them turned out to be Pigeon's blood relatives from the Banuba people. A conflict of tribal loyalties then arose which Pigeon had to deal with.

During the night the prisoners accosted Pigeon and goaded him into turning against the white man. Swayed by their entreaties, he picked up a rifle and shot Richardson to death. This was the man whose life he had previously saved. He unshackled the prisoners and by morning all were gone back into the bush.

From that day Pigeon assumed the leadership of a large force of his fellow aborigines. They roamed the countryside raiding and looting and killing at will. Their new leader trained his force in the ways of the white man and in particular to be skilled in the use of firearms, which he knew so well. This made them a serious threat and they were soon attacking police parties and releasing the prisoners held by them. It was rumoured that this band was gaining in confidence and intended to sweep all white resistance before it and even go so far as eventually attacking the town of Derby on the coast.

Remembering their mistake in using a local black man to fight against his own people, the police recruited a number of black trackers from Queensland to help them. These men were sent to find Pigeon's forces and infiltrate their ranks with the idea of gaining intelligence and therefore weakening his position. Pigeon's suspicions were aroused at the arrival of these strangers and he was urged by his henchmen to kill them. Instead, he released them unharmed. Pigeon's mistake of course was that these men relayed vital information on the location, disposition and size of his force back to the waiting police. This then led to an attack by a contingent of thirty police in a fight known as the Battle of Windjana Gorge. Many of Pigeon's men were killed in this fight, while he himself, although hit three times escaped and disappeared again into his favourite hiding places in the security of the ranges.

Tunnel Creek is a unique feature in the Oscar Range where a river disappears underground and flows in the wet season beneath the outcrops to reappear on the other side. Pigeon used this place as his hideout and also the myriad of caves in the area for two years, until he had recovered from his wounds. He then ventured out and went on the rampage again.

Despite having learned that a number of his followers had been captured and hanged, he once again began to harass the whites, who had assumed that he was dead. They realised their mistake when a raid was made on Lilmaloora Station, which had Pigeon's unmistakable signature on it. He made off with a large quantity of food and some guns and ammunition as well. His next attack was launched at the Oscar Range station homestead where he left one of the stockmen dead.

He was flushed out by the avenging police once again and was brought down by one of their bullets. One of the pursuers, a white man named Blythe rode up, coming across him lying face down on the ground. He fired down at Pigeon with his revolver. With uncanny cunning Pigeon rolled over and fired back simultaneously at Blythe with his Winchester. The revolver shot had taken pigeon in the chest while his own fire had struck Blythe's gun hand, severing two of his fingers. Despite his wounds the fugitive took advantage of this temporary reprieve and sprang to his feet to escape yet again.

Pigeon's determination fuelled now by a bitter hatred, spurred him to attack yet another police camp for the purpose of freeing more prisoners who were being held there. Pigeon's gunfire appeared to come from many different directions as he kept changing position, tricking the police into thinking that a large force had surrounded them. They were however successful in calling up reinforcements from a nearby station. This party moved in behind Pigeon to cut off his retreat. He eluded this threat and headed again for his favourite refuge in Tunnel Creek where he had gone to ground before.

One of the black trackers soon picked up Pigeon's blood trail that led up a stony gully. Bailed up at last by his pursuer, he turned to fight and an exchange of fire ensued. In this final act of defiance Pigeon received the fatal bullet that ended his infamous career at last.

We drove along the track that skirted the Napier Range, while Albert pointed out to me the locations of the incidents where the hunt for Pigeon had unfolded. Lilmaloora Station, now a crumbling ruin, was tucked up close to the range. I tried to cast my mind back and imagine the events that led to the murder of Constable Richardson and which were the precursors for the dramas that followed.

A few miles further on towards Tunnel Creek Albert motioned me to stop and pointed out a cleft in a jumble of boulders at eh foot of the cliffs.

"Pigeon was killed up in there," he said simply.

Finally we reached Tunnel Creek, which had featured prominently in Albert's tale. Large boulders surrounded the entrance, where the dry watercourse dipped into the cool dark interior of the tunnel. Pools of fresh sweet water lay by banks of soft clean river sand continuing further on into the darkness. An ideal retreat, I mused, in which to get away from the troubles of the outside world or from the white man's police patrol intent on taking your life.

Albert asked me to stop at a point on the road close to the cliffs. I followed him as he strode along searching the rocky face for some sign or other. He showed that he was still very agile when he began to climb. By way of a series of narrow rock ledges we reached a point high above the plain when he paused.

The small cleft he showed me in the rock allowed me to squeeze my body in to view a cave that opened out to the size of an average room. I could barely make out in the dim light that there were artefacts in there and rock art on the walls. It was a burial cave in which had been secreted the bones of many of Albert's ancestors. Some were in coolamons - native wooden bowls, while others were laid out on the floor of the cave. Skulls stared with empty sockets from rock shelves and clefts. The walls were the canvases of the ancient artists. I realised that I was looking at the relics of a culture which had thrived there many thousands of years back in time, and I was completely overawed by it.

A feature of this cave that also spoke of its timelessness was revealed by a closer examination of the floor. A layer of brown dust about two inches thick coated the entire surface. The texture of it was that of the finest talcum powder. This fine dust I assumed was the detritus from the roof and walls of the cave, the result of decay of the rock over the centuries. Its smooth surface bore no marks. It appeared to have lain undisturbed and untrodden by man or beast since the day those bones were laid to rest. There was the sensation of being awarded a great privilege when Albert showed me that place. He had given me the gift of his trust, which ensured that I would never reveal the location of the cave with its ancient relics – and I never have. If ever I felt caught up in the ancient history of the land it was on that day. So well concealed was the cave up on the cliff face, that if I returned there tomorrow I doubt I would be able to find it.

While declining the offer to share their meal, I did accept the thanks of Albert's family for the kangaroo he took home to them that night.

On another memorable day I received an invitation to join him at a corroboree to be held by his own people at a remote bush clearing which was their tribal meeting place. He said I should be there about sundown.

The celebration of their tribal culture and the storytelling through music and dance commenced after dark in a large arena lit by two fires. The women and other members of the tribe sat in a big circle on the ground.

It was like a surreal scene from another world as I watched the naked dancers in the flickering firelight playing out the stories of their ancient culture and of the Dreamtime.

There was the portrayal of a kangaroo hunt. It commenced with the quarry, which moved slowly into the firelight bending gracefully to feed and then standing erect with all senses alert to detect danger. The movements of the dancer who was the kangaroo were such a faithful replication of the animal as seen in the wild, that I found the likeness to be uncanny. Two hunters with spears and woomeras at the ready appeared from the bush, and moved into the clearing, stalking their quarry in a stealthy circling movement. When the kangaroo looked around sensing danger, the hunters froze and as he lowered his head to feed they drew closer. The re-enactment of the animal taking fright, the spears flying and the kill were an exciting conclusion to this piece of bush theatre. Despite the realism, I believe this particular kangaroo survived to perform another day.

Anyone who has been lucky enough to witness the ritualistic dance of a pair of the tall native brolgas in the bush would not fail to be entranced by it. The two big birds circle and pirouette and bow deeply to each other in what appears to be a perfectly rehearsed bush ballet. The pas de deux usually culminates with simultaneous leaps high off the ground with broad wings extended. There is perhaps another pirouette in perfect unison and then it is over. A pair of aboriginal performers was transformed into two brolgas that night in a faultless performance, while a didgeridoo boomed a haunting accompaniment pierced by shrill bird cries.

Then two opposing parties of warriors charged into the arena and clashed in a fierce fight, no doubt the re-enactment of some battle long ago. The surging sea of naked black bodies glistened with sweat as the earth was pounded to dust and the intricate white designs of body paint stood out starkly through the haze of smoke. The confusion began to abate and take form as the combatants formed lines and a tribal dance began. The instrumental accompaniment increased its volume and rhythm as the cracking of music sticks and impact of boomerangs clashing one on the other joined the throb of the didgeridoo. Then as if a lightning bolt had struck, all sound and movement suddenly ceased with the simultaneous thud of twenty feet - leaving a silence made more intense by the crackling of the fires and the hush in the settling of the dust. This climactic crescendo of noise and action followed by dead silence is characteristic of many of the corroboree performances of the Kimberley aborigines that I had seen.

As with all well crafted performances there was a memorable finale. The didgeridoo began again to fill the air with its haunting baritone drone, while the circle of women took up a low chant to set the scene for the forthcoming pageant. Four tall black warriors entered and moved to centre circle. Each was crowned with an enormous and elaborate headgear. Their white body paint was startling in its intensity against glistening black skins and dark background of the surrounding bush. Above each dancer rose the tall simulated but unmistakable mast of a sailing ship. Secured at intervals up the masts were the yardarms, the extended arms of the dancer formed the lowest of these. Most remarkable were the simulations of white sails - intricate lacings of white thread filling the spaces between the spars. These tall ships rolled and dipped as a ship would do in the swells at sea and only a short stretch of the imagination would paint a mental picture of what the aborigines must have seen off the coast at the first coming of the white man. It is probably close to one hundred years too, since the last of the sailing ships plied the Australian coastline. Who knows that it might have been Matthew Flinders' ship or that of his French rival, which were spotted by the native inhabitants long ago and the sightings woven into the rich culture of the tribes? I will never forget the sounds and the images of that wonderful night of stories in dance with Albert and his people.

Together with my family, I finally left the Kimberley behind, taking away a store of fine memories. It was when I was later in a comparatively tame and civilised occupation in Perth that I scanned the morning newspaper one day. There, in a black and white photograph was the unmistakable face of Albert, with his characteristic hair and a moustache even whiter than before. The article that accompanied the picture, announced the coming of a troupe of aboriginal dancers from the Kimberley who had been flown down to perform in the Concert Hall in Perth. The dancers would present a traditional corroboree under the leadership of Albert whose photo was adjoining on that page. It would have given me great pleasure to say G'day again and shake his hand, but the opportunity did not present itself. And so far as paying fifty dollars to see the show, well, I could not see that their performance under artificial lighting up there on the white man's stage could in any way compare with that corroboree witnessed in the dark and mysterious wilds of the Kimberley. Sadly lacking would be the swirling dust and the smoke and the smell of sweat and the firelight flickering black dancing shadows on the backdrop of the bush.

From a mutual friend some time later I learned that Albert had died. While I still had some photographs of him of my own from some of our hunting trips, it seemed fitting that I should try to create a more memorable and lasting tribute to my old friend. Always keen on art, I was attending classes in oil painting and felt that as a tribute I could not have chosen a worthier subject than the old fellow himself. From the original small black and white photo in the newspaper and reinforced by the strong images that I had retained from our times together, I painted my masterpiece. It's not a masterpiece really of course, but my misgivings about it were put aside when an old friend from the Kimberley days walked into our house some years later and recognised Albert at once – so the likeness can't be too bad. The picture still hangs in pride of place on the wall at home and I seldom look at it without some recollections of the good times we shared.

It is my special wish that Albert has been well received out there in the Dreamtime world of the aborigine and that all the kangaroos there are big and fat and not too hard to catch.

The ruins of Lilmaloora Station, which featured in the story of Pigeon the outlaw. They can still be found by the ranges in the country where Pigeon roamed

Too Close for Comfort

A group of businessmen based in Derby secured a large portion of land about one hundred miles North in the Robinson River country. The focus of their interest was an abandoned cattle station. It was their intention to bring in cattle to restock the old station and try to return it to a viable state again.

This tract of the West Kimberley landscape is beautiful country. It has ranges of rugged hills, box tree flats and lush grass. The Robinson River flows westwards from the high country down to the sea and is the main watercourse for the area, with a network of tributaries and chains of billabongs. As a result of the heavy seasonal rains received in the wet, there is an abundance of feed and water for stock and for native and feral animals.

One still tropical night I lay in my swag acutely aware of the sounds of movement and activity all around me. Nights in the bush are never silent. The night hunters are out and about and nocturnal animals are foraging. In the distance a donkey brays, a wild calf bawls, or a fish flops to break the still surface of the nearby billabong and rock the waterlilies. At that time of the year with the fertile ground still sodden from the rains and the heat of the sun accelerating its prodigious growth, there is another more subtle but insistent sound if you care to listen for it. From the ground all around a persistent rustling provided the background to other night noises coming to my ears one steamy night and I was intrigued to find its origins. It was as though an army of huge ants was on the move.

My probing torch beam lit the Lilliputian jungle of the growing plants at ground level for a considerable distance, and then I saw the reason for the unusual and continuous noisy activity. As each blade of grass sprouted from its mother stem, reaching upwards to find the sun, it became entwined with other young blades, which were intent on doing this too in the profusion of green and damp. The vigorous shoots were trapped by others as they tried to grow longer and straighten - and then were released with a flick and a rustle like uncoiling springs. This intriguing process was played out continuously and could be clearly seen and heard over some distance in every direction. I got back to sleep then in the swag with images of the wonders of nature coursing through my brain.

The new owners of the station had much to do to prepare the old, rundown property to take their new stock. One difficulty that had to be overcome was the presence of a large number of wild cattle roaming the landscape. These were the descendants of those that had strayed or were left behind when the previous owners abandoned the property. A survey had shown in particular a large number of big scrub bulls, which had never seen a yard, or a stockman, or a dog or a branding iron in their lifetime. On our previous hunting trips in the region we had encountered these ferals and had come away with the lasting impression that they could be wild and dangerous if provoked.

The investors approached my hunting companions and me, to determine if we would be interested in culling the wild bulls for them. Hunting was a major outdoor activity for many of the men in the Kimberley and we welcomed the chance of a little excitement, while doing a good turn for the pastoralists as well.

The hunting went well and we were able to report a good tally after each trip. Firstly finding and then culling the bulls was not without its risks however as we discovered. One very large and angry scrubber that a hunter had failed to drop in his tracks with the first shot set his own sights on one of the Landrovers. His rush brought him headfirst up against the vehicle with his horns locked underneath the chassis. With a hunching of his massive shoulders he lifted the car and rolled it on its side. All was consternation and chaos for a while until the cranky attacker thankfully crashed off into the scrub. This gave everyone present a healthy respect for the strength and bad temper of scrub bulls in general and the big ones in particular.

One of my most memorable heart-stopping situations occurred in these serene and tranquil surroundings one night and with not a bull in sight.

Our hunting party had discovered an old bough shed with rusty corrugated iron on the roof and leafy tree branches for walls and with a primitive stone fireplace. Apparently once used as a stockman's mustering camp many years previously, it appeared to have survived the ravages of time, although the termites were well advanced in the process of consuming it. One obviously knowledgeable member of our party insisted on sleeping out under the stars, with the excuse that inside the hut was not a good place for restful slumber. I ignored his advice however and rolled out my swag on the dirt floor the first night and prepared to surrender myself to a sound sleep. It was not to be however because of the din. A sound like a barely muted stone crusher filled the air throughout the night and streams of wood powder soon covered me, coming down from the rafters above. Its source was traced to the termites on the night shift happily eating the thick bush timbers of the roof. It was not so obvious over the sounds of human occupation, but when stillness and quiet descended however, the noise seemed deafening, with a good sleep out of the question. I joined my friend out under the stars in self-defence, and had a few comments for him also about keeping me in the dark in relation to the disturbing sound effects back in the hut. He was hugely amused.

The outcamp we used on hunting trips to the old station. The billabong where I had the night encounter with the croc is close by.

The site for this camp had been well chosen, as it was nestling in shade at the foot of a jumble of rocky hills with a large billabong close by, decked with a profusion of colourful waterlilies and with its banks fringed with reeds and overhanging river gums.

Following a most satisfying feed of steaks and billy tea, I felt totally at peace with the world there one evening and decided to stroll down to the billabong.

242

Sitting on the soft grassy bank, propped against a tree, I lazed in relaxed contentment soaking up the quiet of the bush and feeling completely at ease in the surroundings. The thought of any danger in this tranquil spot could not have been further from my mind.

It was a glorious moonlit night and so balmy with the temperature at what passes for cool in the Kimberley after sundown. I followed with my eye the moon-track across the still water and marvelled at the way the silvery light picked out the glistening lily pads on the surface of the pond. There was not a ripple to mar the surface.

We found it to be advisable in that country to carry a loaded rifle when alone in the bush. This was not so much to try and project the great white hunter image, but to be prepared to deal with any chance encounter with a snake, a wild boar, an angry bull or a cranky croc.

With the rifle across my knees, I must have been dozing for an hour or so, when I woke with a shiver and began to get to my feet and head for camp and the warm swag.

The event was so sudden and so violent I got the fright of my life. With a lash of its powerful tail, a croc suddenly launched from the top of the bank behind me and in a fountain of spray, crashed back into the billabong almost at my feet. I had the horrible realisation at that instant that the croc had stealthily moved right up behind my tree and was preparing to grab his evening meal and I was on the menu. What really concerned me was the fact that as far as I was aware, all had been silent and still. Even while dozing, I felt sure of being alerted to any suspicious or furtive sounds nearby, particularly just so close by. There was not even the sign of a ripple on the moonlit surface of the pool either, to suggest that such a big beast had left the water and climbed the bank some distance away to steal silently up behind me.

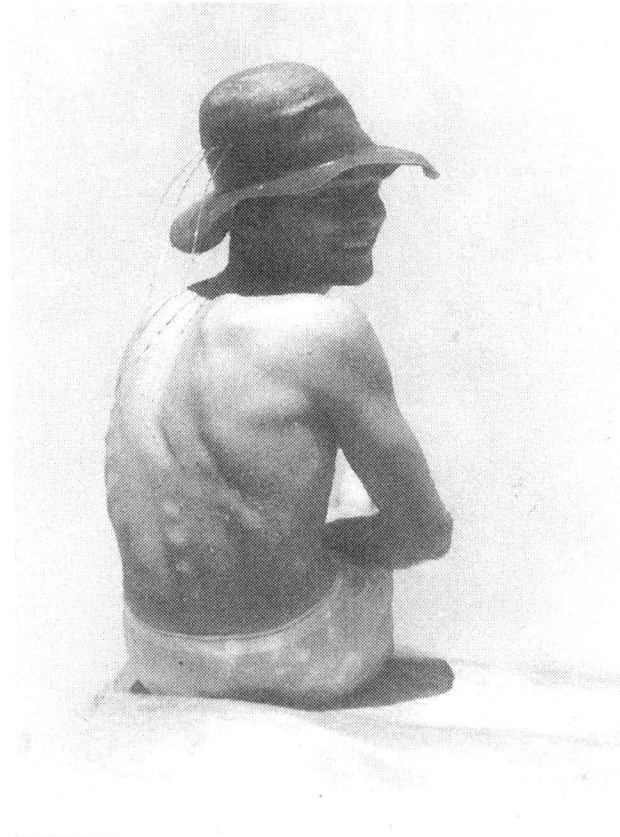

Friend Alan cools off in the Fitzroy River.

The account of my narrow escape from the croc, which was related to my so-called mates after the nerves had returned to a state of some normalcy, did not seem to raise much interest, much less sympathy. Admittedly they had had a few evening drinks by this time, but even so I felt that they could have shown a bit more concern at the fact that Ivan was almost taken by a big hungry predator. These friends made the point that I was well armed and therefore should have been able to deal with the would-be attacker and really had no reason for concern at all.

In retrospect, I shudder to think of him hooking those big yellow fangs into me and dragging me, locked in a death roll into his pool. There would not have been the opportunity to shoot him under those confused circumstances either. All that aside, he was much bigger and more powerful than I was and had a lot more teeth.

It's for the Kiddies

There had been considerable deliberation and debate in the Derby council chambers on the issue of raising enough money to build a new infant health centre in the town. Some funds were available through local government channels, but these apparently fell short of the amount required to build this much needed public facility. The shortfall it seemed would have to be raised within the local community.

Rather than pass the hat around or rattle tins on street corners, it was decided that a public event should be organized on a scale to entertain a large number of people. These would then hopefully feel so grateful that they would dig deep into their pockets for the total of the sum required.

What kind of event could be offered which would attract the requisite number of cash donors, pondered the councillors? What kind of entertainment could be offered that would have the desired pulling power?

The male staff from the one bank in Derby comprised a relatively young and enthusiastic group. Their boss, the manager was both keen and cooperative in community matters. I was invited to attend their meeting with the task of developing a suitable strategy to meet the council's needs to raise the money.

The annual celebration of our traditional Guy Fawkes Night was only weeks away and there appeared to be just sufficient time to organise a function on this date. The final agreed proposal was for a monster fireworks night and bonfire with fun and activities 'for the whole family'.

"So"- said the manager, as he addressed the meeting, "I reckon that a re-enactment of some historic event on a grand scale should be the theme incorporated in the program – what do you think?"

It was agreed that while a re-enactment of the gunpowder plot involving England's parliament house was considered a bit beyond our resources and as Guy Fawkes wasn't allowed to finish the job he started then anyway, there was a need to

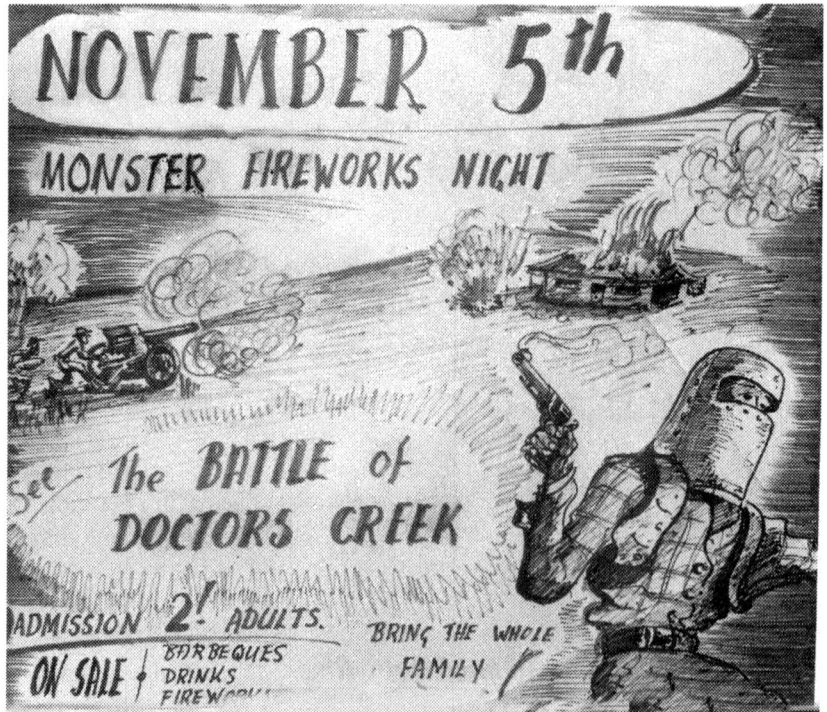

Having made up this poster to publicize our forthcoming event, we displayed it in places all over town.

244

find a suitable alternative theme for our show.

The bank manager continued.

"We need an historic event with an Australian flavour. What about the taking of Ned Kelly? He went down in a great show of fireworks – I suppose we could do that!"

So that was it! - Ned Kelly was going to be our star. We planned to perform it at one of the local landmarks and our public spectacle was to carry the title 'The Battle of Doctor's Creek.' We split up into committees and started work on the project.

Derby is surrounded on three sides by tidal flats. Much of the area of these is covered only rarely by the highest of sea tides. The closer, drier areas are smooth, baked hard and therefore were considered ideal for our Kelly Gang's battle scene. Doctor's Creek was deemed to be the ideal spot as it was within easy walking distance of the town. Posters were prepared and stuck on vacant wall space and in shop windows announcing the event and we got down to the business of preparing for the big night.

To ensure the effect of realism in our Ned Kelly shootout we needed to employ a lot of guns as the historical accounts suggested. This would hopefully provide an atmosphere of high drama with flashes, noise and lots of smoke. For me personally, the attendant effects of a ding-dong battle scene were always music to my ears. If there was a field gun or a cannon banging away as well I was even more impressed. The planning committee therefore decided, though with some reservations, to incorporate cannon in our show. With deference to the historical record, I don't think the police opposing the Kelly gang back in those times actually used cannon. It seemed a deft touch for us however to employ one which would speak with authority and finish off our Ned Kelly in a grand and spectacular style. Call it theatrical license.

There is an effect known to all those involved in oxy-acetylene welding which involves the mixing of the two gases in a confined space. It has been demonstrated - intentionally or not – that if this mixture of gases is ignited, an immediate deafening explosion ensues. This rather startling behaviour was therefore to be incorporated into the operation of our homemade big gun. Though frightening to some, the resultant effects are also known to be relatively harmless, which suited our purpose very well.

It was also revealed that several members of the bank staff had served in artillery units in the army. Here was a golden opportunity to rapidly work up a professional-looking gun crew including the roles of a gun sergeant, gun layer and loader and others comprising a team to attend our weapon with true military discipline.

The fabrication of our field piece was completed late on a Sunday afternoon with not much time left before the big event. With time running short it was decided to have an inaugural run out on the tidal flats that evening. We needed to determine if our piece was effective in turning in a suitably credible and noisy performance.

Repetitive practice drills had honed the gun crew to a fine edge and it was obvious that the training was welding them into a smart and very professional-looking unit. We had every confidence in the success of the whole scheme as a public spectacle, though still remaining aware that it depended on the overall results of the final performance.

This thing was seven feet long with a bore of six inches - a rival for a twenty five pounder - dead set!

The heavy gun carriage was moved to a position on the tidal flats about a hundred yards from the edge of town. With a thought for the townsfolk on this tranquil Sunday evening we quietly went about preparations for the first practice shot.

The night was so still. We could hear very faintly the hymn singing from a church nearby and we were all conscious of a pervading atmosphere of peace and quiet.

The long cannon barrel reflected the failing light as the crew, in low urgent voices, responded to commands from the gun sergeant. The gun layer swung the muzzle seawards away from the direction of the sleepy town. The loading of a substantial charge was completed and the crew was poised with the lanyard in the firer's hand.

The gun sergeant's order 'Fire' was drowned in the very respectable concussion. A big bright orange flash lit up the surrounding countryside and the report echoed back in waves of sound. For some strange reason the hymn singing from the church ceased.

We were naturally very pleased with the success of all our efforts – jubilant in fact. So much back slapping and banter was coming from the excited crew that we failed at first to notice a pair of headlight beams bouncing and weaving towards us from the direction of town.

"What the hell do you think you are doing?" yelled the headlights.

Our resident sergeant of the West Australian Police Force who had driven out to investigate did not sound very amused.

"You have stirred up the whole bloody town!" he bellowed.

"It won't be easy to talk ourselves out of this one," I thought uneasily as we tried to soothe the irate policeman.

It so happened that I had actually already taken the precaution of forewarning our police sergeant of the possibility of some stray noises of the explosive kind to be expected as we practised our routines. It seemed however that he considered that Sunday night was not the appropriate time – and said so!

Our plea was that these activities were all in a good charitable cause with an ultimate benefit to the town and especially to all the little children. Could he not therefore waive his policeman's sworn duty and overlook this minor breach of the peace. He finally reluctantly agreed, fortunately without banning future practice of this kind, but demanded that his rules would be followed in the event of future artillery drills i.e. never on a Sunday.

Anyway, the cannon was voted a resounding success - not in our judgement alone, but in the opinion of most of the townspeople we spoke to - with the possible exception of some parishioners who were in the church nearby. We discovered in fact that it was the best kind of publicity we could have wished for to advertise our project despite the fact that the ancient old church had some of the ceiling come down as a result of concussion from the cannon They said "well, if it's all for the benefit of the kiddies then we will have to forgive you and we'll let it pass."

The Battle of Doctor's Creek

It was dark in the bush and hushed. The outline of the small, rude shack was silhouetted against a star-spangled night sky. There was an air of tension that seemed to foretell the violent events soon to be played out there.

A group of men strung out in single file, warily threaded its way through the trees. Their dark figures were bulky and misshapen by the round iron helmets they wore, by the sacks on their shoulders and by the long-barrelled guns they carried. This was the infamous Kelly gang returning from a stagecoach bail-up, escaping to their safe remote sanctuary in the hills.

I apologise at this point to the original Ned Kelly - in his absence, upon whose exploits the theme of this battle was based.

As they approached the shack the leader rasped out an order. In response, two gang members split from the group and merged into the deeper shadows to stand guard. Their henchmen stacked the loot in the hut, emerging then to squat around a freshly lit campfire. All appeared hungry and tired. Their air of caution seemed to gradually relax and soon the firelight could be seen flickering on the circle of hunched figures as they ate in silence.

We crouched unmoving and quiet in the shadows observing the scene – so absorbed in the scenario that I was scarcely conscious of the odd whisper and occasional cough from those spectators close-by. Behind me in a large semicircle was a crowd of about a thousand townspeople waiting expectantly for some action to start.

Our well-rehearsed programme was running smoothly so far. The quiet voice on the public address system had alerted the crowd to the arrival of the Kellys. It asked for absolute silence lest the outlaw gang be alerted to their imminent danger.

"The gang is settling for the night," said the disembodied voice of the announcer.

"The Kellys are totally unaware that a police detachment has set a trap for them here and they have walked right into it. The police are now ready to go into action, so be very quiet or you may not hear the signal to attack."

Suspense in the crowd of spectators seemed to be reaching breaking point. At last from cover behind some fallen trees came the order.

"FIRE!"

The cannon spoke first, rending the darkness and the quiet with a bright flash and an ear-splitting bang! The gun crew worked feverishly to reload and train the weapon on the hut again.

The whole scene before the stunned crowd erupted in detonations, flashes of gunfire, roiling smoke and chaos.

Six vehicles, hitherto unseen, parked back from the perimeter of the bushrangers' encampment had their full-beam headlights switched on simultaneously. The hut and the frantic actions of the gang were picked out in stark relief as they came under withering gunfire from the tree line.

A frantic gang member in a cloud of steam and a shower of sparks and ash doused the campfire in front of the shack. Scrabbling bodies tangled and tumbled one over the other in their attempts to find cover. Some became wedged in the narrow doorway while others dived in through windows – anything to avoid the murderous volleys of the attacking police.

The watching crowd seemed electrified with the drama of the scene being played out in front of them.

The police detachment maintained their attack from the shadow of the trees. The Kelly gang soon responded with red flashes of answering fire from the darkened windows of the hideout. There were however no whine and thump of bullets in this battle. The guns all fired blank charges. The big cannon too which spoke with such authority was discharging a bag of flour rammed down the barrel for each shot. Realistic? - Most assuredly.

The opposing forces roared abuse at each other as the tide of battle moved to a crescendo. The police appeared to be gaining the upper hand in the ferocious combat.

The imposing figure of a man in colonial police uniform emerged from the shadow of the trees and moved cautiously towards the front of the hut. Hands held clear of his revolver holster; he stood silhouetted in the glare of the lights, casting a long black shadow on the wall of the hideout.

"Ned Kelly," called the officer, "You are surrounded and there is no escape!"

The sounds of the conflict died down to sporadic firing and then ceased. Only the faint cries of the 'wounded' could be heard as all watched in tense anticipation. Silence reined as a helmeted figure filled the doorway of the shack.

"Throw down your weapons and surrender now!" – The order was repeated.

Ned Kelly was a pitiful sight. One arm was hanging limp – shot through. His helmet showed signs of bullet strikes and his long bush coat bore spreading bloodstains.

"I'll never surrender my freedom to the traps – come and take me if you can!" he roared.

Defiant still, Ned fired his revolver wildly at his enemy and dissolved again into the darkness of the hut.

The attack was renewed with greater intensity. And it was obvious now that the terrible punishment was taking effect. Inside the hut a small flicker of flame appeared behind a shattered window. It blossomed, took hold, spread rapidly and soon became a raging inferno. Black figures of the gang members silhouetted in reddened openings, fought for the escape routes from the doomed building. Those who did escape by charging defiantly towards the waiting police soon staggered and fell in contorted shapes.

A woman in the crowd screamed hysterically, heightening the drama.

"They are not all out of there – there's still some inside in the fire!"

There was no sign of the remainder of those hapless bushrangers as the roof of the hideout collapsed in a swirling column of sparks. A thudding explosion of stored gunpowder sounded the death knell of the Kelly gang's retreat. The sidewall blew outwards and scattered flaming debris across the ground. Then all the lights went out, leaving only the blazing hut and glowing fragments and a pillar of smoke spiralling into the night sky.

It was over. The Kellys' reign of terror in the colony was at an end - the most notorious criminal of the time was finally defeated. The roads were now free from the threat of his marauding band, and the stagecoaches would run again in safety.

The crowd was still flushed with visions of the battle played out before their eyes that night on the edge of town. Ears still ringing from the din, they moved away from the arena and filtered, chattering, towards their homes. They were not about to be let off so easily.

Unseen, the vanquished bushrangers picked themselves up from the battle scene. Circling around in the dark they quickly moved to block the departing crowd.

"Bail up! – Stand and deliver! – Your money or your life!"

The townsfolk stopped, confused by the sudden threat from the menacing guns of the helmeted figures standing in their path. Not many of the crowd escaped the hold-up and under the threat of levelled pistols, dug deeper into their pockets and handed over most of their remaining cash to the blackened, beckoning hands.

The funds collected already that night were swelled further as the crowd, startled by the audacity of the unexpected toll, gave again and not unwillingly it seemed. No more resistance was encountered in this robbery than if they had been stagecoach passengers bailed up on a lonely country road.

All those who contributed their time and talents to the making of The Battle of Doctor's Creek, held on the night of November 5th 1961 had reason to feel satisfied with the outcome. A collection and count of the takings - both offered and enforced, exceeded our expectations by a generous margin. The planning and hard work had paid dividends and we were able to demonstrate just what a good community spirit in a small country town can achieve. Some maintained that the fun and camaraderie experienced was reward enough. Any fears we may have harboured that the Guy Fawkes celebration was destined to be a fizzog – a damp squib, were firmly dispelled as it all went off with a bang – most spectacularly so!

Out on the tidal flats the following morning my good friend John and I scuffed through the smouldering wreckage of the previous night's havoc.

"You cut a pretty impressive figure in that Malayan Security Police uniform last night John." - He was proud of that uniform.

"Thanks Ivan, but for a while there I thought I was back in the rubber plantations fighting the terrorists. There are a couple of points I wanted to clear up though," he asked.

"Yes - what?"

"The crowd was pretty worried about those bushrangers who didn't come out when the hideout burned down. What happened to them? Oh and what about the explosion that blew out the end out of the hut?"

I explained. "Well first of all we made it appear that those blokes were being fried in there to work up a bit of excitement. Actually they slipped out the back in the dark on the side away from the fight."

"And did you notice that we evacuated the whole area before the shack blew – it was all arranged. A plug of gelignite was set off by the fire behind the wall there right on cue – one of my mining mates donated that – everyone helped."

"Yeah," said John with a grin, "Nice touch that."

As a postscript to the story of the battle I have to tell you of an unexpected and violent anticlimax to our show.

Feeling pretty pleased with the success of our pageant and congratulating ourselves while moving away from the battleground - police, bushrangers and the rest were suddenly and devastatingly bowled over and left floundering in a sea of mud - this under the power of several high-pressure firewater jets directed from the dark fringes of the battlefield.

Our police party was not the only group hidden in the trees it seems, but also too, was a fire engine complete with fire crew - totally unbeknown to us, which were sneaked in under cover of darkness.

The term used for a debriefing after a fire incident is a 'wash-up' - very appropriate. The explanation for this treacherous act given by the fire chief was that he was only there to kill the hut fire and we just got in the way.

That excuse we thought was highly suspect and just didn't hold water.

Looking Over the Sights

With reference to the title of this offering, I have to plead guilty to the use of a little chicanery. While it may lead some readers into thinking it is about a guided tour of Paris or Rio de Janeiro, they will soon realise that the main topic is not tourism, but guns.

Warning - Should any reader happen to be a SNAP (sensitive new age person) and is likely to experience an ectopic heartbeat or break out in a nasty rash at the mention of these beastly things, I suggest that he or she skip this story.

So it is with gay (original meaning) abandon that I take you to a facet of my life, which has provided great interest and enjoyment and adventure for me since I was a boy. Dad and Grandpa told us many stories of the bush and their hunting adventures, and these fired my imagination as far back as I can remember. Guns were useful tools in those days as providers of food and for personal protection as indeed they had been for many centuries before.

Tucked up in bed at about the age of five or six, the early signs of my interest in guns were emerging. With my brother asleep in the other bed and the house quiet I would pull a little toy pistol from under my pillow together with a torch. Both were presents from a doting aunt and were among my most prized possessions. In the torch beam I projected the enlarged silhouette of the little gun on the ceiling. Turning it this way and that, I admired it until sleep took over or the batteries went flat – whichever came first. This was the beginning of a lifetime affair. Grandpa must have noticed my interest as he began to teach me gun lore and how to use and above all respect them. The lessons soon gave way to actual hunting trips taken with him or my father. We roamed the bush and the paddocks in search of game while the old man was coaching me and also admonished me if I made a mistake or was careless. I was beside myself with pride when I carried home my first rabbit for tea.

Having attended Sunday school and church until the age of fourteen, I am able to recall a reference in the bible relating to our gifts of being able to use other species of animals for food. In Genesis IX verse 2, God said to Noah, –

'Every beast of the earth and upon every fowl of the air, upon all that moveth upon the earth and upon all the fishes of the sea into your hand are they delivered. Every moving thing that liveth shall be meat for you, even as the green herb have I given you all things.'

With the greatest respect for the Scriptures, I reasoned that if it was good enough for Noah to be given permission to eat God's creatures, then it was good enough for me.

We watched the black and white cowboy films that were showing at the local pictures with youthful fascination. Our boyhood heroes were Tom Mix and Hopalong Cassidy. They played havoc with the lawless villains and downed hundreds of bloodthirsty Indians, all with a black powder Colt six-shooter or an antique Winchester, and at incredibly long ranges. Dad, who was very good at picking up on these things, asked me if I had ever seen one of my heroes in a film reload his gun. Conceding that I had not, I shrewdly added that neither did they seem to have a proper job or even go to the toilet.

My Dad asked us if we would like to see and handle a real gun. We were apparently so keen that he brought one home the next day.

In his job as a bank teller he carried the responsibility of not only handling large sums of cash, but also of protecting it as well. To this end the bank provided him with a loaded .45 calibre revolver, which was kept in his cash drawer. The purpose of the gun was to deter bank robbers. The word must have got around because he wasn't held up once. He also carried it in a holster in the street when he was on escort for big cash transfers to banks and payrolls for businesses around the city streets of Fremantle.

I must have been goggle-eyed when he buckled the gun belt and holster onto my small frame, so I could practice a lightning draw like Tom Mix did. Along with all this however came a stern and very detailed set of lessons about gun handling and safety. He was so serious that I sensed the gravity of it all and have never forgotten. The instruction and the lessons were soon to prove valuable - possibly a lifesaver in fact.

It all took place when I was about ten years old. A friend invited me into his house one day after school when his parents were absent. His father was a businessman who often kept cash at home and had a small loaded revolver in a desk drawer as many people did in those days for their protection. With typical juvenile bravado and ignorance – a dangerous combination, my chum picked up the gun and broke the cardinal rule of safety – he pointed it straight at me. I could see the live rounds in the cylinder of the revolver and realised I might soon be in serious trouble. Not too keen on becoming a statistic at such a tender age, I tried to recall the advice my dad had given me about what I should do when faced with this kind of situation.

Holding the wrist of my friend's gun hand I grabbed the top of the revolver. Dad had said that if you could stop the cylinder from turning and get the web of your thumb under the hammer it couldn't be fired. While there was no way of knowing whether or not he would have pulled the trigger, considering him to be a bit of a lair, I would not have put it past him. Either way, Dad's advice worked. I felt like strangling the stupid idiot.

My love of guns was further stimulated when I was sent to the farming town of Beverley to stay with an uncle who was the local doctor. 'Doc' Hodgson was an avid hunter and he owned a pair of fine English shotguns. The one that was his pride and joy had been ordered from a famous gun maker in London. It was a Boswell and was fitted to his own personal measurements the way a suit is fitted to a customer in Saville Row. When examining this piece I was fascinated by the superb craftsmanship in the oiled English walnut stock, the deep blue of the barrels, the precision of its parts and the beautiful scrolled pattern of its engraving.

Doc took me out on a hunting trip on a big duck swamp where my keen interest apparently made an impression on him. He handed me his gun, pointing out a passing mob of blackies. I was standing in water with a soft muddy bottom and it was my very first time at this. After closing the action I made two mistakes – I did not fire until the birds were directly overhead, and I pulled both double triggers together in my excitement, which discharged the two barrels simultaneously. The extra heavy recoil driving straight down through my twelve-year-old frame was pretty stunning and the shot missed the targets too. Doc could not contain his amusement at my expense and was almost choking when he said, -

"You fired the gun straight up Ivan and it was loaded with those big 2_" cartridges."

He was even more surprised when he broke the gun to eject two fired cases and realised there had been a double discharge as well. Still chuckling to himself he waded back to high ground and sat down to wait for me. The delay in catching up with my uncle at that time was as a result of not being able to move from my firing position. The hefty kick of the gun had driven my feet down into the mucky bottom and I was still anchored there waving my arms about trying to get free from the sucking slime. Doc had been laughing a lot before – now he was helpless. In some way that I did not fully understand, this incident seemed to cement the friendship between us and I was his devoted and enthusiastic hunting companion on every safari after that.

Doc tried to pull a fast one on me once when we were shooting rabbits together in a stubble paddock. I was carrying his other gun with pride and care and hoping for a shot when we reached a small wooded hill in the middle of the paddock. He directed me to skirt the hill to the right while he took the left side, with instructions to take a rabbit if I saw one and meet him around the other side. There was the muffled sound of a shotgun in the distance as I approached, so I asked him when I caught up if he had bagged anything. He said he was pretty embarrassed, as he had missed what he aimed at. This didn't ring true to me as he seldom failed to take anything that he wanted to bag.

"Hey there's one over there," he said pointing at a shape in the stubble about forty yards distant. "Shoot him quick or he's going to get away!"

It was a rabbit all right sitting up looking at us, but as I had been taught to positively identify a target before pulling the trigger, I hesitated – its ears were hanging over its eyes making it look funny.

"What are you waiting for?" urged my uncle.

Walking cautiously over to the quarry I soon stood alongside it and gave it a nudge it with my foot. My suspicions were confirmed - the rabbit's soul was departed. Even in my meagre experience I felt sure that dead rabbits were usually found lying down, but this one was sitting up. This, I could see was achieved with the aid of the stick propped under its chin, all the work of my funny uncle, who had taken the rabbit with the shot heard earlier and set the trap for me to fall into. It did make an amusing story at the dinner table that night however as we tucked into a tasty rabbit casserole with lashings of field mushrooms which we had collected on the paddock. These I remember were steaming with mouth-watering fragrance and swimming in hot butter.

The man and boy relationship was further strengthened it seemed when we sat down together on the wide veranda of their house one afternoon. He said that he had the utmost confidence in me with no uneasy feelings about my being at his back with a loaded shotgun out in the field. I was naturally pretty pleased at this vote of confidence but even more; it was a surprise when he said he wanted to reward me. He picked up a fitted gun case and handing it across said that it was mine to use as long as I wanted. The case contained an English classic side-by-side Greener double with 32-inch barrels, with a hammerless action and engraving too.

"You will have to have some ammunition," he added, and emptied two boxes of shells into a cartridge bag for me.

All that was required on my part was in the stipulation that I should look after the guns, keeping them cleaned and oiled at all times. This to me was a labour of love anyway and he must have known that it was too. The feelings that I experienced might be appreciated in recalling that I was only twelve years old at the time. This is five years short of the legal age at which a boy would normally be permitted to even pick up a gun in this present day.

Considerably later in life, our years living in the Kimberley region of the state offered many opportunities to hunt larger and somewhat more dangerous game than rabbits. Along with the challenges involved in stalking wild boars and feral bulls, there was the satisfaction of knowing that we were providing a service for the pastoralists who welcomed our assistance. Wild species were making a real mess and initiating erosion by ripping up some of the ecologically and agriculturally sensitive areas of the station and river country. It was alleged too that the feral bulls that got romantically involved with the station's heifers did nothing for the purity of the bloodlines.

Wild feral pigs were pretty numerous along the banks of the Fitzroy River. We discovered a large mob, which had made its rooting ground at a place called Skeleton Pool. History has it that a tribe of bush aborigines who lived along the river many years ago had contracted one of the white man's diseases soon after he had begun to push out into the territory. With no immunity to protect them and having never been exposed to the disease before, it assumed endemic proportions in their community and wiped out almost the whole tribe. The discovery of their remains led to the naming of the pool and was a sad testimony to their fate. We found many signs around this spot confirming that aboriginals had indeed lived there in the distant past.

When hunting feral pigs you often don't see them, especially in thick scrub country, until suddenly they literally explode from cover and scarper off in different directions to escape.

There was a party of us one day searching for porkers when we came across a whole litter of youngsters with an old sow. Those small pigs can really move fast on their stumpy legs and one of our party decided he was going to run one down and catch it. He reasoned that with Christmas coming up soon he would take one home and fatten it for a lovely pork roast. A brace of the miniature rockets zipped past me zigzagging through the thick grass. My friend Reg was hard on their heels with arms outstretched presumably hoping to grab the first one that lagged a bit. With head down and legs pumping he was actually gaining on his quarry. The picture of what happened next is still as clear as day. At top speed and as near as damnit to success, and with his concentration completely on a capture, he failed to notice the tree. I was watching the piglets, which had been running side-by-side suddenly break left and right in order to pass a stout sapling in their path. They saw it but Reg did not. He fetched up cannoning into the tree with his skull and fell in a moaning heap on the ground. Someone dryly remarked that it was fortunate that he had impacted the tree with his head, as he might have otherwise done himself an injury. That little bit of bush comedy made our day. It was my good luck to catch one of the little blokes myself however and take him home to my kids in a bag.

As youngsters do, my lot took a great fancy to the plump little black and white speedster and he became a house pet. They called him Hamlet. I was still struggling with the dilemma of either dispatching Hamlet to become the centrepiece of the Christmas dinner table, or listening to the pleadings and lamentations of the kids and adopting him as an official porcine member of the family. It was about the time that I was admiring his fat and rounded body, which I could see so clearly in my imagination surrounded by golden roast potatoes and gravy, when he disappeared from the yard. While I cannot normally track a camel in a sandpit, there were unmistakable imprints to be seen of his little trotters outside our fence, heading up the road in the sandy verge. At the front gate of the home of a friend of mine who lived up the road a bit, Hamlet's tracks suddenly disappeared and did not continue anywhere further on that I could see. Enquiries at this house drew blank looks and protestations that no pigs had been sighted, but with the guarded reassurance that if one turned up they would let me know immediately. There are no prizes for guessing what that family enjoyed for Christmas dinner, as it was confirmed later by one of my spies, that which I suspected all along. Our neighbourhood relations soured a bit after this. As any of my friends can testify, I will have no truck with pig nappers.

The guardian of Skeleton Pool near the river was a massive wild boar, which we guessed was the dominant male of that mob. He was as cunning as he was big however, and while we had spotted him several times he was able to disappear, evading us in the maze of watercourses and billabongs in the area. Our new strategy this time was to come in higher upstream and follow the bank of the river down to where he would most likely be foraging. We reasoned that this should put us between him and the river, in line with his usual escape route, thereby cutting him off at the pass as it were.

The plan worked well. The boar broke cover about fifty yards away and came charging down the hill, headed for the sanctuary of the river while we waited in his path. I'm sure anyone in our situation would have been excused for thinking that they were the subjects of an attack, with a good story to tell later of being charged by a wild boar. We accepted that he was only heading for home and we just happened to have put ourselves in his way. I had seen pigs at full throttle before but this one amazed me. Not only did he appear to be travelling at a faster rate than I could have possibly run, but he was hurdling as well – an Olympic pig no less. Large tussocks of spinifex which were in his line of travel and which any other self-respecting animal would have skirted around, he bounded clean over with amazing agility. Being slightly ahead of Lionel, my hunting companion, I looked back to see how he was coping with the situation and froze in my tracks.

He had hinted to me earlier that he wanted to inject a bit more excitement into this hunt and decided that he was leaving his rifle in the car and was going to try and bag a pig with his pistol. Wyatt Earp would have been proud of him as he started banging away as the pig got within range. Bulldozing down the hill in a shower of dust and debris, snorting and squealing all the while, I could see the pig was taking hits. While the .45 automatic is a lot of gun - for a pistol, it proved in this situation to be totally inadequate for the task. Its big round-nosed slugs were literally bouncing off the thick frontal armour and mud-caked hide of the juggernaut. While he was not one to panic unnecessarily, Lionel had emptied the magazine of the gun with little obvious effect and I could see his head swivelling around searching for a handy tree to climb. With my heavy hunting rifle I was pretty sure I could bring things to a conclusion and save my mate from further embarrassment, but I hesitated for a few seconds. I still ponder at this reluctance to act immediately and have done some soul-searching as to why I would have allowed the boar to get so much closer to my mate before taking action. It was hard for me to accept that in some perverse way I was making him pay for his foolishness and was teaching him a lesson he wasn't likely to forget – I still don't know the answer to that one. My solid hit from the rifle, behind the shoulder of the boar, put him down decisively in a big cloud of dust. His snout ploughed a furrow in the dirt to within a metre of Lionel's boots.

Wearing a sickly grin and waving the empty gun he simply said, "I owe you one."

There is a tragic sequel to this story of the hunting safaris with my dear friend. Just a few weeks after this last bit of drama, word came through that Lionel had died. His demise was the result of an accident involving strychnine poison while he was working on one of the stations nearby. He had in all innocence eaten some biscuits contaminated with the stuff - blown by the wind from a spilled packet on a table. He survived less than an hour afterwards. It was a great blow to my family and me and to all those who knew him. Not only did I suffer the pain of the loss of such a fine friend and companion, but also felt the sadness of knowing that he would never be able to pay back the one that he said he owed me. It could be a hard life in the bush in what was sometimes a harsh and unforgiving country.

Further on and in later years I made a devoted friend in Indonesia. He was the local policeman in a nearby village in the jungle on the island of Sulawesi. He really looked the part in his dark green uniform and cap and white webbing, and with a big .38 Special revolver on his hip. He had a problem however, which he revealed secretly in the disclosure to me that his gun 'was broken.' After much cajoling I was able to examine the weapon and could see it was badly rusted and hadn't fired a shot for years. He seemed reassured when I told him that I was in fact a gunsmith back home in Australia and so should be able to get him out of his difficulties. Not only was I able to repair his gun for him, but also gave him some tips on how to use it effectively. This was mainly directed at his ability to loose off shots without causing injury to innocent bystanders. Previously, the safest place to stand would have been in front of him. You should be able to find reference to this bit of drama in the Indonesian stories.

After moving down from the Kimberley to settle in Perth with my family, there was still the urge to burn powder and keep my eye in as it were. Over a number of years I took up the sport again in different clubs and various disciplines. While this was a bit tame and nowhere near as stimulating as hunting in the bush in the North, I resigned myself to be content with drilling holes in paper. At least I had never heard of a club shooter at risk by being charged by a wounded target!

My love of shooting actually helped me with the difficult task of rehabilitation, which I had to undertake after losing my eye. There were no counselling services freely available for help and guidance then and it was a matter of coping as best I could when returning to a normal life. The advice given me by my surgeon was most valuable however, as he basically said that I should get out there and do all the things I had done before and cope with the inevitable embarrassing situations, until many of the skills I had lost were retrieved. That included the shooting sports I had enjoyed.

As a boy at school, it was normal to throw with my left hand and so bowled as a molly duker when playing cricket. From the first day in picking up a rifle I mounted it left handed and continued to do so thereafter. Using my left eye for sighting too for so many years meant that the habit was deeply ingrained and now that I didn't have one, was sure it would be almost impossible to change. Recounting what the surgeon had told me, I was determined that the switch to being a good right-handed shot would be one of my goals.

There was a lot of practice involved to remaster my shooting but there was a day when I enjoyed the sweet taste of success. An old friend invited me to an Olympic pistol match at a meeting of his club in Perth. He was doing pretty well in competition himself and to improve his chances even further, had bought a fine Swiss Hamerli free pistol which he said was a steal at a couple of thousand dollars. The big boost for me on that occasion was achieving the top score in the match. The downside for him was in my bettering his performance using his pistol. That was the first invitation that was offered to shoot with him and perhaps not surprisingly, was the last. This and other minor successes in so many other ways too, contributed to me getting back on the road to a normal level of self-esteem and confidence.

After retiring from a life of travel and activity I still looked for involvement in some worthwhile pursuit. It was a temptation to join a special group of authorised sporting shooters. These experienced hunters collaborated with government agencies in the control of feral animals, mainly in areas where native flora and fauna were being threatened by a proliferation of a range of exotic species. An example of their valuable service to the community at no cost to the taxpayer can be seen in the South Australian example. A large community of feral goats had lived in the ranges there, and perfectly adapted to the environment had exploded in numbers. As goats will eat almost anything, they attacked the fragile native flora and had it stripped back to a brown stony wilderness in a short space of time. The sporting shooters systematically and efficiently hunted the feral pests with great success, and a recent report from there revealed that after long years the hills were showing an even green growth of native plants again.

Entry as a member of this group of hunters had very strict requirements. An applicant has to demonstrate skills in bushcraft and navigation and also demonstrate the desirable attitudes of an ethical hunter. A good level of firearm knowledge and safety performance is mandatory of course and almost without saying you had to be a good shot. I was pleased to be coopted into writing some of the training manuals for this organisation, and also carried out some coaching and assessments of applicants at the shooting ranges.

Some people have suggested to me that the term 'ethical hunter' is a conflict in terms - an oxymoron even, but I would like to give an example to explain my contention that it is not necessarily so.

Take a common Australian countryside scenario - a wild rabbit is nibbling on the farmer's young emerging wheat crop in the morning sun.

Hunting wild pigs in thick bush is exciting stuff

An ethical hunter however is stalking this particular bunny in the wheat crop. A few seconds later without fear or fright, our rabbit is stretched out dead in the paddock. The hunter, using an accurate rifle with a telescopic sight, together with the required skills, has selected and taken his quarry cleanly and without pain or fuss. Did you know that when game is taken in this efficient and humane way, it does not even hear the sound of the report, as the bullet travels faster than the sound does? This then, is ethical hunting.

As many have experienced, a particular sport or other interest can claim one's attention at an early age for no particular reason and remain with us for most of the years of our lives. So it was with me. My father and his father before him shared this same interest, so perhaps it is a hereditary thing. It has taught me many lessons from a young age, in discipline and responsibility and has gained me some respect I am sure, in the company of like-minded people. I have enjoyed a life of sport and recreation with guns, and all in complete safety. The sport of hunting has given me an abiding love of the bush, an understanding and respect for its wildlife and also put many tasty meals on the table. It is maintained that as men have been hunters since time immemorial in order to survive, it follows that for some the desire to hunt is in the genes. That is one part of the nature of man that cannot be extinguished in a few years by government legislation.

Mind you, everybody is different.

Episode Ten

From Oil to Iron

It seems that there has been a lot of space devoted to the accounts of our life in Derby. The main reason of course is the span of time we lived there – long enough to produce our family in fact. We spent almost nine years in that tropical climate. The other reason for my preoccupation with that phase was the number and variety of events that occurred and the remarkable lifestyle we led. It was so different to the alternative, which was to exist in the relative tranquillity and boredom of a quiet city suburb.

Looking at the home front, I took Beverley to the Kimberley as a blushing bride and when we returned it was as a family of six - no evening television. We had Peter, Lise, Richard and Gregory in tow. For a city girl newly arrived from her home in Perth, Bev encountered much to test her spirit. She had babies to look after, a house to run and she helped me an immeasurable amount when I was running the private businesses. This was under the extremely trying conditions of heat and high humidity and the lack of many of the amenities that go towards making life easier for a wife and mother. I am deeply indebted to her for all this.

We found ourselves one memorable day with the whole family and all our worldly goods packed into our utility truck travelling south. The injury I suffered resulting in the loss of my eye had had serious consequences for all of us. I was forced to leave the Kimberley as an emergency medical case and the ensuing long absence from the businesses was the last straw. We knew it was time to pack it all in and go.

There were many unique adventures and experiences and challenges in those years however and I would not have missed them for quids as they say. It seemed that the search for oil – for me at least was over.

Our mode of transport on the trip to Perth was a well worn Holden EK utility. The businesses were sold when we left, along with just about everything else. I bought this car, loaded up for the trip and then we hit the road.

The kids had become used to riding in the back of our vehicles when we made short trips around the country and they always enjoyed it. This was not to be a short trip however as the road ahead stretched for 2,300 kilometres. Neither was it all to be in the warm tropics, as we knew that the further south we travelled, the colder it would be.

Deciding to waste as little time as possible on the trip I continued to push on with only short naps at refuelling stops on the way. The ute had seen better days and among other things lacked a side window. This was not a bother in the Kimberley but became one by the time we reached Geraldton. Travelling in the night on this leg, the blast of cold air whistling into the cab was hard to put up with. At the back of a store in the town I found an empty beer carton. Flattened and cut to fit the window space it worked quite well in relieving the polar conditions. Mind you, without the desirable feature of being transparent like your average window, vision was a bit limited. That was a compromise I had to live with.

The next problem manifesting itself as we travelled further south was that the kids who were bedded down in the back were beginning to turn blue with cold. The EK has a fairly roomy cab, which together with the long bench seat was in our favour. After prising the kids out of the back we proceeded to pack the six of us together in the front of the car. Perhaps due to the fact that children will put up with a lot providing they are warm, we completed our human jigsaw. Bev, Peter, Lise and I occupied the bench seat. This left the two young boys to fit in somewhere. We stretched little Greg out on the shelf along the back of the seat under the window. I knew he would not roll off as I could feel his feet on the back of my neck. Richard solved his own accommodation problem. After we had dragged him inside and shut the door with difficulty, he just slid down into the foot well under the dash. There he curled up like a contented puppy and we set off again. They were all asleep in minutes.

So it came to pass that the great Macmillan family trek was coming to an end as we rolled across the Perth Causeway and on around Riverside Drive on this bright sunny morning.

There is a phenomenon you may have experienced after driving a long distance on empty roads at high speed. We had been travelling over a span of forty-eight hours with my foot virtually on the carburettor as they say, most of the time. Feeling pretty whacked I yearned for a soft bed and it was not surprising that a policeman pulled us up for a chat on the run around the Swan river. It apparently had something to do with the fact that I was still driving at my accustomed speed and was exceeding the city limits by a considerable margin. He was a nice bloke - an understanding bloke. This was in the era when they could be like that. He walked around and ran his eye over our dusty vehicle and must have taken note of the Kimberley licence plates. Back at the driver's door he peered over the beer carton window to count the number of tired and travel-stained faces looking up at him. No one had to tell him that we had been on the road for three thousand kilometres or a bit further north than the nearest suburb. He smiled understandingly and giving me a mild reprimand and graciously waved us on. The employment pages in the newspapers were listing offers of well-paid jobs in the new iron ore mining ventures. Big things were happening up in the Pilbara with the involvement of some very big players. Hamersley Iron was recruiting for an operation at Mt. Tom Price. They proposed to take the top off a mountain to get to the ore deposit, build a rail line to the coast and establish a company town there too. This looked like interesting stuff to me so I said goodbye to suburbia and headed north again.

It was obvious that it really was the pioneering phase of the project when I was shown my accommodation on arrival. Mine was tent number seven. Many units of massive earthmoving and mining equipment soon arrived, construction teams set to work and the miners attacked the mountain. There were many thousands of tons of overburden to be removed before mining could start in earnest on the iron ore deposits. It was a happy day when my accommodation soon changed to more permanent quarters with a door and windows and a roof.

My role was that of Supervisor Heavy Equipment Maintenance. They built an enormous workshop to accommodate the big gear, lined up twelve 120-ton ore trucks together with a hundred or so other pieces of heavy equipment and said go to it.

From that point on I was very busy, very hot and very dusty. There was a staff of 120 personnel attached to the workshops, and I set about planning and organising and scheduling to bring the section to a reasonably efficient level.

Many of my staff were immigrants. While they were carefully selected, there emerged some problems among them. Working side by side in my area was members of those nations that were long standing enemies back in their European homelands. Two men in particular began sniping at each other and although I had them in the office and told them to cool it, the animosity continued. One was a Serb and the other a Bosnian and we are all aware of what happened over there in later years.

There was a noisy disturbance in the workshop by the materials store one morning so I went out to investigate. One of these two combatants had armed himself with a piece of four-by-two timber and looked ready to brain the other with it. While it was never my thing to go around getting involved in fights unnecessarily, I knew it was time to do something. Holding back for a bit at the sight of the weapon, I must have reassured myself that it would be OK as I was wearing my safety helmet. That was a bad call, for while wading in to try and separate the two, the four-by-two completed its arc and caught me on the back of the neck under the helmet. In a brilliant burst of fireworks I found myself pitched into a steel cupboard. The fight stopped as I emerged groggily with the air now full of apologies. They must have realised that there was no future in trying to kill their boss. There was little sense in chatting about it so I just went and wrote out two dismissal notices. Fortunately for me there was no real harm done except for a nagging headache that somebody had for a while.

As most of my prior field experience had been in the oil exploration business I found that large-scale hard rock mining was a bit of a shock to the system. The infernal din and the dust began to manifest themselves as the rock drills started work on sinking the 13" blast holes. These are required for the explosives to carve away the site to form an immense pit. It was awesome to see the effect when a row of shots was fired simultaneously. When witnessing for the first time, a shoulder of the mountain heave and then collapse as thousands of tons of shot rock in a towering cloud of dust, I was mightily impressed. You get accustomed to this display of brute force after a

I inspect a Marion mining shovel in the pit at Tom Price

while.

Big Caterpillar dozers then moved in to heap the ore on the pit floor while the mining shovels scooped it up and loaded the ore trucks for the run down the haul road to the railhead. One of my responsibilities was to continually monitor the machines working in the pit, so there was a lot of time spent in the dust and the noise and the diesel fumes. In the course of the work I was continuously talking to machine operators and checking for faults that we would have to deal with later down in the workshops.

One of my scheduled tasks was the daily visual inspection of the massive buckets on the mining shovels. These take a terrible beating from the abrasive iron ore and I had to examine them carefully to try and detect any excessive wear or flaws in the steel that would require repairs. My written reports were picked up by the nightshift teams who carried out the work in preparation for the next day's continuing operations.

We used to get a lot of rubbernecking visitors in the earlier stages. These had to be escorted around the workings and I was allocated a newspaper reporter one day. He was to accompany me on one of my bucket inspections.

The method I usually employed was to stand near the edge at the top of one of the forty-foot high benches and signal the operator of the mining shovel working below. He knew the routine and would lift and slew the forty-ton bucket up and over my head like a tame dinosaur. He then landed the bucket with a mighty crash about a metre in front of me. This allowed me to climb all over it to make the examination.

This routine was all explained to my reporter companion as he stood next to me so he would be aware of what was coming. The engine exhausts bellowed, the bucket lifted and swung and then crashed down precisely in front of me in a cloud of dust as planned. That operator really had flair - I knew him well and trusted his skill.

My comments to the newshound at that point were not getting a response so I turned and found him gone. The poor bloke had back-pedalled rapidly and was now standing fifty yards away. His body language said it all. To be fair I had to concede that with possibly his greatest danger to date being that of dodging Morris Minors in St. George's terrace, he had cause to be a bit jumpy.

The home leave arrangements were none too generous for staff in the in the early stages of construction jobs in the north. We were committed to working a seven-day week for twenty-eight days straight. Our leave allowance after that was four days at home and then back to work. It was hardly surprising that we were always very eager to fly down to Perth to be with our families.

My day had arrived. I stood on the baking airstrip, with bag in hand together with a huddle of other hopefuls waiting for the Fokker to arrive and take us out of there. Time was well past the ETA as we searched the horizon for the plane that wasn't going to come that day. We got the word the damn thing had broken down in Port Hedland. A dozen very angry and disgruntled men straggled back to the camp and into the bar where they would try and drown their disappointment.

There was a five-seater Cessna parked on the strip. The word was that it was going to fly down to Perth later in the day. Finding its pilot in the bar, I bought him a lemon squash and asked him for a lift to Perth.

"No can do" was his reply, in what I assumed was pilot lingo. "I've got all seats filled to Hedland with a few changes there and then on to Perth."

I knew him to be a really good bloke and as we talked my desperation must have been showing. We made a deal.

Fortune favours the bold they say. I went with him back to the airstrip to help him refuel the plane before his passengers came out to board. They arrived a little later, took their seats and then their luggage was packed into the rear locker. Known only to the pilot and to me of course, was the fact that the plane carried a stowaway – me. Bent over double in the confines of the narrow tail section of the fuselage I was hidden behind the luggage as the plane taxied for the take-off.

If I was ever a potential candidate for an attack of claustrophobia it was going to be determined over the next two hours. It was one of the strangest and most completely helpless sensations I had ever experienced. While curled up in the semi-darkness with my head between my knees trying to convince myself that it would all be worth it in the end, I felt that a distraction was needed to take my mind off this self-imposed situation. Then I spotted the solution. A Phantom comic was protruding from one of the passengers' bags so I eased it out and pulled it along the floor between my feet. By the thin beam of light filtering into my cell from the front of the locker I became totally engrossed in the adventures of The Ghost Who Walks. He did me a good turn that day, as well as saving a lot of the jungle pygmies from the evildoers.

The pilot had put it plainly.

"When we land at Hedland I'll park the plane so the luggage door is facing away from the terminal. When the pax - more pilot lingo, have all gone inside I'll come around and unload the bags and you at the same time."

My impulse to fall down flat and kiss the tarmac was strong, but it would have drawn unwanted attention to us, so I desisted.

"Don't forget, if they spot you I'll do my license."

My aching back told me it was time to stop imitating a travel bag without a handle and besides this there was another reason why I couldn't continue to Perth in the same way. Climbing out of the plane's luggage locker again at Perth airport would almost certainly have got me arrested.

The pilot seemed almost as pleased as I was when he told me there was a spare seat in the cabin of his Cessna for continuation of the trip to Perth. The alternative was to be stuck in Hedland, which was an option that didn't bear thinking about.

At 1:30 am, my sleepy-eyed wife opened our front door at home in Perth. I was feeling sure that the family would be even more pleased to see me, because they had been told I would not be coming home at all that day.

There was another nocturnal homecoming when arriving in a similar unconventional manner and at about the same hour unannounced, to be reunited with my unsuspecting family. For some insane reason I had grown a moustache while away. Pretty big and bushy and dark, it must have changed my appearance quite a bit as even Bev was taken aback at first. In reply to her request for an identity check at the door I declared.

"'Tis your husband" – or words to that effect. Not wanting to wake the kids we quietly went to bed.

During the times of my absences, Bev and my eldest son Peter had developed a routine in which he would come to the bedroom and wake her every morning.

Can you imagine the young lad's consternation when he called from the doorway to wake his mother next day as arranged? In response to his summons this strange bloke resembling a Greek fruiterer sat up in bed next to his mother. I can still see the look of horror on his face as he yelled, -

"What the hell!" and glared belligerently at me.

"It's OK Peter - I'm your father," was my lame response to his challenge.

It seemed to be a while before he ascertained that this bloke in bed with his mum was indeed who he claimed to be. I concentrated on calming him down and reassured him by saying I was so glad he was acting as the man of the house and showed me that he would protect the family. The poor kid walked around looking bemused and embarrassed for days. He must have finally forgiven me, but only when I shaved off the mo. I wondered later what his reaction would have been if I had introduced myself as Uncle Giuseppe instead, but that would have been taking it all a bit too far.

As parents, the things which children come up with often amuse us greatly. Bev and I were lying in late one morning one time when I was on leave. Young Peter in a typical show of enthusiasm asked if he could make us some coffee. He disappeared and was back in a flash with two brimming cups. I felt pretty sure that you cannot make coffee and serve it in the space of two minutes, but started to drink the brew anyway. It was just luke-warm and had a disgusting rubbery taste. Peter knew that you need hot water to make coffee, so in his inventive young mind decided that the best place to get it was from a hot water bottle. Dragging out the one that had been in his bed all night he decanted the contents onto the coffee grounds. Instant coffee made from instant hot water – it's hard to find fault with that!

It was a stimulating time working in these early stages of the iron ore mine at Mt. Tom Price. I enjoyed these jobs where you have to start from scratch and help to build and develop the project, breaking new ground - literally. The variety and the new challenges that were found in Learmonth, in the Kimberley and later in Indonesia were much to my liking. What might be considered by some to be a flaw in my character was that once the operations were up and running, the spark of interest was gone and I wanted to move on looking for another one. Having been involved in the process of carving a mine and a railway and a town out of the wilderness at Mount Tom Price, I began to feel the itch again. When the Company went into full routine production they offered me a permanent job with perks and a new staff house in the town, but there was this feeling that I had done my dash there.

The huge American company Kaiser Steel had just won the contract to construct an iron ore pellet plant on the coast at Dampier. This plant was designed to convert the iron ore fines into hard pellets which were more suitable as feed for the blast furnaces in Japan.

This seemed to have the necessary ingredients for an attractive work situation so I applied and was accepted. One of the drawcards was the Company's offer to have my family housed with me at the site. We all arrived at the fledgling town of Dampier-on-Sea, which was the port that was shipping Hamersley's iron ore product.

The area on which the town houses were to be set was the most inhospitable I could have ever imagined. It resembled a Martian landscape, except for having the sea close by. The base was hard rock and the surface an arid stretch of jumbled boulders stretching down to the ocean. In later years I went again to Dampier. It is a pretty little town these days with a tropical look about it. There are leafy trees and palms and flowerbeds around the houses that line the curving terraced streets. The site had been prepared for building by dumping thousands of tons of soil trucked in to cover the rocky wilderness. A man-made habitable landscape had been created there most successfully.

There was a friend I had in Tom Price. He was an American who had been seconded from the States for the position. He was my superior on the job, being in the American terminology, a Master Mechanic. That is our equivalent for a Maintenance Superintendent. He changed his spots later on as it happened, but he was a friend and colleague then.

While he was younger and less experienced than I, we did work well together and shared a lot of interests. When I moved to the pellet plant job at Dampier, he showed up as my boss over there as well.

Our family housing accommodation at Dampier was small and rudimentary though being insulated and air-conditioned it was well suited to cope with the effects of the hot climate. Set back about five miles from the coast, it was a part of a small village of ten similar houses sited on a flat area, which was spread with scorching blue metal. Behind the houses there rose a high ridge of jumbled black boulders that reflected and concentrated the sun's heat down upon us from sunup to sundown.

Our little settlement was whimsically called Happy Valley.

The family was at most times of the day confined to the house in order to escape the blistering heat of the day and the oppressive humid nights.

Without schools to go to, the kids were started on correspondence lessons, so were in our little home all day. This was bad enough but then I was placed on nightshift with the necessity to sleep during the day. The tension and the difficulties we experienced after that can well be imagined.

There was a lighter side to all this. One day my youngest son Greg aged six, after solemn deliberation announced that he had decided to leave home. He was going to live he said, up in the black rocks behind the house with all the snakes. We went along with his wishes and helped to prepare him for his lone journey out into the unknown. He jammed his hat on his head and then picked up his school case with an apple in it. After saying goodbye to everyone and with a face set with grim determination, he opened the door and stepped outside into the world.

The family waited expectantly after our boy had left us. We are not too sure of the exact duration of his absence – it was between five and ten minutes, before there was someone again at the door. Red-faced little Greg was welcomed back with open arms and much gladness after his epic journey into the unknown.

About four months passed as we battled along. I was determined to see out this contract to the completion of the project, but there were difficulties ahead. It appeared to me that my erstwhile American friend had acquired a bee in his bonnet as the saying goes. This would not normally have bothered me and I sought no favours, but felt I was there to do a job. His inexperience in man management began to show when for some reason or other he started making life difficult for me and naturally I rebelled.

It all came to a head one steamy night at the plant. Dampier is recorded as having some of the highest night temperatures in the country - up to around thirty degrees with streaming humidity, so tempers were often strained anyway. While following some of his written instructions for the night work I saw that one of the job items was to install a new hoist rope on one of the biggest cranes on the site. The only one we had available was under the size requirement so I refused to install it in the interests of safety.

To lessen any unnecessary tedium in this story I will jump to where I was waiting with my family at the airport to catch the plane to Perth. I had been summarily fired.

My boss and ex-friend had treated with scorn my warning that if the job went ahead we could have the consequences of injury or death on our hands. I felt no shame or humiliation after losing the job, only a sense of having done the right thing and the correct thing according to my conscience. The quiet and tranquil and boring suburbs of Perth were beckoning again. We were not sorry to leave Dampier, the kids could attend their essential schooling and Bev would be with her friends and family again. I must have been the most relieved of all, as working conditions had become untenable.

I look from the top of Mt. Tom Price over the mine in its early stages.

My job there could not have continued much longer in any event as they shelved the whole pellet plant project and it was never completed - and the cause for this was not as a result of my departure I'm sure.

Episode Eleven

<u>Good Morning Sir</u>

It was 1967 when I realised it was time to take stock of the family lifestyle and decide where we wanted it to go in the future. We harboured no regrets over leaving the job and the Spartan existence at Dampier. Bev and my young children had been through an uncomfortable period of heat and dust and isolation but now the justification to stay was gone and besides that I had been given the sack, which is always a very good reason for leaving a job. Thinking that perhaps my pioneering spirit had waned a bit, it seemed that I was seeking some stability. It was becoming clearer too, that I had the need for some professional qualifications with which to arm myself, to provide us with a brighter future. Bev seldom complained about some of the rough and ready places where we had gone to live. She was really glad that we were going home to the city though, to friends and family and the prospect of a better life for us all. I had to be responsible and think seriously about the family's health and welfare, not to mention lining up to pay for a big home mortgage as most others did.

It also became clear that my personal assets and attributes necessary to secure a well-paid steady job were pretty meagre. There were not many shots in my locker. With just a Junior high school education, and an apprenticeship - though with a lot of practical experience in railways, oilfields and mining, that was basically all I had to offer. That does not provide the basis for a very impressive CV for a job seeker in the city at age 40.

The Carlisle Technical College had recently expanded its curriculum to include courses for the training of apprentices and technicians working on earthmoving and construction machinery in heavy industry. Mining was booming at the time and there was an increasing need for men with the special skills required for this new discipline. My application was accepted as the first instructor in the college to enter this field and it was very much aligned with my own work experience. It transpired that rather than having to walk in the steps of the established staff in the college with what appeared to be a tired old curriculum, I was promised the freedom to begin writing my own lesson material. This would be enjoyable I was sure and a welcome challenge.

After that first introductory training it was also part of our qualifying process to attend the Perth College weekly for a period of four years. This was to obtain the required piece of paper - the Teachers' Diploma. Described by a few of my colleagues as largely airy-fairy nonsense I took it all at face value and accepted the newfound knowledge. We wrestled with the subjects of Psychology, Philosophy, Statistics, Group Dynamics and Teaching Methodology and others.

As we had all come directly from practical jobs in industry and were unaccustomed to concentrated study, we found it difficult to knuckle down at times. As a sworn pragmatist I looked for value in all this hard work and over time found it to be very relevant to the new job.

The ages of my students ranged from 17 to 22 years with later some mature-age people to teach as well. The head of our department said that in order to ease us into the business a little more gradually we would have charge of the first year apprentices initially. Presumably the shock would be lessened for us and possibly for the students as well. The theory behind this was that the teacher should at all times know more than his pupils do. First year apprentices presumably knew little or nothing, so a new teacher would hopefully be ahead of the game from the start.

There is a clear recollection of my first day standing in front of the class of thirty-five young hopefuls, feeling a bit exposed and vulnerable. It was a worry just what was going to happen next, but it all fell into place somehow. In many ways I felt that I was going to be on a steeper learning curve than they were. This was a pretty accurate assumption as it turned out.

It was a very busy time for me as I was literally holding down three jobs. There were the regular day classes through the week, some additional night school teaching and then on weekends and odd nights I sallied out with a camera. Then on top of these jobs there were the study commitments as well.

After qualifying in all the training requirements for a lecturer in the department, I was able to put some letters after my name - 'Ivan Macmillan, Dip. T.T. - MIAME.' I don't think I was ever called upon to use them though. It has always been for me - not who you were, but how well you could perform.

No one had ever addressed me as 'Sir' before. It was accepted in the classroom although as I moved into teaching more mature students it was relaxed to more informal terms. Something I had to become accustomed to were the times when out in public. When a voice came out of the crowd in the Hay Street Mall with the greeting, -

"Good morning Sir." It gave me a good feeling. This continued for many years afterwards too and I have to admit to always getting a charge out of it.

So there I was with a head full of theory, trying to remember the strategies to use when dealing with the frequent problems of class control. One of the older and more experienced lecturers advised me one day saying.

"The theoretical stuff you have learned is useful and it usually works – up to a point. There will be times however when you feel that you would like to let rip, so do it. The lads will see that you are only human after all and that's not a bad thing."

As we get older and perhaps a bit grumpier, the younger generation seems to be more and more of an aggravation to us. This is pretty normal I suppose but there was a risk in this that you would tend to categorise them and lump them together as being all the same. If they are seen as untidy with long hair, wearing their caps backwards or slouching around in baggy clothes, we tend to automatically brand them all as gormless ratbags. As I began to deal with this younger generation of young men on closer terms and got to know them more for what they were, it seemed that many of my prejudices were exploded. I realised to my surprise that some of my wild things were actually the brightest and the most intelligent and sometimes the most respectful of the bunch. It appeared that there was much to learn too, and there was the need to keep an open mind and be ready to revise some of my own creaky attitudes.

Some of the apprentices were my size and larger so there was no advantage of the physical kind to fall back on in the quest to gain their respect. It seemed that I would just have to try and be smarter than they were.

There was a large gangly lad with a cocky air who had decided to amuse himself and his mates by provoking me. I had heard that many teachers are put to the test by their students especially in the male domain and particularly if they are bigger than the teacher. You see, in those days there were no counselling services for harassed teachers and stress leave had not been invented yet – we had to sort the problems out ourselves. Being relatively new to the game I wondered how I would handle a challenge if it came. One day it did.

To give him credit, Greg had feet so large that they would not park together comfortably under his desk. To stretch his long legs he often extended them sideways which caused his clodhoppers to fill the aisle. I had been made aware of the psychological advantage a teacher can gain by circulating through the classroom. To this end I liked to walk from the front of the room to the back to stay mobile and discuss problems and check the work of individuals.

My classes liked to carry out projects working on live engines. This is a group of first year students

It all began innocently enough when I started to move down an aisle between the desks, which incidentally were obstructed again by Greg's big feet. He would usually pull them in at the last minute as I approached and give me a big grin and sometimes a salute as I passed. Our game actually began in earnest when he got tardier and I was forced to stop while he cleared the way - still with the big grin on his face. This caused all his mates to giggle, which seemed to reinforce his position in provoking me.

One day as the game continued I resolved to act like a human – a man even, and deal with the problem in a more direct manner – in a way the group would recognise and understand. I moved across the back of the classroom and up behind Greg who had his feet obstructing the aisle as usual. I thought that resorting to shock tactics might be effective in this case – it was.

The first thing he knew was that he had been seized by the scruff of the neck. Twisting hard on his shirt collar I put my mouth to his ear. In a rasping stage whisper, which carried around the room, I grated, -

"Pull your **#@** feet in or I will stuff them down your **#@** throat!"

He did as ordered instantly, so I gave him a fatherly pat on the head and continued on my way. Greg never challenged me again and in time we actually became good friends. Crude – yes, unseemly – yes, effective – decidedly!

Saving face and maintaining your position of authority in a class situation is very important in a room full of young men who are 'feeling their oats.' Also too, causing a student to lose face in front of his peers can generate resentment. When a teacher is himself humiliated however it can be disastrous and very hard to recover from.

Technical drawing and cartooning were favourite pastimes and I liked to use them for effect and a bit of variety in the class work. Most of the classrooms had large chalkboards across the front wall. These were the folding type and it was a useful technique to place a diagram on one board and then cover it with another. I went to the room before the class began and drew a detailed technical diagram in colour on the board, which was ready to be revealed at the appropriate psychological moment.

The right time had arrived and I uncovered my artwork with a flourish and then nearly died of shock. It had been grossly transformed by the addition of representations of male and female genitalia and other disgusting enhancements. Should I do my block and throw things and shout at the class, or walk up and erase the offending material - or what? It seemed that neither of these would serve to reveal the identity of the phantom artist, which was what I must try to do.

From the back wall I instructed everyone to face front and look at the new artwork on the blackboard. The laughter died away and was replaced by a general air of expectancy. With still no plan in mind as to how I should proceed I began to discuss the merits and otherwise of the work of the phantom artist. It struck me like a bolt of lightning that I knew without a doubt who the culprit was. Phillip in the sixth row had become the centre of class attention. All eyes were on him to gauge his reaction to my comments and it looked as though I had stumbled on the solution to the dilemma almost by accident. His ears were turning red and I knew then that I had him. Trying to keep it all on a light note I grabbed him by the shirt collar, propelled him to the front and roared in the loudest voice I could muster, -

"Get up there and clean it off!"

He was laughing as he erased his graffiti, as was the rest of the class. So was I - but it was a bit of a nervous laugh and more one of relief.

As others will tell you, I have always harboured a love of machinery and live engines in particular. Perhaps it was hereditary, as my dear old dad had been oiled with the same brush. So working with engines, dissecting them and putting them through their paces and experimenting and tuning them was much to my liking. We also had the old and the new side by side at the college to be studied and compared to show historical advances in design.

With the four years training finished and the diploma under my belt I felt the urge to move on. While I think of it, there is a very true saying that goes 'If you really want to learn most about a subject – teach it.' I learned a lot at the college but after six years of 'chalk and talk' things began to pall.

One of the reasons for my discontent was the lack of action to improve work safety. Young and inexperienced students are vulnerable and require special care and supervision. I was acutely conscious of this and it seemed that I was becoming a bit paranoid. It was affecting me deeply. It seemed to be about the mystery of peoples' attitudes that would cause them to ignore important issues that could be clearly seen by others, and which required some remedial action. There was much to learn about all this and I wanted to find some answers.

It was time to do something about it.

With this nagging need to learn more by entering the world of industrial safety I left the college and joined an industrial safety foundation as a consultant. The Teaching qualifications helped and coupled with industrial experience I was soon circulating throughout the industrial world around Perth conducting safety audits, giving lectures and generally advising company managers on aspects of safety in their workforces. The work was not easy but it was rewarding, especially in the case of one company which burnt one of its employees to death in an horrific accident and then just one year later took off the top safety award for their chain of companies throughout Australia.

This valuable experience then led me to combine training qualifications with safety experience and offer myself again to someone – or anyone who wanted my services.

Episode Twelve

Say Cheese

While working as a technical teacher at the college, I still felt considerable financial pressure and decided that more money was needed than was forthcoming on the teacher's salary. While it was possible to make do with just my earnings to provide for our family of six in 1967, we were not getting ahead at all.

It has been said that the best work of all was that where you are doing what you would normally like to do as a hobby. My consuming special interest for many years had been photography, so I thought - why not?

My application to join one of the largest photographic firms in the city as a trainee was accepted. Their speciality was the coverage of family and social events like weddings and other celebrations. Child photography was also a speciality. Their photographers could often be seen working at the Embassy and Pagoda ballrooms and most function centres.

A typical young bride and groom

As a keen amateur I considered myself middling in the photography game. Over the years I had recorded on film the activities at every mining and oilfield location where I worked. At places like Wittenoom, Learmonth and numerous other job sites, it was necessary to make do with whatever could be converted into a darkroom to process the photos.

It required more than a little imagination to convert the smallest of enclosed spaces to take my darkroom equipment. Conversions were made to bedrooms, bathrooms, wardrobes, pantries and broom cupboards. The possibilities were meagre in one location, as I had to utilise a toilet for the purpose. In this one I fitted a makeshift table over the top of the cistern and sat on the toilet seat – with the lid down of course and facing backwards – a sight to give anyone a shock if they burst in! Most of the prints in this book are the results of my processing efforts in places like these. On hot tropical nights it was necessary to add ice cubes to the developer to bring it down to the right temperature - and then sometimes my sweat would drip on the film or on the printing paper and spoil it.

In the five years with this employer, I was eventually promoted as one of their 'top guns' with responsibility for getting professional results in photographing many of the big-money society weddings and other important functions. My battery of cameras was impressive and filled much of the interior of the little blue Volkswagen.

Some weddings were fun - and some were not. A very amusing incident that I photographed during one speech by the bride's father is still clear. The poor bloke had relief written all over him as he nevertheless made quite a witty speech to the guests. He expressed his happiness with the fact that the couple had finally tied the knot. Turning his attention to the bashful bridegroom, he said that he had a present for him and after reaching under the table, leaned over and handed him a special gift. It was a white-painted shotgun with a big white ribbon tied around it.

"I won't need this any more!" he offered with a big grin.

Another father, who had sired seven lovely daughters, gave one of the shortest speeches I ever heard. Having married off only four of them to date, he intimated that he was feeling the financial burden imposed by all the weddings - this was in the days when the bride's parents almost always footed the bill. He was a little round-shouldered man with specs down on his nose with the facial expression of a St. Bernard dog on a bad hair day. Rising slowly to his feet he peered around at the crowd and said, -

"That's four down and three to go," - and slumped back down in his chair again.

The photo shoot at the bride's home as she was being prepared for the wedding, was a test as to whether that the family had accepted you. On one occasion when the three lovely perfumed bridesmaids lined up for me to zip them into their frocks, it seemed that I was approved in that house at least. On the other side of the coin one day, no sooner had I begun to set up for some nice bride's portraits when there was an invasion. Dozens of relatives flooded into the small room and with happy-snappy cameras started their own photo shoot. Most were portly matrons who shoved and jostled and got in the way and were yelling to the bride to turn this way and that! The bride's mother quickly woke up to the fact that the hired photographer was repacking his gear and heading out of the door. When told in no uncertain terms that I was not happy with the situation and would soon be gone, she cleared the room in seconds.

Being on side with the families was one thing. Winning acceptance by the church minister was another. A good rapport was vital in securing the important shots inside the man's church during the ceremony. Every minister, pastor, clergyman and priest was different in his or her tolerance of the photographer. It was my standard practice to front up to the clergyman before the wedding and ask him what his particular rules were. A promise that I would be considerate and behave usually brought some concessions. One cardinal rule was in the use of flash. This was seen as being a serious distraction in most churches, so I put the gun in my pocket and had to shoot with what light was available there.

My favourite shots in a church wedding are the ones taken from the choir loft high up at the back. Most large churches and cathedrals had these and I was always eager to get up there to capture the overall picture of the ceremony. One particular priest had banned all photography from the loft in his church, due to some previous unpleasant incident, but I went ahead anyway, risking detection. Creeping stealthily up the creaking stairs just in my socks, I ensured also that there was not a cameraman's silhouette to be seen above the rail against the bright stained glass windows. The quietest camera in my collection was used for this job. Some SLR cameras are particularly noisy in operation and in the still hush of a church, when the button is pushed it sounds a bit like a road accident. They say that stolen fruit are the sweetest – stolen photographs are often the best too and I felt justified in bending the rules when the shots turned out well and more importantly sold well too. The good report from the bride later on these church pictures was encouraging.

The demarcation between families when seated in the church was often pretty obvious with neither venturing into the other's territorial pews. Sometimes the tension could be felt between them, betrayed by the belligerent looks and the haughty stares, which were enough to wither the apple blossom on the bride's bouquet. At one particular wedding reception I noticed that the two mums were shaping up to each other like a couple of fighting cocks. They managed to get through most of the evening however without ruffled feathers and without a showdown – until the final minutes.

As the time drew nigh for the bride and groom to leave their well-wishers and depart on the honeymoon, I glanced at the exit from the hall where close family members traditionally gather to say their farewells. The two mums, built like pocket battleships, were manoeuvring and jostling for the best positions. Like Roman gladiators they challenged each other across the doorway and let fly with threats and counter-threats until it all finally turned physical. Just as the bridal couple approached the doorway, the happy scene was shattered. It was handbags at five paces as a right royal melee erupted. All pretence of decorum and good manners disappeared as the two screamed and belted each other with the handbags. Scraps of personal finery and the contents of the bags littered the floor. A little bit of unexpected and impromptu entertainment is welcome especially when to all intents and purposes the party is over. This lot shaded the floorshow by a considerable margin. Yes, I did resist the temptation to film this lively event for inclusion in the pages of their wedding album.

There can be many unexpected hiccups in what otherwise should have been the perfect wedding. Many seemingly insignificant and minor situations are often enough to send the already tense bride into a flat spin. It was obvious though that many of them were doing this for the first time. As the photographer on the other hand is supposed to have his wits about him and due to his vast experience, he can sometimes be relied upon to be an adviser and mentor when things go wrong. In order to cater for little emergencies, I carried what is termed Every

An olde-worlde wedding I shot in 1970. The ceremony and reception were held in The Old Mahogany Inn.

Wedding Photographer's Survival Kit, which had been accumulated over the years and was with me at all times on assignments.

Garden photos can be the most prized in the family collection. Taken in natural and colourful surroundings, there are endless possibilities to show these beautiful people to their best advantage. Queens Gardens down near the causeway was a favourite spot to take a wedding party. Its masses of flowers - to complement the bride, and gnarled old trees - to complement the mother-in-law and pool reflections, are just begging to be used as frames and backdrops. Nature is not always kind however and the weather not always idyllic, so you have to be prepared.

In my survival kit was a half dozen white golf tees. High winds in some of the more exposed garden sites tended to pick up the bride's veil and wrap it around her head and sometimes the train would stream out horizontally like Pricilla the Queen of the Desert's. Having positioned the lady to the best advantage it was often possible to defeat the effects of the breeze by literally pegging her haberdashery down to the lawn with the golf tees. This was most effective and being white in colour they did not show up in the photographs. On that note, the tearing of a frock or the parting of a strategic shoulder strap can be disastrous, were it not for the intervention of the photographer. A card of safety pins could always be found in the kit for running repairs. There were bobby pins for collapsing hairdos too. A variety of elastic bands are useful when you see the groom's hands disappearing up the long arms of his hired shirt. On more than one occasion the ceremonial pen brought out to sign the ceremonial marriage certificates had run out of ceremonial ink, thereby bringing the otherwise smooth ceremonial proceedings to a halt. My own imitation gold-plated Schaeffer was always ready in the left hand inside jacket pocket for such an emergency. The sleight of hand, which produces a big umbrella in a sudden rain shower, gets you a lot of brownie points too.

The ghastly look on the best man's face as he searches his pockets when called upon to hand over the ring is an indication of his misery and panic when he realises it is still on the dressing table at home or still in the pocket of his board shorts. Sidling up alongside the best man in this desperate situation I have been able to palm my Eveready spare imitation gold ring to him on more than one occasion. A full jerry can of petrol kept in my car's boot was called into service once or twice, as was my box of spanners to help get a reluctant car running and down to - or away from the church. The list of useful and necessary items in the survival kit is quite comprehensive and no wedding photographer worth his salt should be caught without it. As a photographer colleague of mine once dryly remarked.

"These things will all come in handy even if I never use them!"

A good man in trouble is a disturbing sight. When the best man called for help at one reception I was able to pass him a copy of 'The Duties of the Best Man' and also make suggestions as to what he could say when he is on his feet in front of all those people. There was also a selection of witty telegrams to be read out, which was especially useful when they hadn't received any. One example of these little gems was the popular message to the blushing bride.

"Take a tip from one who knows – tie your nightie to your toes." Signed – "Mother."

All in all, being a wedding and social photographer was a pleasant and rewarding occupation as you worked closely with the public when they were at their best and happiest - or almost always were.

We took our cameras to big social functions too where our firm had the franchises for the candid photo rights.

A good sense of timing was important too in photographing the partying guests when the drinks were flowing freely. If you started working too early when all the punters were stone cold sober and reserved, you would invite a number of haughty refusals. Shooting when they were becoming carefree and garrulous and you were everybody's friend was more productive. Trying to capture them towards the end of the evening when many were well on the way was likewise non-productive. Many of the pictures you got would portray some of the dishevelled obnoxious drunks in a bad state of disrepair, who would probably have forgotten you had taken the pictures of them and had lost your card anyway. There was a lot to learn in this business apparently.

Many of the floorshows and items of entertainment at big functions were really good subjects for the camera. It was also a good move however to ask the performers before the show if they objected to being filmed in action. Some did and some did not and I respected their right of refusal. Some memorable shots were there for the taking at a big wedding in the town of York out in the farming country. Held in the big historic town hall, the guests were all seated around the gleaming dance floor when the star turn appeared. He was a big, striking Scot

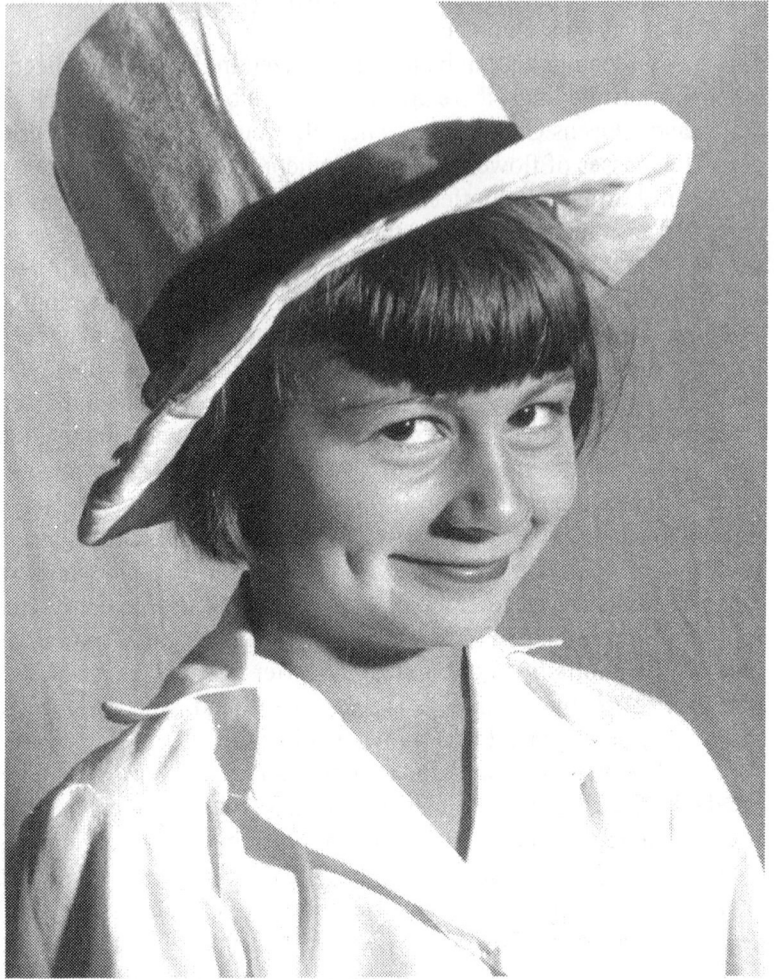

'THE HAT' - photographing children held some of the greatest rewards of the job. This is Roberta.

in full regalia complete with kilt, sporran, gleaming buckles, a tall busby and a set of bagpipes. It was late afternoon and a shaft of sunlight speared down from the windows above like a theatre spotlight. He strode to the centre of the floor to stand in this pool of iridescent gold light and played the haunting piece Amazing Grace to a hushed audience. Those pictures were rated exceptional.

Using the additional earnings from my photographic job, we managed to finance the purchase of a block of land at Mahogany Creek in the Eastern Hills suburbs. Liking it there so much, we decided to build a new home on it and lived there for a number of years. It was even possible to include in the design of the house a purpose-built darkroom. This allowed me to work as a freelance for a time afterwards - these earnings of course also contributed to the payments on the mortgage for the house.

Owning my own darkroom at last certainly beat having to do the photo processing work skulking in a toilet.

Episode Thirteen

Aussie Guru in Indonesia

The need for change had been pressing me for some time so I applied for a position with this big company that was advertising for staff to work in South East Asia. A transfer from my present dusty landscape to a land of swaying coconut palms had a certain attraction and besides, it promised to be another pioneering job, which is the kind I like anyway. Little did I know just how pioneering this new job was to be, but it wasn't a long wait to find out. My application was accepted and my passport processed and I was on my way north – way up north to Indonesia.

This was going to be new territory for me. Added to the other matters of interest, there was a perceived threat everyone was talking about involving their President Sukarno who was rattling his sabre. Having never worked in a foreign country before either and as this nickel project in the jungles up there was just starting, - more pioneering, I welcomed the opportunity for some excitement and a new row to hoe.

This new employer was International Nickel Pty Ltd, or PT INCO, as it was known in the region. Nickel deposits had been discovered in the mountains of Sulawesi, the large island north of Bali. Inco was a Canadian company, which also had mining interests in Venezuela. They were poised to pour big money into the project, an open-cut mine and a smelting works. This would produce thousands of tons of very pure and valuable metallic nickel in the form of matte.

My job title was Senior Technical Training Officer. The allotted duties initially were to train a group of selected Indonesians up to the technical levels required to service and maintain the modern machinery at the mine.

The green coast of the large island of Sulawesi slipped below our small aircraft as we approached the final landing. The ocean was like lead and I could see rain squalls sweeping over the mountains. An interesting aspect of travel from Australia to a remote outpost like this is that the kind of aircraft they keep transferring you to, gets smaller and smaller all the time. This occurred to me when I climbed aboard this one and reckoned they could not get much smaller - so then I must be near the end of the trip. I was right.

Jim, the American pilot, was a friendly bloke and with just the two of us in the plane he put me in the right-hand seat and we talked on the headsets over the roar of the engines. Jim was a mine of information, having flown as a bush pilot for this company since it began its exploration phase. He would prove to be a great friend in the ensuing years and I can still hear his laconic Yankee drawl with all the news of the day.

We flew a lot of kilometers in Jim's Islander taxi

"This is what they call a STOL airplane" explained Jim. "They are designed for Short Take-off or Landing on bush airstrips with bitty runways carved out of the jungle."

He then started to demonstrate the plane's capabilities.

273

As we slanted down to the narrow gash in the dense green mat of trees, the plane's speed dropped off more and more until it seemed to be flying like a kite. Two sliding panels popped out of the leading edge of the wing when a stall condition was approaching. This gave the plane an extremely low landing speed and a very short run after we had thumped down on the gravel strip. Jim then related a story that seemed to suggest that the anti-stall properties of his aeroplane were not necessarily a good feature. It seems that he was landing into the wind one day – not an uncommon thing to do, when the breeze suddenly picked up into a mini-gale and the plane didn't want to land. So low was its designed landing speed that in the stiff westerly it began to just hover there over the strip like a bird. With only one thing to do, which was increase the weight of the plane, he yelled for his amused audience of locals to run over and hang on to the plane. The extra ballast of about twenty Indonesians did the trick and he was brought down to earth with his human ballast laughing their heads off.

"We have to make a pass over the landing strip to chase off any dogs or other wandering livestock," Jim explained. "I took the head right off a mutt one day last week – he just dashed out of the bush when I was on full power taking off - kinda messy."

The map of Sulawesi will show a deep bay extending right up into the middle of the island, known as the Bay of Bone. Sulawesi Island is in the form of the letter K and was originally known in the geography books as the Celebes and a part of the Spice Islands. It was the home of pirates and privateers who plundered the heavily laden merchant ships from their swift Macassar schooners. Examples of these lovely rakish vessels can still be seen in the harbour at Ujung Pandang, the capital of Sulawesi, which was once known as Macassar. Just after the Second World War when the Dutch, the Portuguese colonists and the Communists had been ejected from the country it was declared a republic and they altered many of their place names. Batavia in Java is now Jakarta, Macassar is Ujung Pandang and there were many more places which fell to this name-changing frenzy.

Standing on the tiny airstrip watching my plane take off again was a moving experience and I wondered what I had got myself into. Not many hours previously I had been in Perth's dry temperate climate, while here I was in a land of contrasts only a few miles from the equator. While not exceedingly hot as in our Australian summers, the humidity was extreme. My shirt was saturated within minutes and a dank, earthy smell came from the tangle of trees and vines in the jungle surrounding me on all sides.

The assignment with my new employer PT Inco, which was the operator of the proposed nickel mine, had put me on site ahead of the other staff. As the first expatriate trainer, my brief was to recruit a number of young Indonesian candidates to undergo an intensive program as engineering apprentices. As with many training posts in developing countries the expatriate literally works himself out of a job. When his trainees reach a high enough level of skill and competence to run the industry themselves, he then becomes redundant. They called this 'Indonesianisation.'

Part of my job induction explained that there was only a very basic level of technical skills and experience among my students. Although they would be recruited from senior high schools and colleges, theirs was after all basically an agricultural country with very little in the way of modern technology or heavy industry. The expectation was apparently that prior experience of many of my recruits would be restricted to years growing rice in the paddy fields with probably a total exposure to engineering equipment being that of the family push bike. This was to be a test indeed and I could see interesting times ahead.

When Jim's plane departed the Malili airstrip I found myself to be the only white expatriate there at the Inco base. The small group of buildings adjoining the bush airstrip had been the centre for the exploration teams that discovered and proved the nickel reserves inland in the mountains. Now all these pioneers had departed and I was left to my own devices. It was a pleasure and a relief at last to meet the Indonesian camp manager who was waiting to greet me. I humped my bag up the hill from the airstrip in the valley.

A young neighbour comes to pay me a visit

The camp boss spoke excellent English and flashed me a dazzling smile, which displayed an almost unbroken row of gold fillings. We shook hands and I felt an immediate liking for the man. His name was Andi Azikin. We were to become good friends and associates and I know now that I would have been lost without his support in the times ahead.

"Mister Ivan," he addressed me, "you will live in that Company house on the hill overlooking the village of Malili. You have been assigned a maid who will cook for you if you wish and a gardener and of course I am at your service at all times."

It was obvious that Andi had been expecting me and was determined to make me welcome. The fridge in my new home – the first port of call where I dumped my bags, was stocked with frosty green stubby bottles of Bintang beer. This label proved over time to be the ideal drink to quench a big thirst in the tropics. Brewed to the recipe of the Dutch Heineken beer once made by the now-departed colonists, it is a crisp lager that went down very well after my exhausting day of travel and excitement.

Sitting on the broad veranda I met my domestic staff of two. The gardener was a wiry little bloke with more flashing gold teeth, who carried a big sharp parang on his belt, but it was the maid who intrigued me.

She was a petite beauty who wore gold jewellery set off against a flawless brown skin. Her shiny black hair was drawn back in the traditional bun and her bare feet could only take very small steps under the full-length floral sarong with gold trim. They are small, almost child-like people when compared to the average person of European descent. Unfortunately for me they spoke no English. I would have to learn their language, Bahasa Indonesia very soon.

Lying back gratefully in a big rotan chair on my wide veranda with a cold Bintang I looked at the village of Malili spread out below. A row of thatched houses on stilts lined a muddy thoroughfare, which wound along the banks of a swiftly flowing river. A swaying cable suspension bridge linked the two halves of the settlement and the figures of the villagers could be seen moving about in the fading light.

These latitudes so close to the equator experience heavy rainfall for most of the year. The recent afternoon downpour had wetted the thatched attap of the roofs of the village houses and a red sunset was developing to paint the tranquil scene. The shining rooftops were tinged blood red and the swollen belly of the river was flecked with crimson. A straggle of villagers padded past the house and eyed me curiously. They had obviously heard of the arrival of this white man Australian guru who had come to live in their community. The men waved cheerfully with broad smiles and the women giggled among themselves as they bobbed down the winding path towards the village.

All were very small in stature but good looking and graceful. Some of the women carried loads on their heads, which imparted a sinuous swaying movement to their walk. The flowing lines of the long sarongs they wore enhanced this effect. Most also wore what I was to learn was a kabaya, a short gaily-coloured jacket in richly decorated patterns. The men wore a short skirt or baggy pants and a jacket, usually white. Almost universally the males carried a parang in a scabbard on their belt or slung across their back. Similar to a large bush knife or machete, this was a general-purpose tool used for a multitude of jobs and I don't doubt, in times of need, as a very fearsome weapon.

Hoping to enter this new country and not be found wanting in the language department, I thought it best to learn a few useful phrases in Bahasa Indonesian. To be seen as totally inept and unable to communicate at all would be disastrous, especially as my role there was to be that of a teacher. My initial efforts as a new arrival did seem to spark some interest among the villagers there, as they seemed bemused at my efforts to talk to them and giggled a lot. It would not have surprised me much if the literal translations of what I said were interpreted as stuff like, - "What is the name of your monkey's mother?" - Or, "Where is the nearest railway station?" or "You are standing on my foot."...! While perhaps not quite as bad as that, I think you have probably grasped that while first attempting to speak in this new language I could have been saying almost anything.

Andi, who had greeted me on arrival, came visiting that evening. A perfect gentleman in every way, he showed impeccable manners. I also had my first introduction to the very formal way in which meetings are conducted in this country according to their customs.

"I am of royal blood" he confided. "I am a descendant of the royal family which ruled a kingdom in these parts many years ago before the Japanese invasion and before Indonesia became a republic. All the independent kingdoms were abolished at that time. It was a brutal process and the members of the royal families became hunted fugitives in their own land. I do not make it known that I am a descendant of royalty but all my people here know and they respect me for that. I am one of their leaders and their adviser."

Family transport under the palms on the Malili River

I learned that the population is predominantly Muslim with a few pockets of Christians scattered throughout the country, with all apparently living in harmony. Standing prominently in the centre of each village and town is the high golden dome of the local mosque. I awoke next morning to the singsong wailing of the Morning Prayer that wafted up from below, and any momentary feelings that I was still back at home quickly disappeared.

276

Andi could see that I was quite ignorant in the ways of the local people, their religion, their customs and their culture. I tried to show him however, that I was eager to learn and willing to blend in with their lifestyle, become familiar with the language and generally understand as much as possible about them. If I was to teach some of these people I had to get close to them and be accepted or my assignment would fail.

In answer to my enquiries on the status of the nickel mine Andi explained.

"There is no mine yet Mister Ivan."

I was becoming accustomed to my new title, which stayed with me for the whole of my stay in the district.

"The site is still covered with jungle about fifty kilometres up in the mountains. There will be a large town built by the company on the shores of Lake Motano but there is nothing there yet. None of the staff of expatriate experts has arrived for the construction and operation of the mine. You are the very first Company representative we have seen here. Teachers are highly respected in our country, and especially if they are from Australia. Indonesians like the Australian people and we look to your country to help us with our development. Welcome to Indonesia."

This was when it became clear that I really was a pioneer in the establishment of this project. It made me feel valued, though not a little lonely and isolated, being there on my own in a strange country. It was obvious that there was much to do so I had best buckle down and start doing it.

It was arranged that I should live in Malili on single status to begin my tour. I wanted to clear the way so to speak and ensure that Bev would feel safe and could cope with living conditions without unnecessary stress or discomfort. After six months there alone I began to feel somewhat bereft and lonely so made arrangements for her to join me and bring Lise, my daughter of seventeen too. More 'expats' were arriving. Couples and families from a variety of countries came to swell the ranks – all spoke English thank goodness, as I was having enough difficulty mastering the native tongue, Bahasa Indonesia. Many of the new arrivals were Americans together with a good representation of Australians and Canadians and some Brits. Inco was a Canadian-based company from Sudbury. A Company store opened in Malili, which was stocked with local produce but also with a big variety of Western commissaries, much of which came duty-free from the good old USA. I saw my first Pringles in this store along with Crisco and we could buy large bottles of Kahlua for four dollars American and Jack Daniels Black for five. I could see there would be some compensation in living at this remote outpost.

We had lodged our two younger boys, Richard and Greg in boarding college in Perth and our eldest, Peter was working in a Telecom apprenticeship. We were happy to be a family again, albeit incomplete, but we made arrangements for the two schoolboys to spend their holidays with us in Indonesia. My daughter Lise was seventeen and her big blue eyes opened wide at the exotic sights and sounds and smells of Indonesia. In the way of young people she picked up the local language very quickly. Just a couple of weeks after she arrived I came upon her in the garden chattering away with the housemaid in their local dialect, as if they were old time pals.

The Indonesian students obviously impressed my daughter. I believe she was naturally comparing them with the boys she knew back home. All these were very athletic and clean cut and extremely polite. Many were artistic in one form or another, in music or art and as I discovered in teaching them, their hand skills were exceptional.

My little lady language tutor Sri, didn't let me out of her sight and coached and harangued me to improve my local language skills. She even had me singing school songs and reciting nursery rhymes in Indonesian. This was a bit demeaning for the 'Great Australian Guru' and I'm sure my students would have thought it a hoot had they overheard my faltering efforts. Indonesian is a relatively simple language to learn with no tenses or plurals and is far less complicated than English. Fortunately, English is the main second language taught in Indonesian schools, so I was able to communicate immediately with the better-educated students. The arrangement of words in a sentence is the reverse of that in the English version with the main noun or subject coming first and the predicate following. For instance, in English we say 'The wrinkled, grey, stooped old man.' They say "Man, old, wrinkled, grey and stooped.' They all told me it would be easier when I got the hang of it.

My students adopted Lise soon after she arrived and she seemed flattered. They organised outings and parties and everyone had a fine time. Beverley and I felt she was in good hands as the Muslims have pretty strict moral codes, especially regarding sex, marriage and alcohol and there is safety in numbers too they say.

Drunkenness among the Indonesian people is almost unheard of, although they smoke like chimneys. Their choice in cigarettes is a pretty powerful and pungent weed called Kretek. With the spice cloves mixed in with his tobacco any Indonesian could easily be tracked down if he was having a puff by just following the smell of cloves.

My new friend Andi Azikin turned out to be a good teacher, advising me on the many aspects of Indonesian customs and culture. It was a part of my job to meet with some of the high-ranking officials too. Not only did I need to integrate with the people as soon as possible by observing and understanding them, but wanted to create a good impression also to help me in my job. It felt a bit like being the Australian flag bearer and a kind of ambassador; especially in the early days when I was the only Company staff representative in the area.

A visit to the family home of an Indonesian friend is a good time to observe the protocol and get some practice, to ensure that you don't make some unforgivable social blunder. You know the saying - 'When in Rome'.... To give them credit, they seemed very tolerant of the expatriates' lack of manners and tolerated a lot of ignorant behaviour, especially from the minority of those who saw themselves as superior and who felt they had no reason to try to conform to local customs.

Andi's home was fairly typical of Indonesian middle-class living. In contrast, the people in the villages live in houses of bamboo and thatch without the benefit of running water, electricity or sanitation. They make the most of their primitive conditions however and still maintain high standards in many ways. I was mightily impressed with the immaculate dress and grooming of these people who also showed extremely good manners and were very open and friendly with an infectious sense of fun. I don't recall ever seeing a dirty or neglected child or a drunk. Alcohol was freely available in all of the roadside shops or 'tokos' but their Muslim religion forbade any excesses. I did sample the 'Scotch' from a local shop one day - it reminded me of paint stripper - perhaps this was a strong reason for their abstinence. Liquor licensing restrictions were unnecessary. Very religious, with high morals, they lived by their beliefs and were a pleasure to be with. The placid village scene belies their fierce warrior spirit however, as they had waged bitter guerrilla warfare against their many invaders over the years. Andi had a scar left by a Japanese bullet. He was shot while mounting a raid on an enemy patrol during their occupation of Indonesia during the war in the Pacific.

Always extremely polite and formal, the Indonesian householder is dedicated to making the visitor feel welcome. You do not just barge into an Indonesian's home. First of all, it's shoes off at the door where you are greeted and invited to enter. You can always note how many guests have arrived by the number of shoes in the neat row by the front door – divided by two of course. The reason for the stockinged feet inside is soon obvious as you notice that the floors gleam and everything is scrupulously clean.

You are introduced to the assembled family, including every one of the children. They all seem genuinely pleased to see you and make a lot of fuss, waiting on you hand and foot.

Some of the customs I had to learn so I wouldn't stand out in a crowd. When seated in company you should never raise your foot off the floor as you would if you were in a recliner or using a footstool. To face the soles of your feet towards another person is considered extremely bad form and even insulting. Again, in the case of two men standing talking, you have to try and avoid the habit of placing your hands in your pockets or on the hips. Both of these actions are a form of body language considered to be implying that you are superior and therefore are seen as demeaning. The natural greeting for a small boy in Western countries is a pat on the head. This is bad form also. When offered finger food, you should also remember not to take it with your left hand. This hand is considered unclean and associated with going to the toilet, so it's the right hand only. I had some trouble remembering this one and sometimes found myself lacking circulation in the left hand - from sitting on it when I was in polite company.

While relaxing in a low rotan chair one day, I was intrigued by the sight of our little Javanese housemaid bent almost double with a worried look on her face and scuttling around the room like a crab. She was behaving in the way of all low status servants by observing that when entering a room, the head should never be at a height above that of the superiors present. Not a problem when everyone is standing, but I was slumped in a low chair this time and hadn't realised her predicament. She seemed extremely relieved when I stood up.

A modern Indonesian will offer you a cold Bintang beer on a hot day in his house, but will probably abstain himself. One of the most common drinks seen on the table during preliminaries at meetings and very formal occasions is either weak black tea or fruit cordial, both served hot in tall glasses with a decorative lid. I learned subsequently that one of the best preventatives for tropical diarrhoea is hot

It was a great day when Bev arrived at Malili after I had spent six months of my life as a single man

red cordial. No one seems to know why it works but the Indonesians obviously had cottoned on to it. Come to think of it, I did not ever suffer the pangs of 'Bali Belly' at all as many other expats did – it must have been the complimentary hot cordial partaken on my visits to Indonesian friends. I even developed a liking for it in time.

We were soon enjoying the local food. Only a little red meat is used in a dish as it is relatively scarce and expensive, but they can do wonders with their rice and vegetables and fish. Superb full flavours are attained with the bewildering range of spices used. This country was originally known as the Spice Islands after all. Another benefit the people appeared to enjoy was as the result of living on this kind of a diet and leading an active lifestyle - I never saw a fat Indonesian villager.

279

You will demonstrate good manners and consideration for your host if you leave a little of the food or drink behind when you finish the meal. This is customary and suggests that you are replete - had enough.

Not long after Bev's arrival in Malili, I offered to take her for a boat ride on the river which passed through the village. In retrospect, after this experience I would not have been surprised if she had wanted to board the next plane back home.

The local villagers constructed dugout canoes hewn from the trunk of a jungle tree, prized for its lightweight timber. Being so long and narrow, it was necessary to stabilise them to prevent capsizing. Outriggers were fitted on one or both sides of the boat, mounted on two spars secured at right angles to the hull. These were fairly crude but effective. As these boats were only paddled up and down the river with no need for speed, they served as a very cheap and effective form of transport.

It was decided to bring a little Western technology to the construction of a dugout canoe for myself. I explained the special features of the new boat to the local boat builder. This raised his eyebrows a bit, but with a little more cash incentive up front he began work on it at once.

The hull was shaped and hollowed out from a selected log and dried in the sun for a time to achieve maximum buoyancy. I then painted it to seal the wood, which also made it unsinkable and prevented it from becoming waterlogged as the native canoes usually were. Only about eighteen inches in the beam, it was twenty-two feet long and slipped through the water like a racing scull. I shaped the outrigger spars from a very springy long-grained local timber and formed the outrigger floats in the form of skis. Unlike their native counterparts, these outriggers were designed to plane across the surface of the water. The height of the gunwales on the boat was also raised with a coaming to keep the boat dry in choppy conditions.

The final requirement was the twelve-horsepower Yamaha outboard motor. Normally fixed to a stern transom on conventional boats, this was mounted on the side aft, as my boat was pointed a both ends. A seat was fitted opposite the motor for the skipper and also for weight balance. My dugout speedster was ready.

While talking in reassuring tones to Bev on this her maiden voyage, I helped her into the boat where she sat facing aft among the paddles and the life jackets and the fuel tank. The motor fired and we soon turned out from the bank and felt the tug of the swift current in the mainstream of the Malili River. Heavy rains had filled the big lakes in the mountains and we were now riding the great rush of overflow down to the sea.

As the water in the river was running at about ten knots and we had to power downstream to steer the boat, our combined speed was exhilarating to say the least.

As Bev was facing me rearwards in the narrow confines of the hull, she did not see the big coastal steamer ploughing ponderously against the current dead ahead and blocking the middle of the narrow channel. We were on a direct collision course, which called for some swift evasive action on my part. She must have noticed the worried look on my face and turned to see the big black hull bearing down. She moved to stand up in fright – not a good idea in a dugout canoe at speed, but I was confident I would be able to power across and down the side of the big boat. I had visions of being run down by a thousand tons of steamer, so began to institute evasive tactics. But we were not out of the woods yet...

There was a loud bang, the motor jumped in my hand and then silence, while we drifted with the current. Frantically tilting the motor I saw the problem - a heavy plastic bag was entangled in the prop, jamming it solid.

At the very last minute on leaving the house that morning I clipped the Old Timer lock-blade hunting knife onto my belt – just for emergencies. This situation fitted the definition perfectly and proved to be one of the best decisions I have ever made.

With one eye on the fast approaching steamer, I used the knife to hack through the tough plastic entangling the propeller. It had to be freed quickly and all the time I prayed that the drive pin had not been sheared off. This would have immobilised us completely and left us drifting helplessly downstream and underneath the steel bows looming over us. I yanked on the starter cord – it fired immediately and we powered across the path of the steamer with only yards to spare, heading for the narrow strip of water between the big vessel and the riverbank.

Our final lasting image was streaking down the side of the steamer, ducking under the lowering sago palms and bouncing out through the churning wake of the big boat - to safety. A hurried glance upwards showed a row of grinning passengers lining the rail waving and cheering us on.

Poor Bev did not want to talk about that experience for some time afterwards. As I explained though, everything turned out for the best and after all, what's life without a little excitement anyway. That Old Timer hunting knife takes the credit for saving our skins and it accompanied me on all my trips after that day of action. I still have it today and will never part with it – ever.

My two sons, Richard and Greg had made the flight from Perth where they attended boarding schools, to stay with us in Malili for their holidays. We planned a trip together in the canoe out to Pulu Puloi, our adopted tropical island out in the Bay of Bone.

This entailed a four-mile run downriver to the coast and then a further sixteen-mile voyage across open sea to reach our destination.

Greg, Richard and I made the sixty five kilometer return sea trip to Pulu Puloi Island in this boat.

These two, like most boys, were interested in the technical side of most things, so we had a navigation briefing before leaving. I had fitted a bracket in the centre of the boat where I could mount a magnetic compass to steer by. It lined up accurately fore and aft in the boat. As ours was a wooden craft there was little chance of us suffering a compass needle error.

After stocking the boat with diving gear, food for the day, some motor spares and the compass, we swung downriver with great anticipation. It was a clear sparkling morning as we skimmed over the sea from the river mouth with the wake streaming out astern. The boat ran fast and as straight as an arrow on our set compass course.

Pulu Puloi is a small but beautiful island ringed by coral reefs and clad all over in green jungle. It reminded me of the mystical Bali Hai in the musical show South Pacific. It was uninhabited, visited only occasionally by local fishermen and of course the odd crazy Australian expatriate. It was said that a few wild pigs lived in the bush there.

My two sons were chuffed to see that the bearing we had plotted and followed with the compass was proving to be accurate. The island grew larger as we approached with the central peak now quite visible. On the way out we had discussions about running on the reciprocal compass bearing to find our way back again. It seemed very unlikely however that we would need to steer by compass on the return trip as the weather was so clear and calm and I intended to arrive back well before dark.

My two sons were enthralled with our little bit of paradise. We explored the island, dived on the coral reefs and enjoyed an unforgettable day at our tropical getaway. I had no inkling of just how unforgettable the day would be, as we headed back towards the river mouth, which was somewhere distant on the low smudge of the shoreline.

About half way back I was worried to see a thick and menacing cloud band rolling down from the mountains like a blanket of dirty cotton wool, obscuring everything in its path. These sudden and sometimes violent tropical storms are common near the equator and although not very destructive on land, could certainly pose problems for a small boat at sea. I watched the storm front with some apprehension as it swept down towards the coast and felt very vulnerable in our open dugout canoe. It was going to be a race to reach the river mouth before the heavy clouds and accompanying wind and rain blotted out everything.

To complicate our dicey situation, the Malili River mouth is quite difficult to find even on a normal visual sea approach. The river channel splits up into a delta before spilling into the sea, with many winding streams going nowhere, branching out through a tangle of mangroves and palms. We had to find the one main navigable channel with some accuracy before the storm caught us.

We lost the race to the river mouth and safety.

Suddenly the wind was buffeting us, whipping the sea up into a nasty chop. Our little craft began to ship water and visibility fell to just a few meters. The boat was riding well but I began to lose all sense of direction in the grey murk with no reference points for guidance. The three of us took up our duties in the boat, as it seemed that what we did from now on might determine our chances of survival.

One of the lads baled water from the bottom of the boat to keep us buoyant; the other checked the compass course, while I continued to drive. Visions of being capsized or losing our way miles from shore flashed through my mind like a bad dream, made worse by the responsibility of having my sons to account for as well. We were totally dependent on our calculated compass bearing after a while. What began as a light-hearted navigation exercise had now turned to really serious stuff and we drove on in the

The girl on the steps of the house by the jungle track always waved.

murky half-light and gusting wind squalls.

Like magic, the bucking of the boat became a smooth steady ride and we began to recognise the barely rippled surface of the main river channel. The palm-lined riverbanks were visible now, gliding by on both sides and we felt the tug of the strong current as we drove onwards.

The main channel is only about fifty yards wide at the mouth and we had entered it right on the nose. Dead reckoning and that little compass had brought us back to safety.

After coming ashore at our starting point back up the river, we straggled up the hill to the house, with the boys trying to act nonchalantly and play down our adventure to two very worried females. Realising we had not returned when the storm hit, they feared the worst – which was only natural - for females.

There was another character to meet the next day with whom I was destined to have a good relationship on this project. Pat was the company helicopter pilot and had flown Hueys in the American Army and under fire too during the Vietnam War. He now piloted the Company's high performance French Alouette gas turbine chopper and was a master at the controls, as he was able to demonstrate to me later. Pat's anecdotes of his wartime experiences were pretty intriguing, but the story he related about the tribal battle I think illustrates the calibre of the man.

The Alouette had to be flown the great distance from France to Indonesia to begin work there. Choppers are not the ideal long-distance carriers compared with the fixed-wing planes, nor are they designed for that purpose. Undeterred, Pat pulled out all the seats from his aircraft and loaded as many two-hundred-litre drums of fuel as he could carry and took off from the land of the Eiffel Tower flying solo, headed for Indonesia. It was necessary to land whenever his tanks were running low, to transfer fuel from the drums before heading off again. Being in a helicopter, he was able to land almost anywhere along the way. Pat related this story of a particularly exciting part of his trip. -

The Principal's office in the Malili Technical Centre - an abandoned warehouse

Passing over the highlands of Papua New Guinea at a fairly low altitude, he spotted some activity in a clearing ahead. There was a ding-dong battle being waged between two rival tribes of natives. The warriors were in two lines facing each other, dancing up and down behind their big war shields and hurling abuse and spears and arrows and all manner of missiles at each other.

As he was an avid collector of exotic artefacts, Pat saw a great opportunity here to add to his inventory. There was a momentary lull in hostilities as the big noisy bird interrupted their fun and the plumed champions gaped as this strange beast settled between the factions in the middle of their no-man's-land. Pat left his rotors running for a swift getaway. Leaping out he began to gather up the expended spears and arrows that littered the ground, bundling them into the aircraft. The apparition of this mystical being that came from the belly of the big silver bird and was making off with all their weapons suddenly brought the two armies to life and the hostilities began again with renewed ferocity. Flying missiles were thick in the air again but all were now aimed at Pat! He reckoned that this renewed supply of genuine native artefacts was too good to pass up, but as they were starting to zero in on him, he decided to take a leaf from the warriors' book and protect himself. The big door of the helicopter was easy to remove, so wrenching it off he used it as a shield and sallied forth again. With the missiles of the combatants pinging and clanging off his Plexiglas umbrella, he finished his collection and ducked back into the cockpit to safety. The astonishment of the assembled tribesmen can only be imagined.

Left standing almost empty-handed in the rotor wash of the departing chopper, they no doubt were already working up the story, which they would have to relate to the people back in the village. With all the details of the battle and the strange vision which appeared, told and re-enacted around the cooking fires that night, I have no doubt that Pat's appearance would have been big news. With a little more imagination, you could see it woven into the legends passed down in the folklore of those wild men of the Papuan highlands.

My close association with Pat and his helicopter plays a major role in the memories of my stay in Indonesia. While not being able to demonstrate the exploits that he described to me previously with the Papuan warriors, there were still some memorable feats ahead.

Pat flew me up into the mountains next morning to find the proposed mine-site. Nestling in the mist-covered highlands was a memorable picture. Lake Motano lay beneath us as a gunmetal blue stretch of water twelve kilometres long and reputedly 2,500 feet in depth. Pat pointed out where the mine would eventually be situated. All I could make out was a sea of treetops and thick matted jungle.

"I used the chopper to sling-load a small bulldozer in bits down there yesterday. There's a crew assembling it and they will soon be cutting access tracks out from the site, so construction can start."

This explanation gave me an interesting insight into just how you begin a large project in this kind of country. The adaptability of the Alouette was beginning to reveal itself. Given that the pilot knew his stuff, our aircraft would apparently prove to be a very versatile taxi, a truck, a crane, a general commuter and a fun machine as well.

Back again in Malili, I got down to work. Although a large technical training centre was planned for construction at the new Soroako town site in the mountains, I could not wait for these new facilities. Training takes time as everyone knows and given that my charges would be starting at a very basic level, there was a gaping shortfall that would have to be filled. This would not have been required to such an extent with the average Australian students of my acquaintance. My goal with these lads was to have

My friend Adwar the village cop

them able to demonstrate skills and knowledge of a satisfactory level and begin applying them by the time scheduled for the start-up of the mine project.

A recruiting trip I made to Jakarta identified twenty-four young Indonesian college students who were the cream of the crop in technical achievement by Indonesian standards. Unlike the average Australian lad however, who had grown up in an environment of technology with all kinds of machines commonplace in his life, these people had minimal exposure to technology and therefore little of the right relevant experience. It seemed we would have a long way to go and I have to admit I began the task with some feelings of doubt and apprehension.

It seemed reasonable that if you cannot communicate with someone you are not able to teach him very much, so I befriended the villagers in order to become familiar with the language. The trouble with this approach was that they spoke a local dialect, which was one of the many hundreds of different languages spoken around the country. Not much help there, but salvation was at hand in the form of a petite little Indonesian lady called Sri. She was an English teacher but had been directed to be my personal tutor by Andi and was available at all times, 'so that Mister Ivan will be able to teach us.'

My first walks down through the village created quite a stir. Like kids the world over, the youngsters trailed me in a giggling swarm pointing out my personal physical differences with gales of laughter. The boldest would dash up to me and examine my arms and legs closely, even stroking them, egged on by the others. Collapsing in hysterics they would shout the word "Monyet – Monyet" over and over. I learned later that the Monyet is a hairy monkey! Alongside their smooth brown and hairless bodies perhaps I did resemble one of those forest swingers.

It wasn't long before I began to discover ways with which to bridge the gap between us. Showing admiration for the children always seems to work and proud parents are the same all over the world it seems. Another of my 'conquests' was the local policeman. Thinking that I could do a lot worse than get on the right side of the law, I struck up a friendship with the uniformed one with the big gun on his hip. Indonesians love uniforms, weapons, parades and all things military. Adwar the village policeman was resplendent in his hip-hugging green uniform and cap and especially with the dazzling white webbing he wore on ceremonial occasions. His secret regret when we first met however was that his big revolver would not work and go bang! I was able to help him out with that nagging problem.

My camera proved to be another means by which I could make friends with the village people. Having set up my little darkroom in the house, I was soon ready to take and also print photographs. The village was a photographer's dream, with the people going about their simple lifestyle in such an unspoiled and beautiful setting. Many had never seen this magic before and clamoured for me to photograph them all. This meant that my efforts had to be restricted to recording only special events and ceremonies. These included weddings, deaths, new babies and street parades. While asking no payment for this service, I knew that it would all come back in their friendship and help. This proved to be the case. The villagers were predominantly Muslim and my pictures of the children competing in Koran-reading competitions were highly prized by their families.

A Sulawesi bride

One day I was drawn into the despair and the heart-rending grief of the villagers when a catastrophe occurred. One of the passenger vessels, which plied the coastline and the rivers, was overloaded by a greedy operator and had capsized with a full load of passengers. It happened in the swiftly flowing Malili River, just downstream from the village. This disaster drowned 110 men, women and children. The boat had rolled over, completely trapping most of the people under the awnings stretched across the deck.

The first I heard of this was when the headman knocked on my door, politely apologised for disturbing me and told me of the tragic accident. All the bodies had been recovered he said

285

and would I bring my camera and follow him. We walked in shocked silence towards the sound of wailing and keening down by the river and soon came upon the unforgettable scene.

The request was simple. Would I take a photograph of the face of each one of these dead people who lie out on the grass of the riverbank and make a print of each to give to the bereaved families? I have seen bodies before but never in such number and in such pitiful circumstances. The faces of the children were the most difficult for me to record with my camera. There were some I recognised as those who called me a hairy monkey and laughed at my embarrassment in the village street just a few days before. The most difficult part of the whole grim situation for me came in the darkroom later where it was necessary to closely examine each picture to ensure a good quality print. Those images will be with me for the rest of my life. What made this emotionally charged experience worthwhile however, were the looks on the faces of the grieving relatives as I gave them their precious photo, while they grasped my hand and thanked me through their tears.

Closer contact with the people in the village was a study in contrasts to my Westerner's eyes. Long accustomed to our strict health and living standards in just about everything back in Perth, I was intrigued by the extremes of beauty and squalor existing side-by-side here. These people lived in a situation of open sewers, clouds of flies and black mud churned by the passage of feet over generations. There were no vehicles, so people walked everywhere or rode one of the few pushbikes. Yet I never saw a dirty or dishevelled person anywhere. The women especially spent a lot of time on their appearance and looked perfectly groomed, even when seen picking their way between the pools of black water in the roadway. Walking behind a lovely looking young woman one day I could not help but admire her glossy black hair with a silver clasp and her graceful movement in the colourful sarong. There were no toilets as we know them, in Malili, just the river close by where you go to relieve yourself. The girl who had claimed my attention must have felt the call of nature as she stepped off the side of the road, nonchalantly spread her feet apart and peed on the ground. The simple innocence of it all was in the completely natural way it was done, with the long sarong almost completely concealing the procedure. It seemed that I was imagining all this, until one day in a roadside store, a woman who had been standing quietly close to me, moved away leaving me with a very wet boot. That's a glimpse of life in a remote Indonesian village.

Malili existed mainly on the earnings from fishing. The catch is mostly a kind of small sardine or smelt. These are delicious, especially when eaten with a cold beer, having been fried with garlic and served crisp. My enthusiasm for the delicacy was blunted somewhat when I saw, one hot day, the catch spread out on trays in the sun to dry. The stench was incredible with the trays a moving mass of buzzing flies. No doubt this is the secret of the exceptional flavour. It would seem that a strong stomach is a good thing to have if you live in an Indonesian village. I luckily did not succumb once.

As there was no school building where I could house and train my band of twenty-four students, I commandeered a large abandoned warehouse on the edge of the settlement. I had it fitted out with a good selection of machine tools and mechanical repair equipment imported up from Australia. Local carpenters were employed to make classrooms and offices in the building while I drew up all the lesson plans and bought the textbooks and lots of blackboard chalk.

Needing staff to help in the running of my little college, I hired Andi and a local girl who was to be my secretary. Tati, as she was known, had had secretarial training at a commercial college in Ujung Pandang, the capital city of Sulawesi. She had learned Basic English too, which was a big help as my Indonesian was still pretty woeful. Sri, my personal language trainer scolded me constantly when I made mistakes in the new language. Under this kind of relentless pressure, there seemed to be a good chance I would be competent enough to speak in the classroom with the students when they arrived.

The Indonesian populace is fanatical in its dedication and hard work when given a job to do. One of the reasons for this I found is the business of losing face, affecting either themselves or their employer or superiors. Tati was a good example of this, arriving early when I opened the school office and tapping away doggedly at the typewriter and never leaving her post for anything. I say this with some conviction.

Sri told me that Tati had recently married and was newly pregnant. Bouts of morning sickness would grip her almost every day, but she handled these with aplomb. While sitting at her typewriter at a particular time each morning you could set your watch by, Tati would start to make strange gurgling noises and the rate of her typing would slow. With only a brief interruption in her concentration, she would lean over sideways and chunder into the wastepaper tin most unconcernedly. When finished, she would sit up straight again, look at me sheepishly with her big brown eyes and take up the typing where she had left off. I just had to give her full marks for unswerving devotion and dedication.

In the early stages of staffing up the expatriate workforce our small settlement at Malili had only a few families living there. With no European doctors available the Company moved a young Indonesian male doctor into a small clinic there to cater for our medical needs. He seemed to cope well enough with the workload of minor treatments until one day there was a ripple of disquiet among the expatriate women there. The ladies thought it strange that almost every general examination began in a rather personal manner. Further enquiries as to the doctor's professional background revealed that his training had qualified him as a gynaecologist only and he apparently knew little else. It appeared that when a woman went to him for treatment for any of a hundred medical problems he would begin in the way he knew best. Most of us are accustomed to doctors looking down our throat and poking in our ears, but this was embarrassing – for the girls anyway.

One thing was becoming plain - there was much we would have to get accustomed to in the years ahead.

After a time, it was dawning on me just how honoured I was becoming among these Indonesian people. While many of those at Malili were local villagers, I did have dealings with some of the better-educated, higher officials from time to time. I was forming a genuine affection for them, both the primitive and the educated. They were very cheerful in the face of hardship and I noticed were strongly family and group-oriented. They loved their children and were intensely loyal and hard working with a fine sense of humour. My position among them as a teacher automatically afforded me a very lofty status in their eyes.

Tati was my dedicated secretary

287

They were so eager to learn the ways of modern societies like Australia. We are held in very high regard apparently and being Australian and a teacher too, seemed to push me somewhere up there near God.

While this kind of halo effect was fine, it did have its drawbacks. The perception by my students that I was expert and all-knowing in everything, which of course was far from the truth, meant that they were bound by their customs never to query anything I said. It was a case of - if I told them that black was white; they would agree and never query my word. They believed that if you asked a teacher for clarification on a point, you were questioning his knowledge and his authority and risked causing him to lose face – insulting him in fact. Gradually it began to sink in that not only would it be necessary to contend with their low technical knowledge base but also to try to impart knowledge in the face of enormous and ingrained cultural differences. So, it was back to the drawing board as they say. Fortunately, I did find ways of getting around these problems in time and learned much myself in the process. It seemed that a bit of lateral thinking was the key.

My group of eager trainee recruits had arrived and were billeted in the Malili settlement by Andi Azikin. Andi was proving to be my right-hand man and my invaluable link with the lads.

Our Hornbill made a good school mascot

Perhaps due to his royal lineage or the father figure he presented, he soon showed perfect control and began to take care of their every need.

These lads, aged from about nineteen to twenty four, were a mixture of the products of Java, Sumatra, Bali and Ambon. The evidence of Chinese blood could be seen among them and also some Malay. Some were light skinned suggesting a smattering of the Dutch or Portuguese colonials too. All in all they were quite a mixture of ethnic and racial backgrounds and were predominantly Muslim. I had handpicked them from what was offered as the cream of Indonesian senior high schools and colleges. They were well turned out, fresh and eager to make a career for themselves in this large international mining company.

"We want to be a part of making Indonesia great," they said with conviction.

My little warehouse school at Malili, Sulawesi, Salatan, in the Republic of Indonesia was ready, so we began with high hopes and fingers crossed - mine were.

A description of our little technical college would be incomplete without mentioning the college mascot. As with many groups and teams, an animal of some kind is sometimes adopted, or in this case the animal adopted us. The mascot's name was Rex. About the size of a pelican with technicolour plumage, it waddled out of the jungle one day and took up residence. It seems that the feeds he scabbed at lunchtime confirmed to him that he had chosen wisely. Rex was a red-crested hornbill and though gentle by nature, looked pretty intimidating with a mean look in his eye and a beak like a bulldozer blade.

My lads I noted were all very fit, lean and tough. Their favourite sport was martial arts and they all had a military bearing and were highly regimented. This was not surprising as Indonesia had a military style of government with the generals practically running the country. The students loved parades, uniforms and discipline. I felt that if I could work on this side of their national character it might lead to success.

I decided that our little training establishment was going to be set up and run along military lines - to try and meet their expectations. I had all the lads fitted out with smart and stylish tailor-made coveralls, which looked good on their universally athletic bodies. There were nametags sewn on every pocket and all, including me, wore an ID badge. There was a list of duties drawn up for each of them, and everyone was awarded a title and rank befitting their individual roles.

One aspect of this organised situation emerged quite soon. They became so enthusiastic I had to curb their fire a bit.

Jimmy Yapianto was checking the gates and perimeter fence one night when an armed and uniformed Indonesian Company security guard tried to get in. These guards had formed a little army and had become pretty arrogant. While never challenging me directly, they were always bossing the locals around and looked pretty trigger-happy to me with their Russian AK47 assault rifles. Jimmy apparently felt his responsibility was to do his job in protecting the school compound and so he attacked the guard with his bare hands. From eyewitness accounts it was a very uneven contest. Jimmy disarmed the guard and had him bleeding and powerless on the ground in seconds.

I had to arrange for an apology and a compensation package to be sent to the security group and also paid for the guard's medical expenses, but they never interfered with us again. Jimmy took it all very calmly.

"I was only doing my job as you told me Mister Ivan," he explained simply.

He must have felt pretty confident in tackling the guard, and the possibility of getting shot with an assault rifle had apparently not occurred to him.

This account of our time in the tropics would be incomplete without some reference to the wildlife. While I am sure

Meet Irving, Rod's mad monkey. That's not his tail - he doesn't have one to speak of. It's a garden hose.

289

that the jungle supported a wide variety of animals, they were seldom seen, as much of their habitat was virtually impenetrable. The one animal that frequently came into the lives of the people was the local monkey. He was actually called the Sulawesi Ape and he had no tail to speak of. I learned later that he is a member of the Macaque family. Bev adopted one of these as a pet while it was quite young and began to regret her decision almost immediately. They are the most inquisitive, mischievous and downright destructive of beasts and can be bad tempered to boot. One day Marilyn as she was known bit Bev on the hand, which required that she have a tetanus shot. She was not impressed, having to front up to the local Indonesian doctor – who was really a gynaecologist. Many people there who had a monkey would take it for walks on a lead. An interesting thing here is that you don't put the collar around its neck as with a dog or a cat. They don't have much of a neck, so the next best place is around the middle - the narrowest part.

An Aussie friend who drove one of the large transport trucks had a Sulawesi Ape called Irving. He played up a bit too, but Rod the driver would take no nonsense and would give him a clip on the ear to show him who was boss. To dodge the roaming dogs around the town, Irving climbed a power pole every night and walking and balancing along the live cable he would sit down, lean back on the insulator and sleep there until morning.

Irving came very close to joining his ancestors one day. He loved to ride sitting up on the load on the back of Rod's truck, pulling faces at everyone they passed on the road. He still wore his collar and lead that was fastened to the truck so he wouldn't be lost along the way. The truck had begun the long hill climb to the mine one day when it gave a sudden lurch. Irving fell off and down onto the road. With his lead still attached and the truck still moving, there was only one thing he could do – start running. After they had covered some distance like this a bystander spotted the monkey pounding along the road behind the truck and hailed the driver. Rod's account of it, about which I had some suspicions, was that the witness told him there was a monkey trying to overtake the truck so would he pull over and let him pass. The monkey was so knackered after his marathon run that he just collapsed on the road, apparently deceased. An Indonesian who saw this just shrugged his shoulders and said "Monyet itu habis contract!" (That monkey has had his chips).

Rod left his wallet in the truck cab one day and Irving stole it, and then climbed up on the load as usual when they set off on another trip. A few kilometres down the road a pedestrian yelled and pointed up onto the truck. Thinking there was a problem with his load Rod stopped to look. The wallet had contained a large sum of money and travellers' cheques drawn for his forthcoming holiday trip. These had been pulled from the wallet, shredded and were scattered like confetti as far back as he could see. Irving meantime was jumping with glee up and down with a sadistic grin on his face and with the empty wallet jammed on his head. It seems that the use of the word mischievous when describing monkey behaviour falls way short of the mark.

The village headman called on me one day on a very official and important matter. Having completed the customary protocols and complimentary introductions over a glass of tea, we finally got down to business.

"All of my people here admire your students very much," he said. "We are going to have a big street parade soon and would be very honoured if you would agree to allow your contingent to march in it."

I accepted his offer after consultation with the students. They were full of assurances that they would make me proud of them on the big day.

Before leaving, the headman extended an invitation which protocol dictated I could not easily refuse.

"You, Mister Ivan are the senior representative of your Company here in Malili. I, together with officials of Government and Police will review this parade from a podium and take the salute. I wish that you should join us." I accepted with thanks.

There was a brass band of sorts leading the parade and making a huge din. What they lacked in musical talent was more than compensated by enthusiasm, sheer volume and spectacle.

The main thoroughfare curving down from the settlement to the village was a riot of colour with flags and bunting flying. The crowd appeared to comprise every man, woman and child in the district and never have I seen so much excitement and sheer enjoyment. I stood to attention on the reviewing stand with the other VIPs, trying to look as important as they did.

There is no doubt that I felt immensely proud of my team, as they swung down the road in perfect cadence with the band. They made a fine spectacle in their smart blue uniforms, proudly sporting their badges and chevrons and shining boots all topped off with bright yellow industrial helmets. To a man, they wore their hard hats at the identical jaunty angle, and the crowd cheered.

My role was really only as a big fish in a little pond there in Malili, but I was caught up in their special day. It felt good to be an accepted member of that happy and wildly enthusiastic Indonesian island community.

Training these boys was hard work, but they compensated me with their enthusiasm and dedication. One of worst times of the day was when I was trying to get them all to pack up and go home. They had to be practically thrown out.

It was all great fun though and the schooling was advancing at a reasonable rate.

My students were smartly turned out at the Malili street

Andi Azikin was my liaison with the students.
Here he is with friends - second from the right.

291

Red sails off the Sulawesi shores. Fishing boats return home after a day at sea

Life by the Lake

The marks of class distinction are very obvious in Indonesian communities, with the wealthy and important people being segregated in more ways than one from the lower classes. Built along the shore of our beautiful Lake Motano was the company town of Soroako and in the national style we had segregation of a kind within the group of expatriates too. There were four levels of housing; A, B, C and D. Each group was separated from the other by a considerable distance. The manager was in the only A house, the senior staff where we were occupied the B section, middle management were in the C block and the indigenous employees could be found in section D. Our home was built on stilts on a slope which came up from the lake in front and on up to rise to a jungle-covered mountain behind us. Many times we had seen the dense cloud layer at the top slide down the face of the mountain to seemingly rest on the roof of the house.

I am the Official Company Photographer

As the appointed Company photographer I was asked to film some of the major events that took place at Soroako and at the mine. One very important event had been scheduled, which had all the locals in a tizz. It was a visit by the man himself in Indonesia at that time – President Soeharto and Madame S. too. It was the celebration of the opening of our new nickel mine.

In a big quadrangle, resplendent with flags and bunting, we expats waited, dressed in our best bib and tucker, sitting in a small viewing stand enclosed with railings.

Two big helicopters thundered in and landed in clouds of dust and as we watched, the presidential party, which was quite large, disembarked from one and took their seats in the 'royal box' about forty yards to our front. The reason for having two helicopters I learned was to confuse any would-be assassins. As the music played and the speeches droned on I became increasingly twitchy. It was essential to get some good pictures of President and Madam Soeharto, but from where I was sitting, they in the shade of the canopy appeared just as indistinct blobs. One thing I did notice about the distinguished party was the tight ring of security guards, which stayed close to his personage with loaded automatic weapons at the ready.

Well, it's now or never - so I ducked under our railings and with a cheesy grin walked directly across the square towards the President with my hands both in clear view. I tried to dispel the feeling that I heard the faint clicks of safety catches being released. Despite the shuffling in the ranks of the bodyguard, I kept going and secured some really good close-ups of the great man and his wife. Madam flashed me - a dazzling smile, despite my apparent audacious behaviour in approaching them.

For recreation, the community of expatriates in the new town at Soroako on the lake devised various means for enjoyment and social interaction.

In a workshop under that house I built our family catamaran. Using marine plywood and odd bits of timber filched from the mine site, the boat was completed and we were soon cruising on the lake in decadent style and comfort propelled by a big outboard motor.

Obviously not wanting to be outdone, many of the growing number of expatriates began to assemble a veritable fleet of craft so they could enjoy the beauty and tranquillity of Lake Motano. It is large, about ten kilometres long and has a depth reputed to be 2,500 ft. of clear fresh water. The fish we saw were small and few and far between, so no one was tempted to cast a line in. The range of the expatriates' boats to be seen on the water at weekends was quite diverse and ingenious. Many were catamarans employing two native dugout canoes fixed side by side with a deck built across them. Some even had a frame with a thatched attap roof to keep off the hot sun. Ours was a bit more sophisticated with a sunroof on top and a ladder to reach it. All were powered by outboard motors of various sizes and makes.

On one of the most memorable trips up to the village at the head of the lake we motored into a small cove hidden by jungle ferns. The water of the crystal pool there was so clear, being fed by a fresh water spring bubbling out of the bank. Its main attraction however was that it was the home of myriads of butterflies. They were there in profusion and in all colours of the rainbow. Imagine this serene setting, which Bev, Lise and I enjoyed there that day. I had a small charcoal-fired Hibachi barbecue sitting on the deck with some prime steaks sizzling on the grill and when the warm weather made us thirsty we extracted ice cold drinks from the built-in icebox on board. In that wonderful tropical hideaway it was easy to forget the cares of the world and submerge into the deep blue waters of the lake. Almost without a word we three stripped off, jumped overboard and swam around like a family of seals – in the nick. That's what I call freedom and togetherness.

When Pat, our company helicopter pilot related the story of his involvement in the New Guinea highlands battle I was intrigued. He showed me the collection of native weapons he had acquired during that fight and I began to see him as an adventurous character who promised to be interesting company. It was not only the potential for some excitement flying in his high performance Alouette that attracted me I told myself, but also we formed a great bond and ultimately had some great times together on the Inco project.

Some of the stories I can relate about hairy flights in Pat's helicopter may paint him as an irresponsible larrikin, possibly putting himself and his passengers at risk. He was such a contrast to the usual run of commercial pilots of my acquaintance, who to be fair were mostly bound by their strict flight regulations and aircraft performance limits, but nonetheless seemed pretty stodgy and unimaginative by comparison. Pat was licensed to fly fixed-wing aircraft also but it is an interesting rule he told me, that if you flew one, you couldn't fly the other and switch around at will. It's something to do with the radically different behaviour of one, which can interfere with your reflexes when flying the other. Pat's explanation of the difference between them was simply that a fixed-wing aircraft wants to fly and a helicopter doesn't. The bit of experience I had at the controls of both types certainly bears this out. We were a long way from any commercial airlines flight regulations in such a remote region and given that you have a high performance machine under you, why not indulge in a bit of fun? This place was a helo pilot's playground.

Much of his flying time had been served in the war in Vietnam. Landing and extracting troops in the bush, much of the time under enemy fire, had given him an edge that few commercial pilots would ever achieve, nor would they have need for. It was 'seat of the pants' flying over there, with exceptional skill as the main ingredient that kept you alive.

Pat confided to me that he had left a true love behind in Jakarta. Her name was Kate and he was keen to have her join him in Malili. As we had some vacant houses and I was the senior Company representative in Malili, I was able to arrange this for him. He said that I could count on his undying gratitude.

Kate was born in Monado, a large town on the northern tip of Sulawesi Island. Monado is renowned for its beautiful girls, which are a mixture of ethnic Indonesian and the Portuguese colonists there. They are taller than the Sulawesi locals with very light coffee-coloured skin and are in great demand as dancers and models in the larger cities of the country. Kate was a fine example of these lovelies.

Our pilot's other passion was rainbows. While flying in the tropical rainfall areas of the world he was in a position to observe plenty of them. He asked me one day had I ever seen a round rainbow. I had to admit I hadn't, as all those in my experience were an arch with both ends disappearing at ground level. I even doubted his claim that there was such a thing as a round rainbow any way. He said that conditions were ideal at the moment and would I like to go chasing rainbows for a while.

The 'test flight' was logged and we headed for a likely area of rain showers and sunshine out over the mountains. You apparently cannot observe these circular displays from anything but a helicopter. A fixed-wing aircraft flies right past at a fixed altitude while a helicopter can be carefully manoeuvred to be in the exact spot in space for the display to be seen.

We found a likely rainbow above the mountains but it was the conventional kind and needed some work on it. Pat said, -

"Now just watch this."

He very precisely moved the aircraft in a hover slowly up and down and from side to side watching our rainbow intently all the while.

Gradually the crescent shape began to form into a circle and after a lot of careful flying, there it was - this glorious kaleidoscope of saturated colour completely surrounded the aircraft and drenched the cockpit with its spectrum's glow. It was indeed a perfect rainbow band, entirely circular in shape and it seemed to be about a mile in diameter.

Pat then played with his enormous sky painting for a while, changing its shape and then returning it to the wonderful natural original work of art. I have not met anyone since who has witnessed this glorious phenomenon.

All you need is a tropical rainstorm at about four thousand feet, with a burst of bright sunshine and be in a helicopter with a pilot who likes looking for round rainbows. The mind images imprinted on that day will never be forgotten.

More expatriates were arriving on site, many of them Australians. A number of these were qualified teachers and training people who were assigned to me in the training department at the mine site. Few of these men had ever flown in a helicopter as they were from city-based jobs back home. It seemed too good an opportunity to miss in giving the new chums a taste of bush flying. I must ashamedly admit that Pat and I conspired to give these new arrivals the flight of their lives. His codename for this bit of mischief was 'the two dollar ride.'

Imagine if you can, sitting in a helicopter just droning along flat and level, admiring the view. Down through the plexiglass cabin floor you can see the dense mat of the jungle canopy swirling in the rotor wash as you pass. There is a feeling of relative security and a connection with mother earth, which seems to instil confidence and dispels any apprehension, even in the new arrivals. At this point you are actually flying along just above a high plateau, which is part of the mountain range in the interior of Sulawesi Island.

The jagged edge of a canyon sweeps under the nose and you are suddenly in empty space with all your reference points missing. You look down a vertical cliff face laced with waterfalls and clinging vegetation to a narrow winding ribbon of a river a thousand feet below.

Imagine now that without warning and with a gut-wrenching loss of gravity, your aircraft noses into in a sudden power dive and plummets down the cliff. You hang weightless in the safety harness as the foam-flecked rocks of the river rush up to meet you and sudden death seems sure and certain.

At the point when your life begins to reel off in front of your eyes and the crash seems inevitable, the chopper flares out in a gut-wrenching return to level flight, ramming you hard into the seat and roars off chasing the winding course of the river between towering canyon walls. The deathly silence in the cabin is understandable, and is relieved only by the pilot whistling tunelessly some Yankee song – no doubt aimed at easing the tension among the passengers, but in fact seeming to have little effect on them.

This 'two dollar ride' always unnerved me even though I was able to anticipate it coming and surreptitiously cinched up my harness and grasped the edge of the seat before the dive. The effect on the new chums must have created an unforgettable scar on their minds. The staring eyeballs in panic-stricken faces said it all. Many took the dim view and while conceding to Pat's flying skills, probably considered him to be stark raving mad.

Another, milder one of Pat's little jokes involved the helicopter flight manual. While flying over the mountains on an otherwise routine bus run to Soroako with new people on board, he would take his attention off flying and begin rummaging around the cockpit and under the seat for his flying bible. In a loud voice he would ask me had I seen his 'goddam Flying Instructions Manual' anywhere. When I retrieved the book from under the seat and handed it over, he would read out the section titles from the index, also in a loud voice. Then he reached the bit about landing the aircraft and read that out too. He accounted for this weird behaviour by explaining to the ears straining from the back seat, that he hadn't flown this type of aircraft much before and that he was just revising the correct way to land so he didn't forget anything. The jaded looks and the mutterings from his passengers showed that this little scam was having the desired effect. He was a worry at times.

In the very bad weather conditions of a sudden tropical storm one day as we took off from Soroako, Pat was helping out a fellow pilot. It was common practice for one aircraft to 'talk in' another on its approach, in conditions of poor visibility, until he had the airstrip in sight and could land safely. Our helicopter jock was doing this one day, while looking back over his shoulder as he did so, keeping the other plane in sight while talking to him on the radio. While apparently not looking where he was going, he flew us directly towards a towering rock face just off the end of the runway. This monolith in the path of the planes does not normally disturb passengers, providing the pilot appears to have seen it and is preparing to take evasive action. Pat, to all intents and purposes was oblivious.

As a part of this con, I was supposed to shout a warning to him at a certain point when he would let out an expletive and pull the chopper round in a tight bank to roar along the rock wall with metres to spare. His final comment was for effect, –

"Goddam that was close!"

Well rehearsed, this little bit of drama is a real heart stopper for the uninitiated. Looking over to the back seat to reassure the distraught passengers on one occasion after a typical performance, I saw an amusing sight. One of the men had reached down for his industrial safety helmet stored on the floor and had jammed it firmly down on his head with only two staring eyes visible under the brim. When I asked him later what good he thought it might have been in an aircraft crash he replied that it must have been a reflex action and seemed like a good idea at the time.

Throughout all these shenanigans, while it still scared me a bit at times, I maintained the utmost faith in Pat's flying ability. I knew too that under the assumed ineptitude and stupidity he displayed for effect, he was in complete control at all times. He had nerves of steel I reckoned.

Not to be outdone, the fixed-wing pilots played little games on their flights over the mountains too.

The 'STOL' configuration aircraft that they flew, is reputed to be very strong, is very manoeuvrable with a powerful engine and is favoured by the bush pilots in this part of the world.

While in the seat alongside my other pilot friend Jim one day we were in one of two fixed-wing aircraft that departed Soroako at the same time. The course was back over the mountains or 'down the hill' as this trip was known. I was taking little notice as the two planes separated to fly in different directions down through the mountain passes. It turned out to be a cat-and-mouse game with each pilot trying to ambush the other with a surprise attack in the maze of ravines and canyons.

Jim had managed to outfox his 'enemy' and was in a power dive on his tail as he streaked out of a canyon hugging the trees. The sound of a burst of simulated machine-gun fire broke the radio silence as Jim pounced on his victim. The game resumed in earnest with displays of flying skill, the likes of which I had never seen before or since and the engagement only broke off as we came in serenely and put down on the Malili airstrip.

Later, in the bar you could be excused for thinking you were in a fighter squadron mess during the battle of Britain, to overhear the two pilots discussing the tactics and manoeuvres of their operational sortie in the sky over the mountains.

There was something I learned and found useful. The Indonesians respond well to training providing you can bring them up to a good basic level before introducing the more technical stuff. It was reassuring to find that my lot seldom let me down, but the expatriate pilot who had been assigned a young Indonesian trainee was not so lucky

Bruno had done his mandatory initial flight training but could still not be relied upon in all situations. Jim, our American pilot explained it to me, -

"I had given Bruno control of the aircraft one day from the right hand seat while we were crossing the mountains. Feeling confident, I began to relax and read a newspaper. Suddenly looking up I could see we were headed for a mountainside directly ahead, so had to grab the controls and sort out the aircraft. Bruno was warm and comfy in the morning sun in the cabin and had trimmed the plane for level flight and then gone to sleep!"

This was something I could relate to, as I had seen the locals at various times during the day draped over the landscape in the weirdest of relaxed attitudes, completely out to it.

"He was good at chopping up dogs with the propeller on landings on the airstrip too," Jim added. When I asked Jim how he himself handled the dog problem he said, "Well not with the prop - it's untidy and you can bend it. I try to whack 'em with the undercarriage."

Bruno put on a star turn one day at the strip. He was taxiing the plane over to where the passengers' luggage was stacked on a trolley. I must ask you first – have you ever seen a high stack of suitcases and personal effects with a spinning aircraft propeller going through them like a big circular saw? Oh the mess, and the noise! Parts of cameras and portable radios skittered along the tarmac, while thinly sliced suitcases were tossed high in the air. There were shirts and pants and frilly underwear picked up by the breeze, descending in tatters at the feet of the horrified passengers. I said to myself, -

"Bruno, you have really blown it this time."

My cranky office phone jangled one morning.

"Hi Ivan it's Pat. I've logged a test flight in the Alouette this morning – destination Pulu Puloi. Wanna come?"

Sweeping aside all work-related trivia, I headed for the helo pad.

Some passengers were already loaded and while climbing aboard I noticed the freight locker contained two chainsaws and a variety of machetes and parangs. There was also some beer. This looked like it was going to involve some hard work and it promised to be a hot day.

He revealed his plan as we flew out over the Bay of Bone to the island.

"I want to build a landing pad for the Alouette in the jungle out there," he explained, "so we can go out any time we want."

Our landing on the steeply shelving scrap of beach was pure artistry. The narrow strip of sand was only visible at low tide, which was now. Pat had done his homework with the tide charts it seemed.

We set to and felled all the coconut palms and trees in a rough circle, which was to be the landing area, leaving the trunks an even height of about a metre. These then became the supports for the platform, which was constructed from bush logs. All brush was cleared away and the job completed by the time the tide came in forcing us to leave the island. The chopper lifted off without getting the skids wet.

Needless to say, there were more 'test flights' logged after our construction efforts on the island. Many happy hours were spent enjoying the unspoiled attractions of Pulu Puloi - our jewel out in the Bay of Bone.

Then the construction phase was completed at the mine and the helicopter had to depart, returning to civilisation piloted by my mad friend Pat.

There is little doubt that by now the jungle has swallowed up our private landing site and all is green jungle there as before. I am sure too that the wild pigs over there are relieved with no noisy chopper disturbing their peace.

The Indonesian story for us was one of exotic places and people, with a lot of hard work and much personal satisfaction as the end result. My daughter Lise made many friends, especially among the young Indonesian students who spoiled her with their undisguised care and attention. When asked later about Indonesia, Bev however might say the jury is still out on that one as it was so different for an Australian housewife living there. I do know that we will not forget the sights and sounds and smells and the experiences we shared.

To remind ourselves of those interesting times we have been heard to drop a few of the local words now and then, so – selamat jalan Pak.

Episode Fourteen

God's Refuge

With the Indonesian experience concluded in 1978, my feet were now firmly planted back in Western Australia. It seemed however that the call of distant and remote places was not entirely stilled and I began looking again for more of the same. Having had previous employment with Wapet (West Australian Petroleum) in 1953 in the Pilbara and Kimberley regions of the state, I looked to that Company again for a job. With some of my credentials improved, it seemed right to try for a training position in the field, which was my location of preference.

Wapet took me on as the Field Training Officer to be stationed at their oilfield on Barrow Island off the coast of Western Australia. The offer was, to have a free hand in establishing a training facility for the island personnel, an autonomous position that suited me well.

The Rock Wallaby - curious and unafraid

The island, which lies 70 kilometres off the Pilbara coast in our North, comprises about 230 square kilometres of semi-arid landscape set in tropical waters. Native animals are apparent in numbers and seem to be everywhere, showing little or no fear of man. These native wildlife species comprise 15 kinds of land mammals, 7 marine mammals, 110 birds, 54 types of reptiles and one frog, which I never could find. This indigenous community has flourished in its isolated environment since its home was cut off from the mainland by rising sea levels about 8,000 years ago – or so we have been told. Unlike many of their mainland cousins, which have been harassed by predators, including people, these have survived and flourished. The feral foxes, dogs, cats and rats have pushed many of the native species to the brink of extinction elsewhere, while apart from the effects of the predators in the natural food chain, the island inhabitants have no enemies. Barrow Island is reminiscent of a living and breathing Jurassic Park – in miniature; truly God's refuge for some of nature's survivors.

The government recognised Barrow Island's value as a sanctuary for its endangered native animals in 1908, when it was declared a public reserve. This was upgraded to the highest level of protection as an A-class nature reserve in 1910 and so it has remained since. The island was originally named in 1818 by a Royal Navy Lieutenant on an exploration voyage, in honour of the then Secretary of the Admiralty, who also founded the Royal Geographical Society in England. In 1873 and prior to its declaration as a

299

nature reserve the island barely escaped decimation. It was thrown open for use as a pastoral property. Thankfully the lack of surface water prevented the introduction of grazing stock and so it remained pristine and intact.

Barrow Island looked promising to Wapet's field geologists who were looking for oil. Further work in that direction came to a halt when Britain proposed to detonate an atomic bomb on the Montebello Islands, a scant 16 km to the North. The whole area within a 72-km radius of the blast site was declared off limits and Wapet just had to wait.

Given the green light at last, the Company moved in the drilling rigs in 1963 and oil flowed to the surface in 1964. The record shows that Wapet subsequently posted a production figure of 100 million barrels of light sweet crude in the first eight years of production. We may well have asked - what happened to the wildlife while all this was going on?

The Company must have done some pretty fast talking to persuade the government to allow exploration there, much less to establish an oilfield that extends over most of the land area of the island.

Enter the champion of the wildlife - a well known naturalist called Harry Butler. He dedicated himself to the task of ensuring the indigenous fauna would co-exist with the oilmen and that the oilmen would respect the rights of the wild life too.

He succeeded on both counts

On the first trip to Barrow Island, I sat in a window seat of the Fokker aircraft and watched the island roll over the horizon and under the plane. With no prominent features it appeared as a flat and semi-arid pancake of land with a covering of spinifex clumps studding the sparse red soil and cap rock. Descending closer I could pick out the few buildings clumped together at the airport and further up the coast the main camp complex. Out in the plain the oilfield Lufkin pumps were nodding amongst the tracery of tracks and roads. From the collector stations, production pipe work created spider web designs radiating out to link the individual wells in the area. A group of big black storage tanks squatted on the northernmost tip of the landmass and randomly, the red flame of a flare rose and fell marking the separator stations. All this was ringed with a necklace of creamy surf and white beaches that contrasted with the incredible blue of the Indian Ocean.

My little empire on the island was centred in a long building on a hill with a fine view of the island oilfield. The training centre comprised a large lecture room with all facilities, with an office and a workshop at the back. The aim was to settle in quickly and design the courses that would be needed for the workforce and prepare training material with which to do it. Most of the needs were for technical and skills training together with the important areas of safety, first aid and fire fighting. Unlike most projects in my past experience however, the topics of natural conservation and the preservation of wild life played an important part too.

The complement of people operating the field numbered about 120 men with the occasional female filling the role of medic at the clinic. To add variety and extra technical content to my courses, I would often co-opt one or other of the experts in specialised fields to give lectures. An obvious and popular choice was the medic who put the trainees through the procedures of first aid and CPR. Having the skills to perhaps save a life had a lot of appeal for the troops who could never know when it might be required.

Burrowing Betongs, an indigenous species, extinct elsewhere which are seen along with kangaroos, wallabies and others

We had a male medic whose field nickname was 'Sailor', for some reason I was never able to determine. He was a regular lecturer at my induction courses for new recruits. His favourite subject and a very important one for that location, was entitled 'Bities on Barrow.' This had to do with all the precautions you should take to avoid being bitten, stung or otherwise suffer injury to your person, with possibly very painful consequence or even death as a consequence. Prominent in this list of villains were the stonefish, the blue-ringed octopus and the stingray. The sea snake was awarded the guernsey for having deadly venom too, which could kill, but it had a very small mouth, apparently finding it difficult to fasten on to a larger animal like a person. At least that's what the pundits said. Even the humble kangaroo tick got a mention. There were a lot of them around in the scrub and if the site of a bite on a person was not attended to, it could lead to a fever or a nasty infection. Sailor took a delight in removing ticks from people, especially when they were found buried in the nether regions of the victim. He would get the tick in the sights of his ever ready spray can and freeze it to death. With the tick deceased and the victim/patient numb with the cold, the scalpel would come out to exhume the offender. Sailor was the recognised and established and revered authority on all bities there on Barrow Island.

While walking to the airport one day to catch a light aircraft flight to the mainland I was surprised when the field ambulance pulled up alongside me in a cloud of dust. While accepting the lift offered by the driver, I sensed that something was amiss. Everybody knew that driving the field ambulance was Sailor's job, but this fellow was a stand-in. I enquired as to where Sailor was at this time.

"The silly bugger is the patient in the back of the ambulance and we are going to medivac him to the hospital onshore." I asked had he suffered an accident.

"I suppose you could call it that," said my informant. "He was walking across the reef at low tide this morning in his rubber thongs and a bloody great stingray in a pool speared him in the foot. The barb went right through under his Achilles tendon. What a bloody nong!"

The reason I suspect for the disparaging comments made by the driver, was that in his lectures, Sailor always laboured the point that sturdy boots must be worn at all times when one is walking out on the reefs around the island.

"Yes," I had to agree, "What a bloody nong."

The injured medic was grimacing and groaning with pain as the poison was doing its job and his foot was swelling to the appearance of a soccer ball. As the only other passenger, I sat alongside the stretcher on the floor of the plane and assumed the role of medical escort and reassured him that I was there with him and everything would be all right in the end. I struck up a cheerful conversation.

"You are a medical expert and know about these things Sailor, is it likely that you will lose consciousness?" Between groans and grimaces he replied, –

"That's possible, I suppose I might. Why did you ask me that?"

Actually, I was imagining myself administering CPR by pumping his chest heroically and giving him the kiss of life to avert an untimely death and replied, –

"Well, I would like to practice those CPR techniques you taught me and perhaps even save your life." His face appeared to take on a more sickly pallor after this offer.

Having a beer together at the bar later when all had returned to normal - including his foot, I asked how he was feeling and if he remembered the plane trip.

"I was a bit disappointed when you didn't pass out after all when I would have had to give you CPR," said I.

"No bloody fear," he replied. "The thought of you getting stuck into me kept me awake – I may not have survived that! Anyway, thanks for the offer."

"Well, that's gratitude for you," I said to myself.

In referring to the medic who was known exclusively by his nickname Sailor, I could open a veritable can of worms. Any man known simply by his correct full name was rare indeed, with most being awarded a pseudonym of some sort that stuck for his full term there. Many of the nicknames were coined to commemorate some monumental stuff-up they had made, while others were labels awarded according to some unusual or outstanding physical characteristic. Few of these titles were complimentary and once earned were carried like a brand.

To illustrate, here are a few examples. I will not include all of the true given names in order to protect the innocent, and also because I never found out what many of the real names were anyway. So there was, –

Tail Light - he wasn't bright enough to be a headlight.
Alice - Alice in wonderland – a dreamer.
Cindy – wore the latest in sports fashions.
Hammer - had an accident with a hammer on a vital job.
Jibber - talked a lot and very fast in a high pitched voice.
Maps - describes how his eyes looked after a heavy night.
Sludge - describes his less-than-sylph-like figure.
Teach - me.
Pygmy or Pig - undersized team supervisor.
Bib and Brace - two inseparable workmates.
Arfa - his surname was Sleep.
Goda - his brother.
Big Brown (bottle) - surname Brown.

Stubby - his son.
Mongo - see 'Blazing Saddles' for the likeness.
Pooh - big fierce looking Maori.
Bubbles - remarkably developed stomach.
Chainsaw - had rows of very uneven teeth.
Mule - his surname was Skinner.
Muddy - abbreviation for the unkindly term of mud guts.
Rags - not the best dressed in the workforce.
Kodak - used to frequently appear in other people's photographs.

While the list is not exhaustive, I must apologise if there are any deserving contenders left out - but we must not exclude the Ball family.

Keeping the real name for this family anonymous, I would like to introduce you to, -

Beach ball - the father with a lovely rounded potbelly.
Pinball - the son who was rather thin.
Blackball - the son with a dark complexion.
Mothball - the smallest son of all.

Actually, while they had never worked on Barrow Island, these are in fact also members of the same Ball family. Please meet, –

Maball - the wife of Beach ball and mother of the Ball boys.
Noball - their daughter.

There are unconfirmed assertions that the Ball family had a dog and a cat with Ball names too. My informant I am sure was making too much of a good thing and stretching the bounds of truth when he told me their names -

Hairball – the dog and, –
Furball - the other one.

It is interesting to observe over the years, that groups of men, particularly when living in isolated places, tend to use their sense of humour to carry them through, especially under trying or extreme conditions. Prominent characters tend to emerge as natural leaders in different ways as we know and the camp comic and the camp organiser of entertainment are highly valued members of the group.

Sitting in the big recreation hall on Barrow Island in the evenings watching a film, was an experience in itself. Many of the plots could not be taken seriously and were often pretty dull or very dated. A bad film always encouraged the wags to heckle the heroes and the heroines as they played their parts out on the screen, and the more puerile the story line was, the more spirited and humorous and rude were the comments. As the audience was almost always composed entirely of men, the heckling which came forth as the actors struggled to do their best had to be heard to be appreciated. A love scene in a particularly unremarkable tale would be turned into a riot of laughter – much funnier than the best of the intended comedies we saw there. There were references to members of the audience too in different ways linked to the plot, which usually had the theatre in an uproar.

A full day's strenuous work in the tropical sun around the oilfield gave most of the men prodigious thirsts. I can vouch for this in recollection of my days working as a roughneck at Rough Range in earlier years. I learned quite quickly that it was not a wise move to sit down in a school of six rig crew workers for instance, which had just come in from a day out in the tropical sun. As men do, one ordered a round of drinks for the group, followed soon by another order from the bar and so on, usually going clockwise around the table. This is according to standard male protocol, but these rounds were for six – not glasses, or schooners, or pots, but jugs of beer.

I too was pretty dry on my first experience of involvement in this ritual of theirs, but after one round of six jugs, I returned to my quarters legless, and the last thing I remember hearing from the group was, -

"It's your buy Stumpy - let's go round again."

Harking back to the end of the day drink at Rough Range back in 1952, the beer was rationed to two large bottles per man per day and that was it. As the equivalent of six middies of beer, this would seem adequate for most, but extreme thirst can have strange effects. I found that I could skol the first ice-cold big brown bottle without taking my mouth from the top. The cold beer seemed not to touch the sides as it went down, and far from quenching the raging thirst I had built up in the sun that day, I was reaching for the second bottle almost immediately. But it tasted so fine and really hit the spot and it was heard that they had the recipe just right.

Fishing was one of the main leisure-time activities on Barrow. The rocks and beaches and reefs around the island were well stocked with the larger northern varieties of snapper, coral trout, sea mullet and the one I favoured as the real delicacy, which was blue bone cod. We were thoroughly spoiled with the quantity and the quality of what seafood was available and with no competition from any other anglers or commercial fishermen, we had it all to ourselves - served up on a platter as it were.

The Company had provided for each man, in a special storage area, a refrigerated locker where he could store his catch over the period of the working hitch. If you had been successful in having some prime beauties in you locker, come the time to fly home, you would retrieve your fish, wrap them and take them with your bags to the airport. It was a common and an amusing sight to see a line of men homeward bound, filing onto the aircraft with a big fish tucked under each arm. The odd ribald comment could be heard too relative to the special favours that could be anticipated from the wife or girlfriend when the delectable prize was presented to her. So large were most of these trophies that they would still be frozen solid and fresh when they were carried in through the front door of the family abode in Perth.

There was a large concrete pad provided at the main camp with solid cleaning benches for the fishermen. When the big haul arrived, a dozen men set to with knives to reduce the catch to a great mound of succulent fillets. You will find the name Pooh in the list of pet nicknames in this story. He was a giant Maori with a taste for fish and meat and offal, which may not seem too remarkable until I tell you that he preferred to eat most of it raw!

Pooh stood alongside me as I rather inexpertly filleted a mullet on the bench and tossed the innards and bone and the head into the scrap bin.

"Hang about," he roared, "You are throwing away the best part of the bloody fish!"

Plunging his arm into the mess of bones and guts he retrieved one of the heads. Splitting it open with one stroke he probed with a big finger and retrieved a circular organ with a hole in the middle. Dragging me over to the water tap Pooh demonstrated how to slice open this rubbery ring and wash the sludge out of it.

"It's his mud filter," said Pooh. "They are bottom feeders and the mullet is the only fish that's got one of these."

Later in the mess kitchen I followed my big friend's instructions and deep-fried a dozen of these in a beer batter. They tasted like a scallop, and washed down with a glass or two of the leftover beer from the batter, they were a gourmet's wet dream. I reflected that I had gone to Barrow Island to teach, but soon found that I was the one who seemed to be learning the most.

There was a rocky part of the coast on the southernmost corner of the island that dropped into very deep ocean. The serious fishermen told me that this was the territory of an old man groper, which lurked in a hole under the cliff. Should the big fish take the bait of anyone using standard tackle he would power away into the depths like a nuclear submarine, taking the tackle with him.

Every so often a few men would go to his lair and drop in a big bait on gear especially designed to hold his great bulk. It was steel stranded cable with a chain trace and with the free end then secured to a Landrover towbar. Being no match for this arrangement he was hooked and then hauled up on shore. The purpose of this exercise was not to eat him, but to remove all the hooks and sinkers and broken lines that he had been carrying around for a considerable time. With the job done and the big fellow now clear of his encumbrances, he was returned to the sea and sank back down into his deep hole under the cliff.

We were out from the shore in a small boat one night in the calm waters of the lee of the island. One of our group who must have been holding his mouth right, was really hauling in a lot of fish to pile them in the bottom of the boat. He developed a method of putting his sandalled foot on each catch when it was landed, to hold it still while he freed the hook. He was grunting with the strain of pulling in another big one and dragging it over the gunwale and then moved to step on it. Someone flashed a torch on the fish just as his foot was coming down and I saw him go rigid with fright. We were looking at the biggest and ugliest stonefish of all time. In the session 'Bities on Barrow' in the training centre, the stonefish was described as a really dangerous villain with wicked spines on its back, slimy with potent venom. It was said that the resultant agony suffered by a victim of this fish could either see him go mad or perhaps even die if he grabbed one or stepped on it.

One member of our party yelped and sprang backwards as if confronted by a black mamba. The fellow who caught the nasty fish calmly swung it back over the side and cut his line to let it escape. A look around the boat revealed the bloke who had panicked, to be seen crouched with staring eyeballs in the torchlight – perched on top of the outboard motor. The big laugh we had from that episode calmed the nerves somewhat. The outboard motor-sitter had to take a lot of good-natured flak for his panic attack and also for his athletic prowess. Later I was surprised to note that he had not been awarded a new nickname to add to my list, in honour of his feat. This was a bit disappointing.

The happy coexistence between the wildlife and the oilmen was quite remarkable. Some of the staff made a hobby of studying the habits and the habitats of the animals and became quite knowledgeable – amateur naturalists in fact.

The smaller beasties provided some entertainment for us too. Having little fear of humans and not being threatened in any way, they blended in with our social life. The smaller hoppies infiltrated into the big mess hall and recreation area and mingled with the crowd especially in the evenings. Their curiosity was fuelled by hunger I suspect as they were always looking for scraps or handouts from the men. In the darker areas around the camp you had to be careful you didn't trip over or step on one.

At about the middle of a film show one night there was a commotion in the audience. A large driller was standing and yelling accusations at all those seated around him. The effect on the rest of the crowd was predictable as they complained and told him to sit down and shut up so they could follow the movie. It was bedlam for a time. It seemed that his large family block of Cadbury's Old Jamaica had disappeared from where he had parked it under his chair earlier. The culprit was in fact a very skilful artful dodger golden bandicoot, which was seen later tucking into the chocolate in the darkness outside. The bandicoots and the boodies were invariably sleek and fat due to the easy pickings of chocolate, potato chips and cheezels, which they harvested from the unwary on movie nights.

There is a fearful predator roaming the spinifex and the beaches in the form of the Perentie lizard. Larger specimens have grown to two metres in length and can be seen hunting most of the other animals up to quite large size. They are the biggest of their own species in the world, second only to the Komodo Dragon of Indonesia. While similar in appearance to other large native lizards, these however are equipped with a fearsome set of very sharp teeth. Smorgasbord time for the Perentie is after the sea turtles have laid their eggs in nests that they have excavated in the dunes. The lizard digs for the eggs and gorges himself on the great feast. Should a baby turtle be lucky enough to actually hatch, dig his way out of the nest and head for the sea, it is very likely a Perentie will be waiting. The little bloke's first perilous journey will be cut short, as he becomes another bite-sized morsel for the big lizard.

Meet the Barrow Island Perentie lizard. Feared and fearless, his methods for catching a meal are often quite spectacular.

The most spectacular catch for a Perentie is when he fancies seagull for lunch. I observed a group of unsuspecting terns fossicking around the base of a sand hill one morning. Events took a really bad turn for them when a hungry perentie burst from the dune grass and launched himself up into the flock of screaming seabirds. The jaws snapped shut on an unfortunate gull as it was brought to earth in a cloud of white feathers. The leap in the air must have been at least two metres high.

The birds must have been worried sick after this, wondering whose tern would be next. Almost every species of the small marsupials there too were listed on his menu.

One of my friends on the island was a keen blue water yachtsman. He was looking forward to joining a crew in England for one of the classic ocean races to Australia – in fact he could talk of little else in the months leading up to his departure date. He had hidden under his sleeping hut a store of pretty seashells, which he had collected to take home to his kids. The night before he was scheduled to leave he went out in the darkness and reached under the hut to collect the shells. A very large perentie was lying in wait for any unsuspecting little animal that might be passing. My friend's groping hand was mistaken for a ready meal and the predator latched onto it. Teeth sank into his flesh all the way up to the wrist. With the man recoiling in shock and the lizard scrabbling in the opposite direction the hand came free, but not without causing deep lacerations, with the skin peeling off like a shredded glove.

Our man did travel to England after all to be with his yachting shipmates for the start of the race, but because of the injuries, he was unable to participate. I learned later that despite his disappointment at the disruption to his plans, there was one bright spot in the whole saga for him. Reporting to a London hospital to have his wounds dressed, he was treated by a gorgeous nurse there. His version as he related it to her was that his injuries were the consequence of wrestling with a big crocodile back in Oz. This bravery must have really impressed the girl. Our colonial hero enjoyed a lot of sympathy and tender care under the nurse. - His words.

According to the natural seasons and cycles, the big green and flat back and loggerhead turtles come out of the ocean to drag themselves up the beach to lay their eggs. At other times as determined by nature's plan their partners joined the females. I crept up behind busy pair which was coupling in the shallow water one day, and was intrigued at the obvious disregard they showed for the world around them. Normally shy and seeking to avoid close contact with humans, these two could not have cared less that I was standing close and watching them. The male - top position, was wearing a dreamy glazed look on his face with his eyes closed. The female on the other hand - lower position, appeared to be a little more wary. She may have had a headache, but appeared to be suffering the indignity of it all out in the middle of a sunlit beach purely in deference to the male ego. All went pear-shaped however when I picked up a piece of driftwood and gave Romeo a smart rap on his shell. Juliet panicked and with a thrashing of flippers exploded from under her swain. Regardless of the potentially painful consequences of the sudden separation, she took off and headed for deep water. Romeo reacted with shock and having been literally left up in the air as it were, was nevertheless stroking strongly when he hit the water in a cloud of spray to frantically head for the breakers close behind Juliet.

To many of the seabirds, Barrow was a man-made amusement park. It was not uncommon to see an Osprey or a majestic White-bellied Sea Eagle sitting on top of and enjoying the smooth rise and fall of the walking beam on a Lufkin well pump. Looking invariably smug and disdainful, the birds would enjoy the ride and the sea breeze for hours. By the number of squitters you could see on the pumps this was obviously a very popular pastime.

Lufkin oil well pumps are the principal method of lifting the oil to the surface.

Due to the highly flammable nature of the products from an oilfield, the risk of a serious or even catastrophic fire or explosion cannot be ignored. My point is that when working with petroleum products this danger is ever present. In my training role in different branches of the oil business it was necessary to insure that every man was capable of tackling and putting down any fire that might occur. Barrow Island was no exception and to meet this need the Company constructed a special-purpose fire-fighting complex in a remote area of the island. The interesting part of this training was seen in the great number of different flammables that may be ignited and consequently the variation in techniques and equipment required for controlling them.

We had blazing wood fires, sizzling oil fires in pits, and roaring gas fires to extinguish at the fire training ground. There was even a mock-up oil well to ignite. All were extremely hot and potentially dangerous. After each fire fighting technique was demonstrated, every trainee was required to knock down each type of fire successfully and safely to qualify. One day a young hopeful had to be hauled back and physically restrained, as he was about to walk right into a fire, which would have surely immolated him. The product burning was methanol, or methyl alcohol. Ignited in a large pit it was impossible to see that it was alight. In daylight it burns with an invisible flame and is therefore a hazard that any fire fighter would need to be aware of.

One of our lads had just succeeded in knocking down a crude oil fire in a large pit using a single hand-held extinguisher. This is no mean feat and it was obvious that our hero had listened carefully to the instructions and achieved what few others had. I can still see this fellow as he turned his back on the fire pit and began to stroll back to the group with a big cocky grin on his sooty face. His jubilation turned to fright as the smoking hot oil re-flashed behind him with a roar and was an inferno again in seconds. To his credit he did accept the criticism he got for turning his back on a fire, and also my explanation that the fire was rigged. Some pieces of scrap steel in the middle of the pit were intended to retain enough heat to cause the re-flash and get it going again. That kind of incident tends to instil a healthy respect for fire in everyone. It was all great fun too!

The young injured Osprey was rescued and has his wing in a sling.

After a two-year term I left Barrow Island with some strong feelings and many pleasant recollections about the place and all its inhabitants. If the definition of the word 'unique' describes that which is remarkable, rare or unusual, then this place fits that picture. It was considered to be a rich field for study and research in so many disciplines, for those interested in wildlife species, geology, marine biology and most of all in the social dynamics of the Barrow Island Clan.

A journalist who visited Barrow Island wrote 'The Island is one of the most remote places in the world – but it is never lonely.'

Juliet Turtle still has her headache

308

Episode Fifteen

<u>Working for the Gas Man</u>

In 1980 a likely contender appeared on my horizon of potential employers.

A giant energy enterprise was being developed in Western Australia. It followed many years of quiet but intense activity by this company exploring the oil and gas resources potential of our state. Production testing had proved that further development was viable and they set the wheels in motion to tap into the reserves and sell the petroleum products on the world markets.

Harking back, I remember a staff meeting held when I was with Wapet during their exploration drilling Exmouth in the North of WA back in 1954. The company had poured considerable resources into a wildcat drilling program in the area, but despite a promising start with the discovery of just one good well, had come up with only dry holes and disappointment.

Our group of people sat in the mess in the Wapet drilling camp at Learmonth after a particularly hot, gruelling and fruitless day. I sat across the table from some of the prominent experts of that time in the petroleum exploration business. All were despondent, as the latest hole drilled had come up with no traces of gas or oil. In oilfield jargon a dry hole is called a 'duster.' We had had a lot of experience with dusters. The whole explorations project in that onshore area around the Northwest Cape was drawing to a disappointing close.

Someone asked the question, -

"Given that we have had struck oil here but not in commercial quantities, where do you look next?"

One of the petroleum geologists replied.

"I believe that there are big reserves under the sea bed north of here - we will have to search offshore along the continental shelf."

So now, in 1981 I recalled that informed prediction made some twenty-eight years before, that petroleum deposits might be discovered deep under the Indian Ocean. This had ultimately proved to be correct and I was now standing on the threshold of what proved to be a successful bid to find it there. Unlike the Learmonth oil find, these leases promised to yield mostly gas and that in enormous quantities. I had no way of knowing in those early days that our one-well oil strike was the spark and the trigger which encouraged wider exploration in the north and the ultimate development of a vast and rich petroleum industry for our country.

The job application contained all the shots that were in my locker and the Company accepted me for the initial staff position of Senior Training Advisor.

There did not seem to be too many fears about being up to the job, as I was pretty familiar with all the training disciplines it was likely to cover. As Western Australia's first offshore oil industry however it seemed that there would be a lot of specialised knowledge I didn't have and which would need to be obtained pretty quickly. Most of the trainees for whom I would be responsible would not have worked on an offshore platform before. Come to think of it, neither had I.

The Company had established its headquarters in Perth in the Terrace. My new office on the fourth floor had a nice view of the Swan River reaches and the green esplanade. I wore a suit and tie to work every day.

My first job was to establish a training facility and staff it. Well-trained people were required to operate the platform when it was ready to go. Naturally these were hard to find in Western Australia, as offshore drilling was still only a fledgling industry here.

The physical size and scale of the offshore structure itself required for drilling the ocean floor and producing an offshore oilfield is enormous. The North Rankin-A platform was to have a mass of about sixty thousand tons and contain a maze of technical and electronics wizardry. This first offshore facility was to be the largest gas platform in the world at that time and I was going to be there to see it materialise.

There was a serious problem for me in the prospect of working in this new industry, which was offshore drilling and production, and that was the lack of the required level of specialised technical knowledge that I had to offer.

Something needed to be done about this.

At this early stage, the section modules of the platform were being manufactured and assembled in many parts of the world. Major steel sections were being fabricated in Japan and Australia. Other services and process components were being built in Europe and USA.

Management accepted my request and a proposed itinerary for an information-gathering world trip. Airline tickets, passport, a generous spending allowance and an American Express card were soon in my pocket and I boarded an international flight, with my first destination - Holland. The heart of the Shell Company, our major technical partner is in The Hague – a good place to start.

The main goal for this trip was to gain a much better understanding of offshore platforms

The drilling rig on a production platform in a thick North Sea fog at night

and how they worked. It was planned to meet with representatives of the major companies involved in designing and also fabricating our platform. These links were scattered around several countries in other parts of the world and I set off to locate them on their home grounds. The major useful contacts were in Holland, in Scotland - the land of my ancestors, and in England and the USA.

One major highlight of the Scottish visit was a helicopter flight from Aberdeen, via the Shetland Islands to one of the huge offshore platforms operating in the North Sea. This involved flying just above the waves through a dense, swirling fog. There were two eventful days spent on a Chevron oil production platform. This was my very first experience of the life offshore. In retrospect I probably drove the staff and the operators mad with all my questions. I climbed all over the structure, soaking up the technical details and enjoyed every minute.

One black spot during that visit was the crash of a Wessex helicopter on an adjacent platform, with the loss of several lives. As Bev knew that I was supposed to be out in the North Sea somewhere, it proved to be a wise move to ring home and assure her and the family that I was alive and well.

After crossing the Atlantic Ocean, I began my round of USA visits from New York. Among the memorable experiences was flying from the airport to attend a meeting in Manhattan. On that flight I met the Statue of Liberty at more intimate terms than probably a lot of New Yorkers had. The shuttle helicopter manoeuvred within a few meters of the imposing statue and circled her head for a personal introduction to the great lady. This was certainly a good way to appreciate her massive proportions, and I was mightily impressed.

An itinerary item stated that I was to attend a meeting in an office of one of the large buildings near the ferry terminals in Manhattan. The helicopter landed me on the pad on top of the building and shut down to wait. It was a glorious September morning, which bathed New York City in bright sunshine. Although I was standing twenty stories above the busy streets, it was necessary still to crane my neck to see the top of the tallest buildings. Most spectacular of all were the twin towers of the World Trade Centre. So tall were they that the morning sun cast their shadows like two long black fingers across much of Manhattan.

Newark New Jersey was the home of one of my contacts, where I was royally entertained. While on a tour of the local shopping centre, on rounding a bend in the mall I saw a policeman there straight out of the film 'Smokey and the Bandit.' Standing with legs astride in the middle of the mall he just rocked back and forth in his tooled cowboy boots and glared at everyone. He was complete with Boy Scout hat, a bright blue uniform with yellow striping and pockets, big mirrored sunglasses and a big revolver slung low on his hip. His thumbs were hooked into the sagging gun belt under his big belly, which severely tested the buttons of his shirt. There was a very large shiny sheriff's badge on his chest to complete the picture. They warned me that I could have very quickly found myself in the local slammer if he had heard my involuntary chuckle when suddenly coming face to face with such a colourful character.

"They take themselves very seriously these cops," warned my American host.

They brag about their beaches there in New Jersey too but I didn't see one to even come close to the quality of an average one that we enjoy here in Australia.

I then headed west in my jet-engined wagon train to Texas, where, I was told; the real oilmen are to be found. The climate, the countryside and many other features of San Diego, which was another stop on the itinerary, were so much like Perth, even to the gum trees that I was quite taken with that city.

Among my impressions of this business trip around the world was the feeling of being disconnected and without things or people I could relate to. Waking up each morning in a different bed with a new and strange view out of the window is pretty novel for a small town boy like me. Walking out into a different city or landscape with crowds of strange people around you each day is unsettling, but intriguing. It reminded me of the comment made by a tourist in a film I saw once that was, -

"If today is Tuesday then this must be Belgium."

It was in all however, a great educational experience and an unforgettable adventure, though with a serious purpose. I would like to go back to some of these countries one day with the time to travel and see the sights – but I would not want to live there. The technical knowledge that was accumulated and the personal contacts made were to prove invaluable in my new offshore job. The many wonderful attributes that my hometown of Perth has to offer only become evident when you have seen some of the rest I think.

The demands on me in the main office became heavier. Soon I was involved in the preparation of job descriptions and position specifications for a wide range of the platform jobs.

My previous varied industrial experience branded me as a kind of resource person, able to assist a number of different departments. It was a time to enjoy feelings of usefulness and satisfaction at having my finger in a number of very interesting pies.

My relative naivety and lack of experience with corporate protocol and office politics must have been showing however, although some commented on my direct approach and called it 'refreshing'. It seems I had been at the 'sharp end' of industry in my past experience, which in office speak refers to someone who has actually been doing some of the work.

The people who did win my respect however were those in another department to me. The people in the Engineering and Production group, were direct, spoke a language I could understand and were generally practical and good to work with. One of these was an Operations Manager, who was at the top of the tree in the drilling and production area. He was an inspiring bloke who got the best out of his staff and whom I found I could really relate to. I found myself more and more working together with him and his staff in the big planning job required in bringing the new production platform on line. Life was good for me then, but this condition was not set to last it seemed.

My new boss and I had little in common. As time passed the rift widened and it seemed we were incompatible to put it mildly. He set me an impossible task to complete over a weekend and when it wasn't done on Monday he said we were going to have a parting of the ways. I responded that I didn't know he was leaving!

Was that the final nail in my coffin? - It was! It seemed he was stubborn too, but he was the boss after all.

Soon after this I stood alone on the footpath outside the building. With my briefcase in my hand, there was a severance notice and cheque in my pocket. Reflecting upon shattered hopes and a battered ego, I wondered with a sense of disbelief just what the hell had happened.

The phone rang at home the next day and jerked me out of my dark cloud of despondency. It was the Production Manager, who said, -

"Stuff that lot Ivan, take a holiday. We need good experienced men on the offshore team and we will be recruiting soon. I'll be in touch."

He was as good as his word. This therefore was a sharp lesson for me in the hazards of corporate office politics, with me as the victim. There was time to think about my fall from grace and the indignity and shame of it all, though I doubt I would have changed my approach, but just watched my back more carefully.

Following the advice, I took the holiday that had been recommended. With time on my hands I took up leatherwork as a kind of therapy - there's a nice decorated stubby holder I can show you. Though still smarting from the humiliating defeat I welcomed his next phone call.

"A number of the good positions for the offshore gas platform will be advertised in the papers next Saturday. I would like to see you apply."

I did as he suggested and won the position. I seemed that an ongoing career in the oil industry was assured.

While I was relatively inactive waiting for the new job, there was time for me to reminisce a bit. Back in 1952, for some very strong reasons, I felt the urge to join in the search for oil in this country. It wasn't just the images of the oilmen in the films I had seen, creating a desire to be like them, as much as the very nature of the oil business. The technical nature of it, using fine equipment, the work in remote places, the unknown factor of new experiences and above all the massive gamble of wildcat drilling in unknown territory all had appeal. Playing with the enormous stakes of work and resources against the likelihood of failure had a special excitement I could feel, while the riches to be had in a successful strike would be like a dream come true with rewards for the whole country.

With no lifts, ladders or stairs I had to make the trip 30 meters to the deck above in the Billy Pugh

The Platform

I reported back to the main office after my ignominious dismissal. There was the spot in the foyer where I had stood, feeling defeated and rejected a couple of months previously. There was now a new job to look forward to with better pay and conditions on the North Rankin offshore production platform, which was the hub of the offshore operations. This was the beginning of a whole new experience, of the type I always seemed to be looking for in my life. The title the job carried was Safety and Training Advisor, a staff job and the first of its kind in the company at that stage.

The time dragged while the anticipation mounted, as along with the others waiting in Perth for the offshore posting, it was hard to contain myself while the construction program continued offshore. Not that we were short on things to do. There was a mass of planning and preparation to ensure that the operation got away with a smooth start. This would be a brand new experience for most of the West Australian staff especially, which tended to heighten the feelings of expectancy.

There is a state of readiness in the construction of a new offshore facility known as Life Support. This is the stage when eating, sleeping and other facilities are in place to provide for personnel on board. I pestered the management to the point where they agreed that I could visit the platform ahead of this stage to orient myself and climb all over it to check out its secrets.

The French Puma is the workhorse of the offshore operations

Here I was at last in Karratha, which is the land base for the operations. This was my first day trip, just off the Ansett flight and queuing to pass through the security checkpoint. There was a big helicopter being serviced on the tarmac outside. The Company used a number of different types; the large eighteen-seater being the Puma and then bigger still there was the Super Puma, both from the French aeronautical stables.

Our aircraft rolled down the runway, lifted off the undercarriage, dropped its nose and climbed into the sky.

We banked and circled and headed westwards over the ocean, with the islands of the Dampier Archipelago slipping away and back beneath the rear rotor. Flying steadily on the course to the platform there was nothing to be seen ahead. Guided unerringly by a radio beacon we flew over the empty ocean for 130 kilometres until the speck ahead resolved itself. Flying closer and lower our aircraft at last flared and squatted on the deck of a ship moored alongside the structure.

North Rankin-A platform after it went into gas production. The flame at the end of the flare bridge is the clue

My eye followed the massive steel legs of the jacket from where they broke the sea surface and travelled up to the main deck twenty-five metres above, where all the construction work was going on. There were no ladders leading up through the steelwork to the upper deck that I could see, but that problem was soon resolved. A man basket came spiralling down on the end of a fly line from the tower crane working high above. Looking a lot like an upside-down parachute, the man basket is a large flotation ring suspended on interlaced rope suspension lines. There is no carnival ride which can compare with the rocket-like lift, clinging with locked arms and gritted teeth hanging on for dear life and swinging out in a wide arc high over the ocean. Later, as I came to be more familiar with the terms used on platforms, I learned that this contraption is known as the Billy Pugh. I also learned later that to insure you would have a smooth ride each time it was necessary to be on good terms with the crane operator. Anyone seen to be having a hairy ride was assumed to be in the operator's bad books.

I was destined to meet Mr Billy Pugh again on many occasions after this.

There is a distinct advantage for anyone fated to work on a structure like this, and that is to be able to go on board during the construction phase. Seeing how it goes together gives you a better understanding of how it will operate I have found. The platform was a dead thing then, but would soon come alive with power surging through its systems and the gas screaming through its flow lines. Feeling decidedly minuscule standing there I stood and surveyed the wide ocean from the top of the structure. That feeling was there again in the rush of anticipation at being involved in the start-up of another big new enterprise. Pioneering stuff again I thought.

In a few months, the life support phase was established and I was rostered to report on board to take up official duties as the Safety and Training Advisor. The standard working schedule for offshore staff is two weeks on board and two weeks at home in Perth. It was a pretty busy twelve-hour day with a lot of extra work thrown in. My counterpart, the other STA was to take over on the alternate swing to provide full coverage. While the even-time working schedule seemed generous I have to point out that it was a twenty four hour a day commitment and the company owned you while you were on board.

Broadly speaking, my role was that of a watchdog, investigator, trainer, adviser and resource person all rolled into one. This covered all issues related to the safe working of an oil production facility stuck out in a hostile marine environment. There was a very sobering statistic I gleaned from one of the experts about the heat potential in the flow of the product we would be dealing with, which was natural gas - mainly a mixture of methane and propane. In terms of thermal energy released in the case of a disastrous fire emergency on board, there would be theoretically, sufficient to reduce the sixty thousand tonne structure to a molten mass in the space of twenty minutes. I resolved to keep this in mind.

A message which I always hammered home to men in training sessions offshore was the importance of attending to every small safety detail in their work procedures. When the production phase began, we lived and worked twenty-four hours in the day cooped up in close proximity to enormous volumes of high-pressure flammable gas. After a time most people become accustomed to the conditions and the potential for a catastrophic accident, though I doubt that many were blasé about it. The ever-present muffled scream of high-pressure gas is a constant reminder and everyone goes to sleep at night with this sound as a constant background. As one man put it, -

"When you are offshore and everything goes pear-shaped, you can't run away from it as can do on shore. You're stuck with the problem and have to stay and fight it."

This massive 1000 ton floating crane was used to lift the modules onto the platform jacket.

The reality of the worst-case scenario taking place on a platform is that the personnel, as a last resort, would be evacuated by survival craft dropped into the sea. While these craft and the methods used to deploy them are designed with all the best intentions, there are no guarantees of everyone's survival. When training inexperienced people I aimed to give them a healthy respect and awareness of where they were and what they were dealing with, and hopefully not scare the hell out of them in the process.

Permanent Company personnel usually did not pose a high risk as they had all undergone a very intensive induction and training program. This included helicopter survival exercises, fire fighting and first aid and all the other good stuff, which makes every man useful in an emergency. In my term on the platform however we employed hundreds of inexperienced workers who had been sent offshore to carry out scheduled project work for us. These people were our Achilles heel. Most had never worked on a live gas facility before, much less one that is stuck out in mid-ocean. Much of this project work involved steel construction with its attendant risk of ignition from the 'hot' work processes they had to employ. The first six months of this work on board required only the usual controls and precautions needed to stop the men falling into the sea from a great height – and the procedures to fish them out if they did. When the first wells were drilled and began producing gas however, the whole complexion of safety precautions changed overnight. We became very strict and serious and breaches of safety rules were dealt with severely. Everyone had to realise that any carelessness or shortcuts could lead to having us all blown out of the water – and no one wanted that. Every work crew unfortunately had its cowboys who were hard to keep in line and took unnecessary risks. They were a safety officer's headache. One of these characters got his comeuppance one day with no help from me.

One of the automatic systems installed on Rankin was a network of very sensitive detectors, several hundred of them in fact, which were scanning all areas of the platform. They were set to detect the tiniest indication of a flame or spark and also a gas leak, and electronically report it. They were also set to trigger other major reactions in the fire fighting system if they were tripped.

My home base offshore was in the accommodation module.

Prior to the start of work one day when the crew was ready to begin a big welding job they were standing by with orders not to switch on and begin welding operations until I had inspected the site and cleared it as safe. One of the items in this check was to ensure that the flame detectors in the area had been switched off from the control room. One particular smart Alec in the crew decided he wasn't going along with all this unnecessary rigmarole and prior to my all clear he switched on one of the welding machines. All hell broke loose. There was a short electrical spark from the welder, a detector saw it and triggered a series of automatic actions. The emergency alarm blasted, which called all personnel to muster stations, warning lights flashed all over the platform, a one-thousand horsepower diesel fire pump kicked in and pressured the water-flood pipe work and all the overhead nozzles were activated. This deluged an area with hundreds of tons of seawater per minute, at such a rate a man can scarcely breathe if caught in it. Watching the scene unfold I was very glad to see our culprit in the middle of this seawater deluge staggering blindly around looking for a way out. I am sure he will never forget it, as I am also sure that the rest of the crew would not have let him, as they too were caught up in the flood along with all their gear. You could imagine it to be a bit like the wrath of God bringing down a mighty deluge upon the sinners.

It did make a lovely story to tell my future trainees on the perils of breaking the rules. There is nothing quite like first hand experience to make an account ring true.

There was my share of running battles too numerous to mention with many levels of the staff and workforce, mostly on matters of principle in the field of safety. Notable among these was when I clashed with a particular manager. There were three in this position rotating in the position offshore. Two we got along famously with, but Neville - we clashed all the time and I dreaded being on the same shift as he was.

As an illustration, it is worth relating the incident in which a half-tonne steel bracket was ripped off the top of the drilling derrick by a crane boom. It fell forty metres and narrowly missed killing two men below. Such a serious condition must not be allowed to continue I decided and drew up plans for an engineering modification to this big lump of steel to render it safe before it was reinstalled. For some perverse reason, due to him having a bad hair day or something, this manager vetoed my recommendation and said that the faulty part would be reinstalled on the derrick without my safety modifications. I saw red. We shouted at each other over his desk on the issue, but he refused to concede my point and authorise the modification, which would have eliminated this risk I could see so clearly.

About midnight I returned and woke him in his cabin - he was not amused.

"If you carry on and refuse this recommendation" said I, "I will go down during the night to where the piece is located and personally lever it over the side down into the sea!"

A strange lump of steel this size on the sea bottom four hundred and twenty feet below would necessitate mounting a costly salvage operation. The real problem for the manager however, as I pointed out to him, would be in the subsequent high-level inquiry as to how the thing got down on the bottom of the ocean in the first place – and as the Safety Advisor I would be telling the story. I'll swear there was steam coming out of his ears when he finally grudgingly agreed to my proposal and authorised the engineering modifications. I tossed in a bit more as well reminding him that this wasn't about personal victories but was about saving lives, as this one had almost cost us dearly already. On that basis I felt there could be no compromise. What do they say about more ways of skinning a cat?

One of the things I was made to feel well aware of on North Rankin was my advancing age. It is mostly a young man's game, especially the really hard physical stuff, which I didn't have to worry about so much any more, having done my share over the years. I didn't fail to miss the odd remark or inference though, but they were mostly good-natured jibes and hell, I was only sixty-two or so. In my particular function I made sure I was doing my bit climbing stairs and ladders and masts all day and getting into all the places where I was needed at the double. A situation developed one day where I could not resist the temptation to carry out a reflex action. This was not to make points, but because I felt the urge to do it.

We had a group of our managers and also some heads from Perth rubbernecking on the section just below the helideck. There was a rule that when in the vicinity of the helipad and with a chopper landing or taking off you have to remove your hard hat or use the chinstrap lest it becomes a flying missile. It was part of my safety talk with these people on their arrival, and sure enough a big chopper fired up on the helideck just above us. Before the buffeting downdraft from the rotors hit us, I signalled our entire group to remove their helmets. One did not, and just as I had predicted, the helmet was ripped off his head and went spinning across the deck towards the sea railing. I guess everyone has done things on the spur of the moment without stopping to think. Just like Modesty Blaise, I dived after the flying helmet, hit the railing with my shoulder and grabbed the hat as it started spiralling out and down towards the sea. Getting up from the deck I returned the helmet to its owner with a word or two about him not observing the rules. I think they had the opinion that mine was a really foolhardy act for a man of my years. The success of the catch I must admit was definitely attributable to the big coil of soft rope that I had noticed right there on the deck, and which made a good place to land after hitting the railings.

I have been told that one of my failings has always been the lack of attention to fine detail. This is not a good thing in the safety business so I had to pull my socks up. I knew from my training that it is the small and seemingly insignificant aspects of a work environment, which if neglected can bring you down. The report on the findings from the Piper Alpha disaster in the North Sea brought the whole thing home to me and converted me into a real nitpicker when it came to enforcing the rules of safe working, even the most minuscule and seemingly unimportant ones. The official inquiry into Piper Alpha had tracked down the cause of this shocking disaster. It seems that a work permit, which is a document controlling a hazardous work operation, was lost in the shuffle between shift changes. A pump was started, crude oil escaped and ignited. This led to a chain of fires and explosions and ultimately, total destruction of the platform. The families of one hundred and sixty men were left to mourn the loss of their loved ones. The initial contributing cause of all this was the careless handling of that one piece of paper - so attention to small details is now my goal in life.

Most Australians and especially those living in northern latitudes are aware of the threat from our tropical cyclones. It was a privilege to have witnessed the power of five of these giant rotating storms at close quarters over the years. Most reports were of those affecting areas on land close to the coast, causing varying degrees of destruction and flooding in the towns and surrounding districts. A vessel at sea has an advantage in being able to evade the full force by sailing off away from the storm track or seeking shelter. If you find yourself in the sights of a

Son Peter was a production technician on NRA. It was good to find myself rostered on the same shift as he was.

cyclone on an offshore platform on the other hand, you just brace yourself and hang on. If you have faith in saying a prayer, you do that too.

The design and construction specifications used in building the North Rankin platform required the structure to stay intact and vertical under the battering of the theoretical 'one hundred year storm.' This is a kind of a benchmark describing the most destructive storm that you are likely to experience in those tropical northern latitudes. Cyclones are ranked in severity from category one up to five. The most severe represents a storm with wind velocities in excess of 280 km/hr. Our platform was allegedly certified as being able to survive the destructive force of 300 km/hr wind speeds.

Faith in the strength of our steel island was about to be tested.

On the 23rd of April 1989, a powerful cyclone was reported bearing down from the north in our general direction. The track of these storms is totally unpredictable, so you just batten down and prepare for the worst, hoping that it will shear off on another course. They can stop altogether, change speed suddenly, or as one did when I was on Barrow Island, mark time for a while and then pounce during the night.

The cyclone watch radio broadcast from Karratha told us we could expect a visit from a cyclone codenamed Orson. This was confirmed by the changes in the appearance of the sea and the sky around us. The ocean was flat calm, looking like a sheet of lead and the dark line of the storm front on the north-westerly horizon was sending little puffballs of cloud scudding over us - an ominous sign. The platform was seething with activity. There was an almost continual thump-thump of helicopters from the helideck as all available aircraft shuttled non-essential personnel to the beach and safety. The wind was gusting to 30 knots as I watched the last chopper lift off with the wind up its tail.

We were just a handful of men, a skeleton crew in fact, left to ride out the cyclone.

Bigger and more tortured seas rolled past from the direction of the blow like messengers sent to announce the arrival of the main force. Watching the legs of the flare bridge was a good way to gauge the sea conditions and I winced as the wave impacts exploded higher and higher and the wind began to slash off the crests in clouds of spray. Our standby vessels were sturdy and seaworthy craft that are normally required to be in close attendance to the platform at all times. They are in continuous contact and are equipped and ready for fire fighting, sea rescue and general cargo duties. It is a comfort to see them standing off every day just a few hundred metres distant - ever watchful. The Lady Elizabeth at this time was pitching so violently in the taller and steeper swells; she was almost standing on her tail and was obliterated by sheets of spray from the big green ones tossing her like a cork. We saw her turn, and rolling and pitching alarmingly, head for a sheltered anchorage somewhere in the lee of one of the Islands clear of the storm's destructive force.

We found ourselves all alone after our attendant vessel left us – an eerie feeling. I was brought back to our real situation by the sound of the wind now shrieking and clawing around the platform. The windsock on top of one of the cranes was rigid and crackling with the gale and everywhere gear rattled and banged in the gusts. Just after sundown we had everything battened down and secured in the best possible way. With one last look at the boiling ocean and the flying spume we went below to the accommodation module, checked that all were present and accounted for then locked and barred all the doors behind us.

The radio told us that Cyclone Orson was a category five storm with wind speeds gusting to 280 kph - and we were right in its sights.

An offshore platform is enormously strong but has to be flexible to a certain degree. This was very evident when the big seas started booming and pounding their way through the lower levels of the steelwork. The platform in normal weather is like a rock, stable and apparently unshakeable – now it was like a live thing. The whole structure shuddered and twisted and recoiled from every hit by the enormous seas. Inside the accommodation, crockery was skidding across mess tables and splintering on the floor. In my office the clipboards on hooks on the bulkhead swayed in unison and the furniture began to rearrange itself.

Normal speech was almost impossible at the peak of the storm, with the men shouting to be understood above the shriek and the roar of the wind outside the module. Access doors leading to the outside decks were clamped shut with warning signs fixed to them. Any man who ventured out at that time would be plucked off the structure and lost in seconds.

Even in the most frightening moments you can sometimes find a touch of humour. The heavy steel doors leading out to the deck are designed primarily to withstand great forces from the outside. Their main designed strength is in resisting extreme wind pressure. The wind however had now created a massive low pressure or vacuum area on the lee side of the accommodation module. This tore one of the door locks free and sucked the heavy door wide open. Our six-man emergency team using a stout manila rope lassoed the door from the inside and pulled it shut again. Seeing that the lock was smashed, they looked for an anchor point on which to tie the rope to hold the door shut. With nothing suitable to be found in close proximity, the end of the rope was passed into an adjoining sleeping cabin off the corridor and tied to the leg of a steel-framed bunk. This bunk had a recumbent figure in it, just off a long working shift and dead to the world. The wind blew harder, the door strained outwards and put more tension on the rope. The bolts fastening the bed to the deck sheared and the bunk with its sleeping occupant was dragged across the floor and fetched up against the doorframe. Cinching up the rope again, the team left him there still snoring. His mumbled comment next morning was that he was vaguely aware of something going on in the night but didn't know what it was until someone told him at breakfast.

There were many there that night that professed their faith in the strength of the platform structure and its ability to withstand the battering without toppling into the sea. By the looks and actions of a few others however, it appeared they thought we were all in our final hours.

A very large and tough-looking man was huddled in the foetal position on the floor in the corner of the mess hall around midnight. He looked a bit odd and then I realised he had strapped on a life jacket. Lifejackets had not been specified at this time, as anyone out in the storm would be lost with or without one. Sitting on the floor next to him I asked if he was OK. Seemingly scared out of his wits, he just nodded. Above the continuous racket I shouted in his ear the question - why was he wearing the jacket? I just caught his mumbled reply that it made him feel better - like a security blanket I suppose.

A few of the external floodlights were still functioning. Wiping off the moisture I peered out through the armour-plated window to try and catch a glimpse of the situation outside. It was 2300 hrs, which was about the time when the peak wind velocity was recorded. What I could see outside was very intimidating. It was as if the elements were trying to tear the platform to pieces around us to get inside. I watched a large dish antenna and two floodlights being ripped from their brackets and carried spinning past the window. Our Whittaker capsules, which are the platform's lifeboats, could just be seen in the flickering light, suspended outside the railings of the main deck. These had received special attention in the storm tie-down due to the important role they played in our security. They were bucking and slamming back and forth against their lashings, an effect that did not appear to be caused by the wind. Then I spotted the reason. Wave crests were impacting the boats as they swept past from beneath the platform. I could make out the waves' foaming tops racing by at high speed. The incredible thing about the waves hitting the boats is that they are suspended at a height of 24 metres above normal sea level. This put the wave heights at 24 metres or 75 feet! If the deep trough of the wave is added to this height, which is about one third again I am told, the seas must have had a magnitude of 32 metres or 100 feet from crest to trough! I was privileged to see nature's ferocity in one of its most violent and frightening forms that night and at very close quarters – right plumb in the middle in fact.

At 0100 hrs the noise of the wind abated slowly to near silence, although the big seas still pounded the platform structure. An announcement on the PA speakers to the sleepless crew explained that we were experiencing conditions often found in the eye of the cyclone as it passed directly over us. Despite this apparent calm, all doors were to remain secured and personnel were prohibited from venturing outside, as the wind would pick up again very shortly with the same risks as before.

It occurred to me as a cool idea to form up and invite membership for something like 'The Eye of the Cyclone Club.' It would be very exclusive and all members would have good yarns to tell. I had been in the core of a cyclone some years previous on Barrow Island. That was pretty frightening, but it was a pygmy compared with Orson. That earlier experience promised to be unforgettable and I am glad I didn't miss out on this one either.

One of the privileges of my job was that I was free to go anywhere at any time throughout the length and breadth of the platform. With the (valid) reason given that I needed to check the security of the doors, I climbed the stairs to the top level, which is under the helideck.

Unbarring the exit door I stepped out into a still and tranquil night. As if my mind was a camera, I have the impressions of the next few minutes indelibly imprinted on my memory.

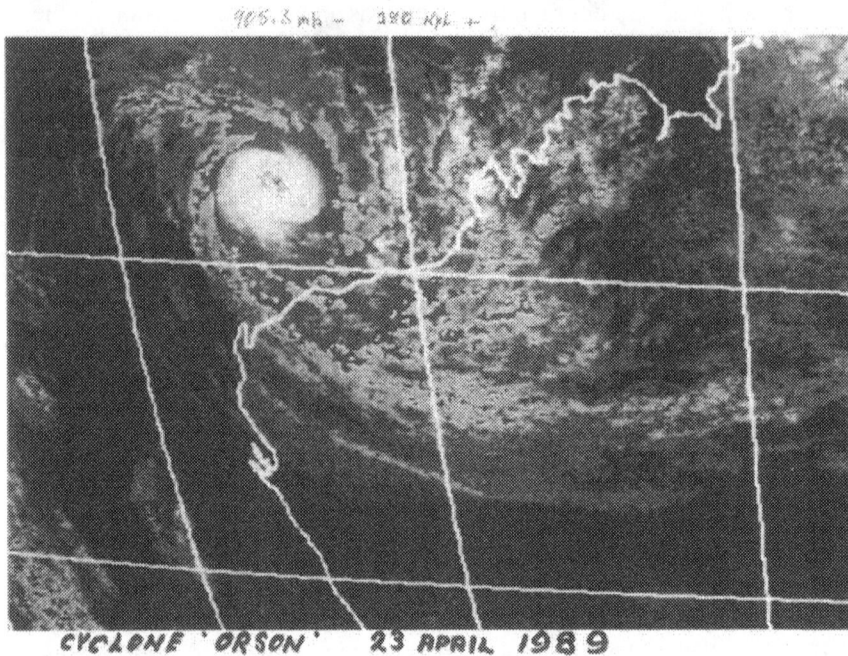

The weather satellite photo shows Orson just a few kilometers north of the platform about to strike. The eye can be seen as a black spot, in which I stood that night

Above me, a great circle, like an enormous black skylight, had appeared. This seemed to be perfectly circular and was studded with twinkling stars and the ghostly images of circling seabirds picked out by the platform lights and trapped in the calm epicentre of the storm. Coming steeply down from the edge of this window was the wall of cloud that defined the edges of the eye. The effect was eerie, with the uncanny silence of the still air and the feeling of breathlessness giving the impression that I was standing there in a vacuum. With eyes more accustomed to the dark, I was able to pick out the turbulent storm ring rotating around the platform a few hundred metres distant. The serried ranks of the great green swells still charged out of the gloom until the lights from the platform picked them up. Their streaming crests rapidly bore down, to go thundering through the steelwork and emerge foaming on the other side. The shock of the impact of every one could be felt even at that high position, from where I was watching in awe and wonder and apprehensive exhilaration.

My dalliance in the eye of cyclone Orson high on the helideck lasted about fifteen minutes. Soon I sensed the beginnings of short insistent puffs of air as the back of the storm funnel rolled closer. These changed to stronger buffeting wind gusts, which signalled it was time to retreat to safety. The back half of a cyclone doesn't usually prove to be as violent as the leading segment, though it can be equally destructive, but with the winds blowing from the opposite direction.

After Orson had done its worst and the gale had abated sufficiently, men began to come tentatively out on deck. The emergency crews were first to inspect the conditions on the exposed and damaged areas of the outer platform decks to ensure the safety of the rest of those coming out behind them.

The repair costs for the cyclone damage on Rankin platform that night ran into many millions of dollars. The large construction work site below the main deck was a shambles. It looked as though a giant hand had swept the decks clean of all equipment and steelwork. Only buckled handrails and the sheared ends of heavy chains formerly securing the gear remained. One incongruous sight was a complete telephone cabinet swinging back and forth high above the sea at the end of its cable. The phone handset was still in place. Evidence of other violent phenomena

322

was found too, which were a result of the intensity of the storm. Many fluorescent light covers had been ripped off. The bare light tubes remaining were part full of water, driven into the tubes by extreme pressures that we could only imagine. A survey of the sea floor with the underwater cameras later prompted the remark that it looked like a metal salvage yard down there.

The ferocity that Orson unleashed can be gauged by the recorded strength of that storm. The central depression was measured to be as low as 905.3 mb, while the wind velocity exceeded 280 k/hr. My partner, with whom I rotated in the two-weekly shifts offshore made his usual complaint afterwards. He grumbled, -

"How come you are always here when the exciting stuff happens?" What could I say? "Just lucky I suppose."

One of the important duties of the Safety Advisor was to enter and test the insides of all the big steel process vessels where men will be required to work. Once the covers had been removed I was the first one to enter, wearing a self-contained Drager breathing set. This inspection was to detect any explosive or toxic conditions. It was interesting and rewarding work giving me a sense of being a useful member of the operations team. Most staff members had to be multi-functional. Due to restrictions on accommodation, we had to take on a number of different responsibilities besides the prime one. I found myself as team leader of a muster station, where a group of men would be assembled and accounted for and then evacuated by sea in a survival capsule if the need arose. We didn't do this in consequence of a real emergency, but had to run many drills to get it right. I trained to be a Whittaker Capsule skipper too and participated in exercises in charge of one of these amazing craft when it were lowered twenty-five metres on a single rope into the ocean with fifty men confined in the hull.

Safety equipment had to be tested from time to time to get my approval for its use. Drilling crews climbing the mast wore a safety belt with a link to a running rail designed to save them in case they slipped. The tests for this apparatus were to climb the ladder about fifty meters above the deck, let go your grip and literally fall backwards. Guess who had to do this one? Yes the belt worked. One safety feature on the drilling mast provided a very quick emergency evacuation by the derrickman from the monkey board about thirty metres up. I reneged on testing this because the evacuee had to loop one arm through a little pulley system and scream 30 meters down the rope at a 45-degree angle. My reluctance to test this gear was excused I think as no one else was game to do it either.

We had some interesting visitors on board for a while. A group of six SAS soldiers under the command of a captain were running anti-terrorist exercises all over and around the structure. They mostly slept during the day and exercised at night time. All dressed in black, they climbed the steelwork, sneaked around the legs in their rubber boat and frightened the tripe out of a few people as they appeared out of the shadows and just as quickly disappeared again. They asked if it was OK to carry their M16 rifles loaded on manoeuvres - they said it was protection against the sharks, and reckoning them to know what they were doing, as they were an elite force after all, we agreed. I insisted on securing their weapons and ammunition in the explosives magazine when they weren't using them however. They accepted this and all went smoothly. They had appeared on board like ghosts, and then disappeared like ghosts as well.

Helicopters are the lifelines to offshore platforms, transporting passengers and freight and doing all sorts of jobs. The machines servicing the platform were the French built Pumas. Originally designed as a military support and attack aircraft, they were in the words of one of the pilots 'built like a truck.' Twin gas turbine engines powered a big four-bladed rotor and they could carry eighteen passengers with two in the crew.

One morning the turnaround flight had just lifted off the platform helideck and was dwindling out of sight a couple of kilometres into its return flight to the beach. I was watching it from a high point on a walkway outside the manager's office. The radio burst into life.

"Mayday! Mayday! Mayday!"

The pilot identified himself and reported that he was experiencing severe vibration and was preparing to ditch in the ocean. He had a full shift change of eighteen passengers in the cabin of his aircraft.

The twelve power binoculars brought him so close I could see all the action. He had lost his flight altitude and was in a hover just a few metres above a choppy sea. The rotor wash was picking up so much spray it was now difficult to detect any further movement. He was just hanging in there. The pilot was reporting extreme vibration throughout his aircraft.

"Like it's shaking us to bits." His voice quavered, due to the severe pounding he was getting.

He had to reduce power to minimise the stresses on the main rotor, which seemed to be at the centre of the trouble. When asked if he thought he could return to the platform rather than ditch in the sea, he did not sound too confident.

"If I pull on more collective and try forward flight, this damn rotor blade could shear clean off. Then we will really be in the drink! There is a good offshore breeze though so I'll stay in hover and see if I can sail back."

This he proceeded to do in a truly remarkable feat of flying.

Almost the whole workforce lined the railings of the platform as the crippled helicopter drifted slowly back towards us. Many of the men had mates on board and I'm sure that there was no football grand final crowd that could have topped the barracking for those in danger. It took ninety minutes to complete the painfully slow trip. Maintaining my watch through the binoculars, I followed the progress of the stricken chopper swaying just above the sea with its white circle of spray churned up by the rotor wash. Landing on the platform was denied him due to the risk of a possible crash and fire, so the pilot opted to try for a landing on the barge alongside. The barge deck was cleared of obstructions – I think a lot of good gear went overboard in the scramble.

It appeared as if the pilot was lifting the aircraft by his own sheer brute strength up onto the barge deck. We could see white faces at the windows as it thumped to the deck with rotors still spinning. The helicopter was vibrating and bucking like a live thing due to the effect of the damaged rotor. The pilot was the last one to leave after he killed the switches and dived out after the rest. As rotor speed decayed the aircraft shook even more, slamming up and down and pigrooting on its undercarriage until everything came to an agonising stop. The onlookers awarded the pilot three mighty cheers. His skill and tenacity saved a situation that could well have ended with tragic consequences. Another wry comment I heard from a pilot one day was that their work was ninety nine percent sheer boredom and one percent sheer panic. I could well understand his comment after this episode.

The cause of the malfunction, which was discovered later, was that one of the bolts securing a main rotor blade had sheared, throwing the whole rotor out of balance. Reports later confirmed that it was a very close thing and only a hairsbreadth from causing a catastrophic failure.

An emergency situation occurred on the platform during an otherwise ordinary day. One of the enormous hollow legs of the main jacket which doubled as a diesel fuel tank, overflowed while being filled. Two welders were working on the structure below the spill and the heat from their operations ignited the gushing cascade of fuel. To exacerbate the situation the overflow could not be stopped immediately.

These two men dropped their tools, grabbed fire-fighting equipment to battle the flames and quelled them successfully. Acting as promptly as they did could well have prevented serious damage and loss of lives. This brave act, which was way above the call of normal duties, had to be recognised in an appropriate manner. I called a special meeting of all the men in their section, and described their prompt act of courage and highlighted the fact that we all owed them a great

debt of gratitude and it was a fine example to others. There was a cupboard full of gifts in my office including fancy belts, radios, tee shirts and caps etc., held for special safety performance awards. I presented an armful of these to each of our heroes to take home. In addition, contained in a big envelope, they were given one of my drawings as something with which to remember their eventful day.

If I were a real action junkie I am sure that I could not have chosen a better occupation than this. It was my last piece of eight years' active service offshore before going into retirement mode. It left me with more vivid memories and images than I could possibly have written into one book - or two or more.

Copies of the cartoon were given to the two welders who killed the fuel fire.

Taking a break

325

Episode Sixteen

Autumn Years

At age sixty four it was time to accept that my full-on full-time occupation as the Safety and Training Advisor on the offshore gas platform had come to a close - and so it did in 1989. The Company made me an offer too hard to decline so I bit the bullet as they say, accepted the windup gold watch and retired. Actually I'm winding you up – no gold watch. One of the managers took me, together with Beverley to an expensive restaurant, gave me a cheque and a big album of photos, shook my hand and cast me adrift.

We had long harboured a secret envy for those living in the Eastern Hills, up in the Darling ranges and vowed to move and live there when I retired from a full time occupation. The lifestyle appealed to us and being part of a village community environment seemed to offer tranquillity and freedom from the increasing incidence of road trauma, home invasion and assaults which were report daily on the media in the city. After a very active and adventurous life the prospect of the odd petty opportunistic burglar upsetting my family life was far from my mind, so with few worries for our security and safety we settled down to retirement in the place where we wanted to be. Forty kilometres from Perth it is a very old settlement called Mundaring.

One frosty night this all changed.

Due to his annoying habit of howling during the night hours, I was driven to locking our ageing Siamese cat Fella out in the backyard workshop where he was free to sing to his heart's content. In common with many male backyard retreats, my workshop was home to a good range of tools and gear. There was also a securely locked inner room where I cared for and maintained my own small gun collection. All this was protected by a series of security lights, locked doors and steel cabinets.

It was really chilly in the early a.m. when Bev woke me with a note of alarm in her voice that suggested I should not go straight back to sleep. Through the bedroom window I could see - nothing. It was black as tar outside which is common for three o'clock in the morning.

Sitting on the side of the bed in a befuddled sleep fog I listened to what she had to say.

"Fella is sitting outside the bedroom window. He woke me up howling his head off."

This scenario was common before we started putting him to bed in the shed and my first reaction was predicable I think.

"You're waking me up in the middle of the night to tell me this - it's happened before and that's why we've been locking him in the worksh...."

Then it dawned on me, that if the cat was free he must have escaped from the shed - and if he escaped from the shed then the shed door must be open - and if the shed door was open now when it was previously locked, there was someone in there.

"Bugger!"

Bev explained that she had started to walk down the path in the back yard to return the old moggie to the workshop again. She was stopped halfway by light coming from the partly open doorway and by strange sounds coming from inside. It was then she decided to retreat to the house and wake me.

Sure enough, there was a light on inside the shed but the security floodlights were not functioning as they should have been. There were noises too of someone busy inside with little effort being made to hush up the activity.

Coincidentally just that evening, I had been watching a television feature dealing with the effective use of negotiation when dealing with criminals in standoff situations. It described the techniques used by police in siege and hostage scenarios at crime scenes and how effective they could be in defusing a nasty situation. Lo and behold it seemed that there could be one such example here in my own back yard. Armed with this new found knowledge and little else, I crept stealthily down to the scene of the apparent break-in with just a shaving coat thrown over my underpants and a failing torch.

While still trying to collect the scattered wits from the deep sleep I had been enjoying, a chilling thought occurred to me. Should the situation turn nasty with the intruder, there was the responsibility of preventing any harm coming to not only my wife in the house, but my two little granddaughters who were staying with us. That thought steels a man's resolve a bit.

With teeth chattering from the cold and a nasty sense of apprehension, I stepped into the pool of light cast by the workshop doorway.

"Hold it right there!" snapped the silhouetted figure.

Blinking in the light I tried to make out some details. He had a very big revolver held steady in his hands pointed straight at my head. Behind him on the floor I could see a pile of my tools and equipment that he had apparently pulled out of the cupboards ready to be taken with him on his getaway.

It seemed I was just in time.

The total surprise of the sudden confrontation and the threat of the gun were a bit of a shock and my mouth went dry and the knees threatened to sag. While trying to examine his face over the barrel of the gun, my brain kept urging me.

'Talk, talk, start negotiating '– This worked for the NYPD on television as I recalled, so I began.

"Why are you ripping off my stuff? What do you need the money for? Are you on drugs? What family do you have and where do you live? If you are in trouble I might be able to help you...."

I thought it best to shut up about then and let him answer some of the questions.

"I'm Bill and I live in Greenmount. My wife is pregnant and we've got no money."

The conversation was progressing quite well I thought and was buying some more time, giving me the chance to study him more closely.

The gun was still pointed steadily at my head and it looked real and dangerous.

The TV program I had seen about these situations said you needed to size up the offender. The eyes looking down the barrel I am happy to say did not seem to be those of a desperate drug addict or those of a man who was likely to shoot me to continue with his burglary work or affect his escape. He had the look of a loser I decided and more like a man who was more defeated than desperate. Also I felt much better after taking the time to study the gun in detail. It was recognisable as one of mine from a locked cupboard in the workshop. I knew it to be harmless.

Knowing for certain that the gun was not a threat to me was a relief. Whether the gunman still thought he had me to rights did not really matter now, but I wasn't about to tell him that he didn't hold all the cards after all.

As we talked and I reassured him that I was not much of a risk with my plastic torch against the fearsome weapon in his hand, I also pointed out that he was now in serious trouble as he had threatened me with a firearm in the commission of an offence. I mentioned too in passing that a gaol term was a certainty after his threatening action and that I would most certainly testify using this against him.

My offer was this, -

"If you put the weapon down and give yourself up voluntarily, I won't tell the police the bit about the gun."

He slowly lowered it and tossed it down on the driveway at my feet, mumbling and nodding his agreement. He backed into the workshop then and I followed him through to my small office inside.

"I had a fair bit to drink earlier on tonight," he confided, which I thought was pretty poor excuse for committing an armed robbery.

"You must be a bit dry after all this,"- it was now 3.30 am. "Would you like a drink?" I offered. He was more than a little surprised I think.

I had this vague plan that the longer I could keep him there, the better the chance of the police coming. The problem was there was no way of knowing if they had even been called.

In the little fridge under the table in the office I was able to retrieve two cold cans. After I had offered him one, we sat down like a couple of old pals to have a yarn. Nobody quite believes that I shared a beer with my erstwhile robber, but it seemed to be a good thing to do at the time. Although obviously much younger and fitter than me, there was no body language to suggest that he was a potential threat to me or my family or was even planning to make a bolt for it. I sat in the doorway just in case he changed his mind though and with one eye, watched the dark street outside through the opening.

Time seemed to drag on interminably and while I didn't feel too concerned, I was running out of things to say. He said he liked Emu Bitter beer, which was reassuring as that was what we were having. This accounted for a bit more polite conversation and a bit more time, but then the supply of beer began to run out.

Bill was showing signs of agitation and I was trying to dispel the thoughts of what was going to happen next and where it would all finish up in the end.

I could only hope that the deal we made about the threat with gun was going to work.

Visible through the crack in the doorway I thought I could see through the gloom a set of vehicle lights cruise slowly past the house. Willing desperately that they could only be those of a police car, I waited and sweated and watched Bill closely. Hope faded as the car continued slowly and turned around the corner of the street and disappeared.

Bill was still unaware of their presence and only realised that he was about to be nicked when the two burly armed constables appeared in the doorway. When introducing Bill to the policemen I was able to assure them that it was him that they should be interested in and left them to it. He was cuffed and marched to the paddy wagon and went without a murmur or a backward glance. While this was going on I stepped outside into the dark again, picked up the gun from where Bill had dropped it and pitched it out of sight into the shrubbery. As angry as I felt about his violation of my property, he came across as a sorry loser and after all he had kept his side of the bargain, so I felt that I had to keep mine. It was decided that I would leave the subject of his threat with the gun out of my statement to the police – no gun, probably no gaol for Bill! I thought he deserved a break.

Anyone who has reported a crime realises that the job's not done ''till the paperwork's finished.' Bev and I dictated our accounts of the night's events to the police in the kitchen and signed the statements required. I walked the senior constable out to his car with a great feeling of relief. Halfway back I was surprised when he stopped suddenly, turned to me and said, -

"Ivan, where's the gun now?' Feigning ignorance I replied.

"What gun?"

"We saw it lying on the driveway when we came in – where is it now?"

Realising I was caught out I had to tell him the whole story, concentrating on the reasons for my memory lapse and the deal made with my burglar to get him to stay when he could have absconded. I also pointed out that I considered we had achieved a good outcome. There was no loss of property, no one was hurt and the police had made an easy arrest with the minimum of fuss. Something that had not occurred to me before however was the disturbing proposition then put to me by the police officer.

"If we had got there first Ivan and he had confronted us with the gun we almost certainly would have had to shoot him. The other thing is that although you recognised the gun and knew it was harmless, he didn't know that and reckoned that what he was pointing at you was real and loaded. He had even cocked the hammer – he is a very lucky man."

Having now realised the more serious potential of the situation, I think I mumbled something about being fortunate to have had a good result all round. A different turn of events, I realised, with different timing, could have seen a vastly different ending with even a body in my back yard. Just retired and blood on my hands - that did not bear thinking about.

Bill eventually appeared in the Midland Court on charges of breaking and entering and burglary and admitted his guilt. He was given six months – sentence suspended! This meant he walked free. After all the trouble he had caused, I considered the sentence given him made it all a wasted effort and in hindsight regretted not slapping the firearms charge on him – so much for moral duty and keeping your word.

It was revealed later by a friend down the street that the police attendance was in response to the call he had made. Bill had raided their place too, earlier in the night. The patrol car was just cruising in the hope they might pick him up, and seeing all the activity at my place in the middle of the night, decided to investigate. In retrospect, these events could have resulted in any one of a number of different endings, but we were all glad of the eventual one.

An interesting side to this situation which I wasn't aware of at the time was that Bev was actually on the point of calling the police when it all started. After watching me and the burglar at some distance from the house however, engaged in seemingly congenial conversation and then moving inside the workshop together, she thought he must have been a friend who was visiting, so didn't raise the alarm. She said she was not able to see the gun from there.

By the way, the reason our two big sensor floodlights were not illuminating the scene was that Bill disliked drawing attention to himself apparently, so he dragged up a stepladder and unscrewed all the globes. He also defeated a padlock and two door locks and broke into a cupboard and then had the gall to tell the police this was his first time – first time being caught I think he meant!

As a postscript to the account of all this excitement and something that made me feel even more foolish, was the discovery next day that Bill had also attempted to steal my car as well earlier that night. In the attempt to hotwire it he had smashed the dashboard and left a big bunch of stripped wires hanging down.

So he was a man of untidy habits too.

Now about the gun - some months ago an old fellow from down the road had heard I was a gunsmith and gave me this revolver. It looked very realistic even to the trained eye, but was only a Dirty Harry Smith and Wesson replica air pistol that needed repair. He asked could I fix it for him. Sadly, he died shortly after and so it remained in my keeping. It was easy to recognise it as an air pistol during the hold up, when I saw that the size of the hole in the muzzle was nowhere near as big as the one you would see in the Clint Eastwood's Mr Smith and Mr. Wesson's' .44 cannon. It was good to know a bit about guns I reckoned.

I often reminisce about the youthful years, the action years and the years of adventure that came my way. I am so thankful to Beverley for putting up with my shenanigans, my absences from the home fires and the stresses and strains of living and raising our fine family in what she termed 'a man's country' for many of our years together.

The heady excitement of the oil strike at Rough Range and the kudos surrounding the discovery that the Last Roughneck is alive and well and writing a book about it all gives me much pleasure and refreshes so many memories. The stories of the triumphs and disappointments experienced in the search for oil must have influenced my three sons who now work in the enormously rich oil and gas industry in my beloved North.

Sorry, I must go - I think my cocoa is getting cold.

Our retirement sanctuary in the hills - the Stoneville house.

With the book writing completed I was reposing peacefully in retired mode in our beloved hills when I saw in the daily newspaper an old photograph I instantly recognised as taken in the pioneer drilling camp at Rough Range in 1953. The oil exploration company had published the photograph in the hope that someone who worked in the camp fifty years ago would see it and recognise it for what it was. The caption indicated that they were looking for people who could help in the preparations for their proposed grand commemoration of the first oil strike, due to be celebrated on the fourth day of December 2003, exactly fifty years since that historic day.

My proffered contribution was snapped up, this being an eye witness account of the first oil flow as I saw it on the drill floor in the company of the others of our five man crew at that time in 1953.

A world-wide search by the Company for the other members of my crew to join the celebrations drew a blank, as the intervening fifty years had apparently decimated our team to the point of leaving me as the sole survivor – the last roughneck.

This discovery led to my being pressed into joining the celebrations. They had me making appearances at company dinners with VIPs, having interviews with journalists and photographers, talking on radio and having my story with photographs in a double page spread of the weekend newspaper.

When the dust had cleared from all this hype I was happy to accept the offer to be an honoured guest and join a tour of the old exploration field with its abandoned wells and most importantly the wellhead at Rough Range No.1 where my story begins.

As part of the commemorations my hosts had constructed a plinth from local stone on the very top of the Cape Range. The site has a commanding panoramic view of the blue Indian Ocean on one side and the Rough Range hills on the other. A brass tablet fixed on the plinth was inscribed with the date and location of the historic oil strike and commemorated the men who contributed to this feat in conditions of fierce summer heat and dust and sweat in this remote area.

It was my privilege there and then to unveil the tablet in front of the assembled press and guests and to relate some stories of life on the oilfield fifty

My return to the oilfields in 2003 after fifty years. The Rough Range 1 wellhead today.

years ago and share some of the unforgettable experiences of those good old days with them.

It is all tranquil and quiet there now under the sun, though evidence of our endeavours can still be seen. Flat drilling locations where the rigs once worked are tucked away among rocky hills, each with its black steel wellhead surmounting the bore holes penetrating deep into the limestone strata below. The roads we carved out of the stony ridges which once were a scene of frenetic activity and dust and diesel smoke now carry the cars of city folk and tourist buses. The black scorched scars are still on the rocks around the site of the big flare where the flow of oil was burned in the testing program to evaluate our discovery and put us in headlines around the world.

BOOKSHOP

A copy of this book may be ordered from most good Bookshops.

If you experience any dificulty in obtaining a copy please visit our website at

www.wapublishing.co.uk for further information.

www.ingramcontent.com/pod-product-compliance
Lightning Source LLC
Chambersburg PA
CBHW080509090426
42734CB00015B/3013